Hands-on Signals and Systems Theory

Kajetana Marta Snopek

Hands-on Signals and Systems Theory

Springer

Kajetana Marta Snopek
Warsaw University of Technology
Warsaw, Poland

ISBN 978-3-031-56080-4 ISBN 978-3-031-56081-1 (eBook)
https://doi.org/10.1007/978-3-031-56081-1

This Springer imprint is published by the registered company Springer Nature Switzerland AG
The registered company address is: Gewerbestrasse 11, 6330 Cham, Switzerland

Paper in this product is recyclable.

Preface

If there's a book that you want to read, but it hasn't been written yet, then you must write it.—Toni Morison.

As an academic teacher in the field of signals and systems for over 20 years, I have always noticed scarcity of student exercise books with solved problems. Evidently, the theory is presented in many books, but often readers can encounter difficulties in finding solutions to attached problems. On the other hand, there is a great availability of books with problem solutions in Matlab or Python, but a question arises: how can results produced by a computer be verified or commented on? This book gives an answer to these problems.

This handbook is dedicated above all to undergraduate engineering students who are interested in analytical methods of solving real technical problems concerning signals and systems. Also, graduate students can treat the content of this book as a helpful revision material. Moreover, I hope that it can serve as a comfortable educational help for academic teachers. The content is related to analysis of signals and systems in time and frequency domains. In nine chapters, the following topics are presented in detail:

- Parameters, operations, and models of signals in time domain
- Orthogonal signal representations including Fourier series
- Convolution and correlation of signals
- Fourier transformation in signals and systems
- Ideal and real signal sampling including aliasing and stroboscopic effects
- System properties, impulse and step response, block diagrams
- Ideal and real signal filtering
- Laplace- and $\mathcal{Z}$ transformations in system theory
- Transfer functions of analog and digital filters
- Chosen methods of analog and digital filter design
- Stability criteria of analog and digital filters

For better understanding of practical aspects of signals and systems, a theory is presented in the form of a concise "lecture" accompanied with solved original problems that are explained and illustrated with necessary tables, graphs and schemes. An inquisitive reader will also be able to check her/his knowledge by solving some additional exercises from the end of each chapter. Some hints and answers to all these exercises are included at the end of the book.

During preparation of this textbook, I have always had in mind its future readers. Therefore, my efforts were to keep a logical structure, to select representative and original problems, and to explain them in a concise and comprehensive way. I would be happy if this publication was well received by a wide audience.

Warsaw, Poland Kajetana Marta Snopek

Contents

Abbreviations

AM	Amplitude Modulation
BIBO	Bounded Input Bounded Output
BP	Band-pass
BS	Band-stop
DTF	Discrete Fourier Transform
DTFT	Discrete-Time Fourier Transform
FIR	Finite Impulse Response
FFT	Fast Fourier Transforms
FT	Fourier Transform
HP	High-pass
IIR	Infinite Impulse Response
LCM	Least Common Multiple
LP	Low-pass
LTI	Linear Time-Invariant
THD	Total Harmonic Distortion

Symbols

$*$	convolution
$\odot$	cross-correlation
(x_1, x_2)	inner product of two signals
$(\cdot)^*$	conjugate of a number or a function
$(\cdot) \overset{\text{DFT}}{\leftrightarrow} (\cdot)$	DFT pair
$(\cdot) \overset{\mathcal{F}}{\leftrightarrow} (\cdot)$	complex Fourier transform pair
$(\cdot) \overset{\mathcal{F}_d}{\leftrightarrow} (\cdot)$	DTFT pair
$(\cdot) \overset{\mathcal{L}}{\leftrightarrow} (\cdot)$	Laplace transform pair
$(\cdot) \overset{\mathcal{Z}}{\leftrightarrow} (\cdot)$	$\mathcal{Z}$-transform pair
$(\cdot)_{\mathrm{e}}$	even component of a one-dimensional signal
$(\cdot)_{\mathrm{o}}$	odd component of a one-dimensional signal
$\mathbf{1}(\cdot)$	unit step function
$\mathbb{C}$	algebra of complex numbers
$\mathbb{N}$	set of natural numbers
$\mathbb{R}$	set of real numbers
$\mathbb{Z}$	set of integer numbers
$A(\omega)$	magnitude response of an analog filter
$A(\Omega)$	magnitude response of a digital filter
$\arg(\cdot)$	argument of a complex number or a function
E_x	energy of a signal
d	duty factor
f	frequency variable
f_0	fundamental frequency of a periodic signal
f_c	cutoff frequency of a filter
$f_{\max}$	maximal frequency of a signal
f_{N}	Nyquist frequency
f_s	sampling frequency
F	normalized frequency
$h(\cdot)$	impulse response

$H(\omega)$	frequency response function of an analog filter
$H(s)$	transfer function of an analog filter
$H(\Omega)$	frequency response function of a digital filter
$H(z)$	transfer function of a digital filter
$\mathrm{Im}\,(\cdot)$	imaginary part of a complex number or a function
j	imaginary unit
$k(\cdot)$	step response
n	discrete-time variable
N_0	fundamental period of a discrete-time periodic signal
P_x	average power of a signal
$\boldsymbol{r}(\cdot)$	ramp function
$\mathrm{Re}\,(\cdot)$	real part of a complex number or a function
R_x	auto-correlation function
$R_{12}(\cdot)/R_{21}(\cdot)$	cross-correlation function
$\mathrm{Sa}(\cdot)$	sampling signal
$\mathrm{sgn}(\cdot)$	signum function
t	continuous-time variable
T_0	fundamental period of a continuous-time periodic signal
T_s	sampling period
$\delta(n)$	unit delta impulse
$\delta(t)$	Dirac delta function
$\delta_{T_0}(t)$	comb distribution of the period T_0
$\Lambda(\cdot)$	triangular pulse
$\Pi(\cdot)$	rectangular pulse
ω	angular frequency
Ω	normalized angular frequency

Chapter 1
Analysis of Signals in Time Domain

This chapter introduces the notion of a continuous-time– and a discrete-time signal and presents different signal classes: periodic and nonperiodic, finite- and infinite-duration, finite energy and finite power, and distribution signals. Basic signal parameters (average value, energy, average power) are also defined. Elementary and advanced operations performed on signals in time domain are described. A separate section is devoted to convolution and correlation of two signals. The reader also will learn how to express different signals in orthogonal bases.

1.1 Basic Models of Continuous- and Discrete-Time Signals

A *signal* can represent any physical variable changing in time, for example, temperature of an object, amplitude of vibrations of any element, velocity of a moving body, etc. If time is continuous, that is, $t \in \mathbb{R}$, we deal with *continuous-time* signals denoted with $x(t)$. If time is discrete, $n \in \mathbb{Z}$, we consider *discrete-time* signals $x(n)$. Signals can be classified according to their characteristic properties as *periodic/nonperiodic*, *finite-/infinite-duration*, *finite energy/finite power*, *deterministic/stochastic,* and so on.

1.1.1 Periodic and Nonperiodic Signals

Definition 1.1 A continuous-time signal $x(t)$, $t \in \mathbb{R}$ is called *periodic* if and only if there is a constant $T_0 \in \mathbb{R}^+$, the smallest one from all, such that for every $t \in \mathbb{R}$: $x(t) = x(t + T_0)$.

The periodicity condition means that the values of $x(t)$ repeat every T_0 called the *fundamental period* of $x(t)$, expressed usually in seconds [s]. Its inverse $f_0 = \frac{1}{T_0}$ is the

K. M. Snopek, *Hands-on Signals and Systems Theory*,
https://doi.org/10.1007/978-3-031-56081-1_1

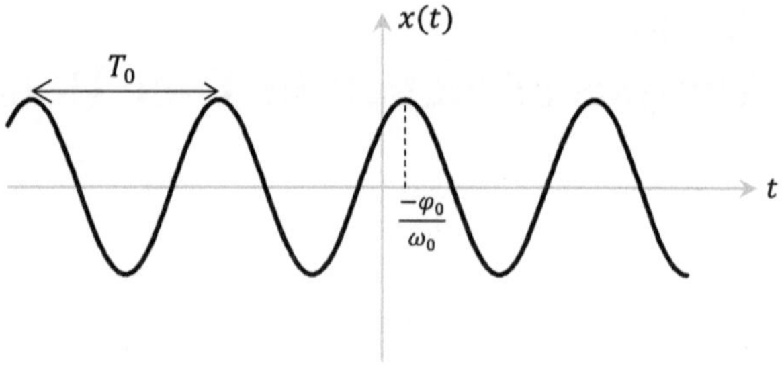

Fig. 1.1 A continuous-time harmonic signal

fundamental frequency (in Hz), while $\omega_0 = 2\pi f_0 = \frac{2\pi}{T_0}$ is the *fundamental angular frequency* (in rad/s).

The classical example of periodic signals is the continuous-time *harmonic signal* defined as

$$x(t) = A\cos(\omega_0 t + \varphi_0) = A\cos(2\pi f_0 t + \varphi_0), \omega_0 > 0, \tag{1.1}$$

where A is the amplitude, $\varphi_0 \in \mathbb{R}$—the initial phase, ω_0—the angular frequency and f_0– the fundamental frequency. The fundamental period equals $T_0 = \frac{1}{f_0}$. Evidently, we can rewrite (1.1) as $x(t) = A\cos\left[\omega_0\left(t + \frac{\varphi_0}{\omega_0}\right)\right]$. We see that if $\varphi_0 > 0$, then a cosine curve is shifted to the left, or if $\varphi_0 < 0$, it is shifted to the right, in both cases by $\frac{\varphi_0}{\omega_0}$. Figure 1.1 shows the plot of the harmonic signal shifted to the right.

In Problem 1.1, we show how to check signal periodicity and compute the basic parameters of a periodic signal.

Problem 1.1 Given the continuous-time signal

$$x(t) = \cos\left(\frac{1}{2}\pi t\right) + \cos\left(\frac{1}{5}\pi t\right), t \in \mathbb{R}.$$

Prove that $x(t)$ is periodic and determine its parameters. Plot the graph of $x(t)$.

Solution We apply Definition 1.1 to prove that the periodicity condition is fulfilled. Let us write

$$\begin{aligned}x(t+T_0) &= \cos\left[\frac{1}{2}\pi(t+T_0)\right] + \cos\left[\frac{1}{5}\pi(t+T_0)\right] \\ &= \cos\left(\frac{1}{2}\pi t\right)\cos\left(\frac{1}{2}\pi T_0\right) - \sin\left(\frac{1}{2}\pi t\right)\sin\left(\frac{1}{2}\pi T_0\right) \\ &\quad + \cos\left(\frac{1}{5}\pi t\right)\cos\left(\frac{1}{5}\pi T_0\right) - \sin\left(\frac{1}{5}\pi t\right)\sin\left(\frac{1}{5}\pi T_0\right).\end{aligned}$$

Now, let us compare the above expression with $x(t) = \cos\left(\frac{1}{2}\pi t\right) + \cos\left(\frac{1}{5}\pi t\right)$. The equality is true if and only if

$$\begin{cases}\cos\left(\frac{1}{2}\pi T_0\right) = 1 \\ \sin\left(\frac{1}{2}\pi T_0\right) = 0\end{cases} \quad \text{and} \quad \begin{cases}\cos\left(\frac{1}{5}\pi T_0\right) = 1 \\ \sin\left(\frac{1}{5}\pi T_0\right) = 0\end{cases}.$$

In general, the above conditions are met if $\frac{1}{2}\pi T_0 = 2k\pi, k \in \mathbb{Z}$ and $\frac{1}{5}\pi T_0 = 2l\pi, l \in \mathbb{Z}$. Since T_0 should be real and positive, we can write $\frac{1}{2}\pi T_0 = 2k\pi$, $k \in N$ and $\frac{1}{5}\pi T_0 = 2l\pi, l \in N$. Solving these equations, we get $T_0 = 4k = 10l$, $k, l \in N$. The smallest $k, l \in N$ satisfying the above condition are $k = 5$ and $l = 2$. Then the fundamental period of $x(t)$ is $T_0 = 20$ [s], the fundamental frequency $f_0 = 0, 05$ [Hz] and the fundamental angular frequency $\omega_0 = \frac{\pi}{10}$ [rad/s]. Fig. 1.2 shows the graph of $x(t)$. Let us notice that values of $x(t)$ repeat every 20 s.

Important Since the fundamental frequency of $\cos\left(\frac{1}{2}\pi t\right)$ is $f_1 = 0, 25 = \mathbf{5}f_0$, we call this component the *5th harmonic*. Analogously, $\cos\left(\frac{1}{5}\pi t\right)$ is the *2nd harmonic* since its fundamental frequency $f_2 = 0, 1 = \mathbf{2}f_0$. The signal $x(t)$ is a sum of its 2nd and 5th harmonics.

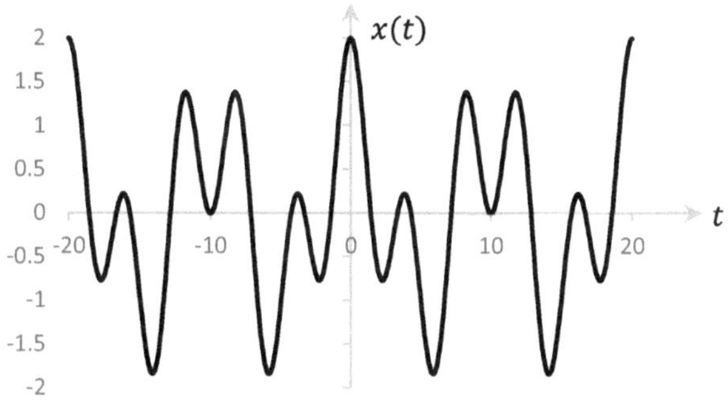

Fig. 1.2 Sum of two harmonics from Problem 1.1

Important The simplest method of computing the fundamental period is calculating the *Least Common Multiple* (LCM) of periods of signal components. Here, we have $T_1 = \frac{1}{f_1} = 4$ and $T_2 = \frac{1}{f_2} = 10$. Then

$$T_0 = \text{LCM}(T_1, T_2) = \text{LCM}(4, 10) = 20.$$

Calculating the LCM of two numbers means finding the smallest $k, l \in N$ satisfying equality: $4k = 10l$ (compare with solution of Problem 1.1).

The *complex harmonic signal* often applied in signal and modulation theory [20, 21], is a complex-valued signal defined as

$$x(t) = e^{j\omega_0 t} = \cos(\omega_0 t) + j\sin(\omega_0 t). \tag{1.2}$$

It is periodic with fundamental period $T_0 = \frac{2\pi}{\omega_0}$.

In signal or circuit theory, other periodic signals are often applied, for example, different kinds of rectangle and triangle waveforms [5, 16]. *Rectangle* and *triangle periodic waveforms* with fundamental period T_0 are characterized by a special parameter called the *duty factor d* defined as $d = \tau/T_0$. For a rectangle wave, τ is the duration time of a pulse with positive amplitude A (see Fig. 1.3). The special case of a rectangular wave is the *rectangle pulse train* consisting of unipolar rectangle pulses repeating every T_0, as shown in Fig. 1.4. This signal model is often applied in sampling theory (see Chap. 6, Sect. 6.2).

For a triangle wave, τ in d represents the *rise time*, that is, the period of time needed a signal to go from its minimum to the maximum value (see Fig. 1.5). If the duty factor $d = 0$, we obtain the *saw-tooth wave* displayed in Fig. 1.6.

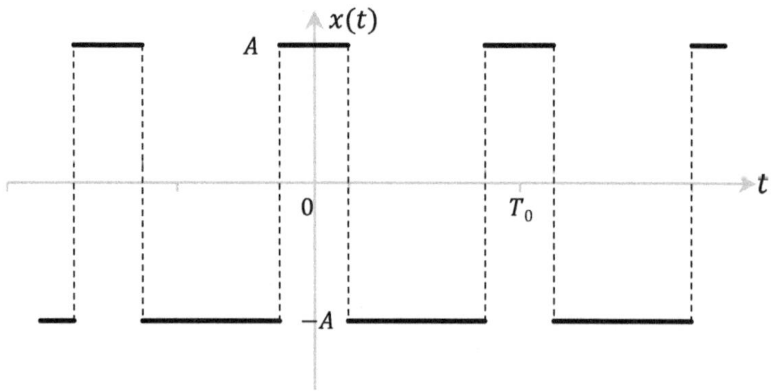

Fig. 1.3 Bipolar rectangle waveform with period T_0 and duty factor $d = 1/3$

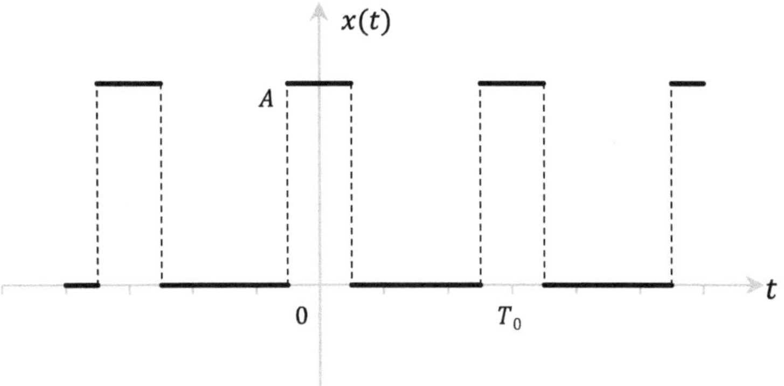

Fig. 1.4 Unipolar rectangle pulse train with amplitude A, period T_0 and duty factor $d = 1/3$

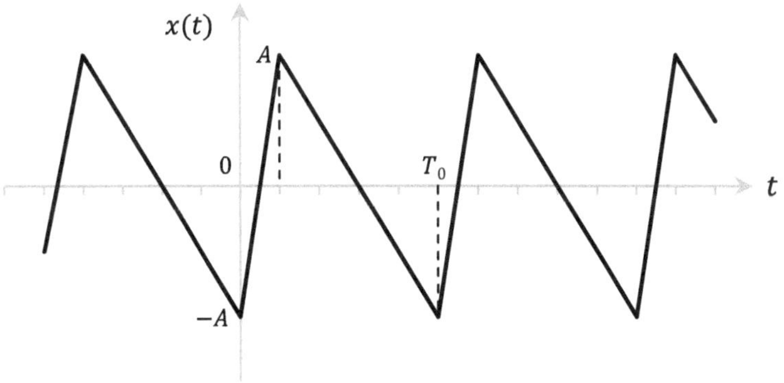

Fig. 1.5 Triangle bipolar waveform with period T_0 and duty factor $d = 1/5$

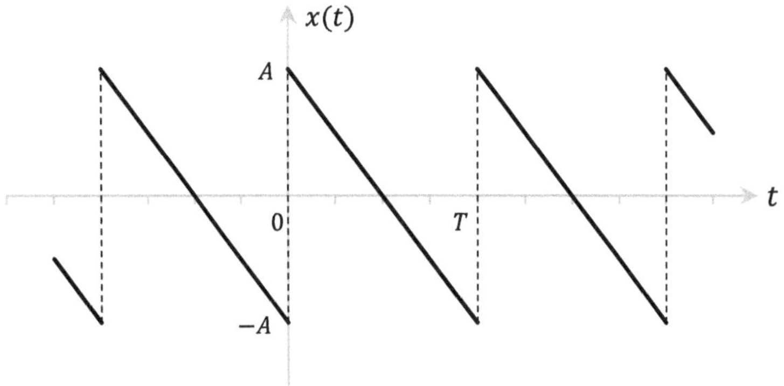

Fig. 1.6 Saw-tooth waveform with period T_0 and duty factor $d = 0$

Let us formulate the periodicity condition in the discrete-time domain (compare with Definition 1.1).

Definition 1.2 A discrete-time signal $x(n)$, $n \in \mathbb{Z}$ is called *periodic* if and only if there is a constant $N_0 \in \mathbb{N}$, the smallest one from all, such that for every $n \in \mathbb{Z}$: $x(n) = x(n + N_0)$. The constant N_0 is called the *fundamental period* of $x(n)$.

Problem 1.2 Given the discrete-time signal

$$x(n) = \sin\left(\frac{n}{6} + 1\right), n \in \mathbb{Z}$$

Check periodicity of $x(n)$ and plot its graph.

Solution Similarly as in Problem 1.1, let us verify the periodicity condition from Definition 1.2. We have

$$x(n + N_0) = \sin\left(\frac{n + N_0}{6} + 1\right) = \sin\left(\frac{n + N_0}{6}\right)\cos 1 + \cos\left(\frac{n + N_0}{6}\right)\sin 1$$

and $x(n) = \sin\left(\frac{n}{6} + 1\right) = \sin\left(\frac{n}{6}\right)\cos 1 + \cos\left(\frac{n}{6}\right)\sin 1$. The periodicity condition $x(n + N_0) = x(n)$ is met if and only if

$$\begin{cases} \sin\left(\frac{n + N_0}{6}\right) = \sin\left(\frac{n}{6}\right) \\ \cos\left(\frac{n + N_0}{6}\right) = \cos\left(\frac{n}{6}\right) \end{cases},$$

which is equivalent to $\frac{N_0}{6} = 2k\pi, k \in \mathbb{Z}$. Since N_0 should be positive, we have $N_0 = 12k\pi$, $k \in \mathbb{N}$. We see that N_0 is not a natural number, so $x(n)$ *is not* periodic. Looking at its graph displayed in Fig. 1.7, we can easily notice that values of $x(n)$ *do not* repeat periodically.

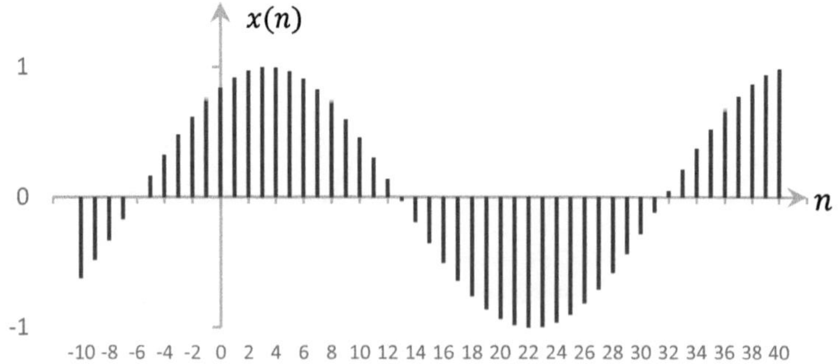

Fig. 1.7 Nonperiodic sine signal from Problem 1.2

The *discrete-time harmonic signal* is defined as

$$x(n) = A\cos(\omega_0 n). \tag{1.3}$$

The following question arises: when (1.3) is periodic? To answer it, let us apply once again the Definition 1.2. We are looking for the smallest $N_0 \in \mathbb{N}$ satisfying the periodicity condition. We have

$$x(n + N_0) = A\cos[\omega_0(n + N_0)],$$

which should equal to $x(n) = A\cos(\omega_0 n)$. It is met if, in general, $\omega_0 N_0 = 2k\pi$, $k \in \mathbb{Z}$. However, the fundamental period N_0 should be a natural number. So, finally

$$N_0 = \frac{2k\pi}{\omega_0}, k \in \mathbb{N}, \tag{1.4}$$

where k is such a natural number that N_0 is the smallest natural number.

Problem 1.3 Given the signal $x(n) = \cos\left(\frac{\pi n}{10} + 2\right), n \in \mathbb{Z}$. Check periodicity of $x(n)$ and compute its fundamental period. Plot the graph of $x(n)$ for $n \in [0, 39]$.

Solution Similarly as in Problem 1.2, we start from $x(n + N_0)$ developed as

$$x(n + N_0) = \cos\left[\frac{\pi(n + N_0)}{10} + 2\right] = \cos\left(\frac{\pi n}{10} + 2\right)\cos\left(\frac{\pi N_0}{10}\right)$$
$$- \sin\left(\frac{\pi n}{10} + 2\right)\sin\left(\frac{\pi N_0}{10}\right)$$

and compare it with $x(n) = \cos\left(\frac{\pi n}{10} + 2\right)$. We can see that equality occurs for

$$\begin{cases} \cos\left(\frac{\pi N_0}{10}\right) = 1 \\ \sin\left(\frac{\pi N_0}{10}\right) = 0 \end{cases} \quad \text{which is equivalent to} \quad \frac{\pi N_0}{10} = 2k\pi, k \in \mathbb{N}.$$

So we have $N_0 = 20k$, $k \in \mathbb{N}$ and the only possible k is $k = 1$, then $N_0 = 20$. If we compare this result with (1.4), we immediately get: $N_0 = \frac{2k\pi}{\omega_0} = \frac{2k\pi}{\pi/10} = 20k$. Figure 1.8 displays two periods of $x(n)$.

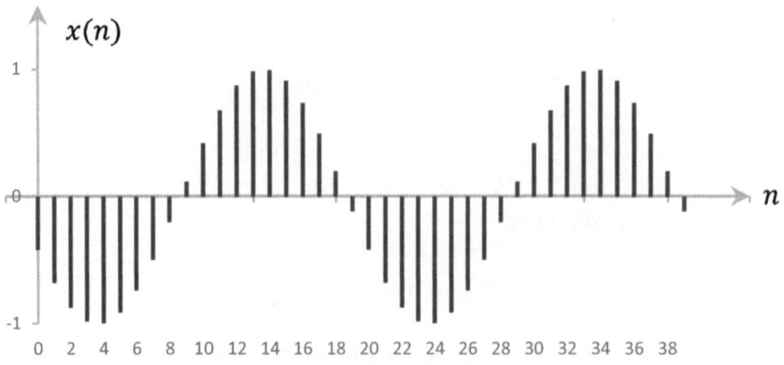

Fig. 1.8 Plot of the periodic cosine signal from Problem 1.3

Fig. 1.9 Rectangular pulse signal with amplitude A and time duration τ

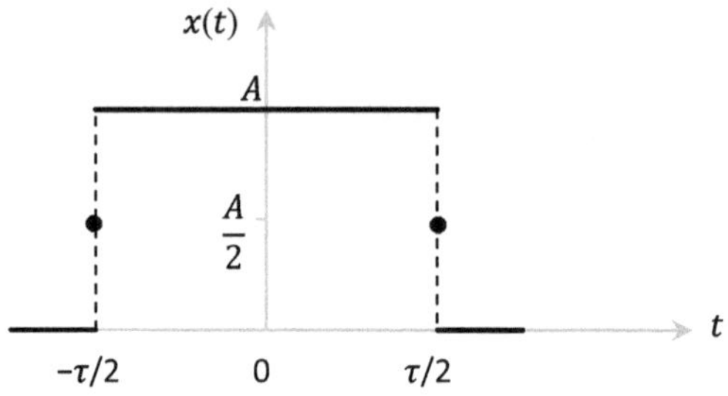

1.1.2 Finite-Duration Continuous- and Discrete-Time Signals

Definition 1.3 The signal $x(t)$, $t \in \mathbb{R}$ is a *finite-duration (pulse)* signal if there are two real numbers $a, b \in \mathbb{R}$, $a < b$, such that $x(t) = 0$ for $t < a$ and $t > b$. We say that the support of $x(t)$ is bounded and limited to the interval $[a, b]$.

The most popular models of finite-duration continuous-time signals are rectangular and triangular pulse signals. The *rectangular pulse* signal (Fig. 1.9) is defined as

$$x(t) = A\Pi\left(\frac{t}{\tau}\right) = \begin{cases} A & \text{for } t \in \left(-\frac{\tau}{2}, \frac{\tau}{2}\right), \\ \frac{A}{2} & \text{for } t = \pm\frac{\tau}{2}, \\ 0 & \text{elsewhere} \end{cases} \tag{1.5}$$

where A is the amplitude and τ is the time duration. This signal is even-symmetric.

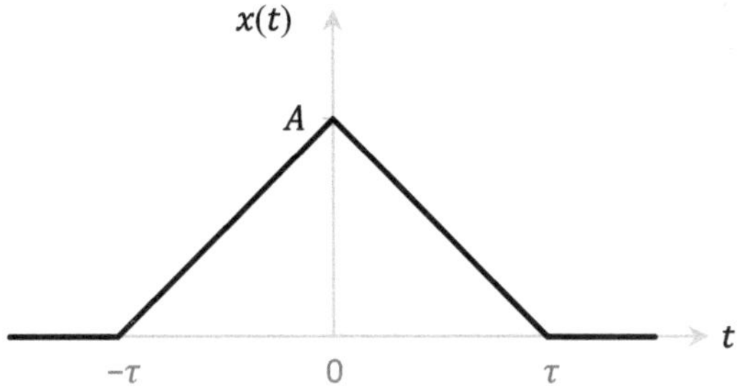

Fig. 1.10 Triangular pulse signal with amplitude A and time duration 2τ

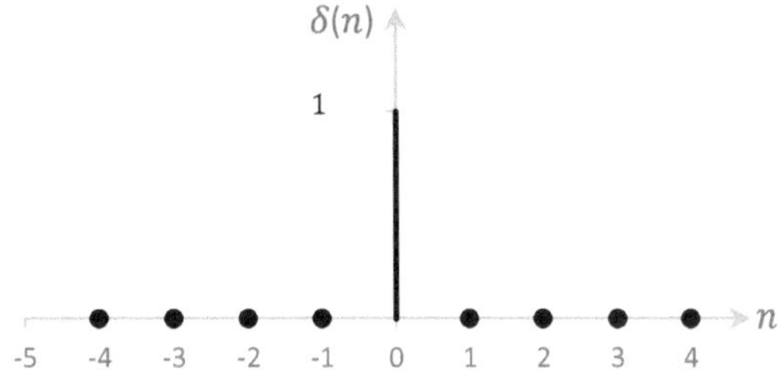

Fig. 1.11 Unit delta pulse signal

The *triangular pulse* signal (Fig. 1.10) is defined as

$$x(t) = A\Lambda\left(\frac{t}{\tau}\right) = \begin{cases} \frac{A}{\tau}(t+\tau) & \text{for} \quad \tau \in [-\tau, 0), \\ -\frac{A}{\tau}(t-\tau) & \text{for} \quad \tau \in [0, \tau], \\ 0 & \text{elsewhere} \end{cases} \tag{1.6}$$

where A is the amplitude and the time duration equals 2τ. Notice that though the notations (1.5) and (1.6) look similar, in the second case the duration time is twice as long.

Let us now present the definition of a finite-duration discrete-time signal.

Definition 1.4 The signal $x(n)$, $n \in \mathbb{Z}$ is a *finite-duration (pulse)* signal if there are two integer numbers $a, b \in \mathbb{Z}$, $a < b$, that $x(n) = 0$ for $n < a$ and $n > b$. We say that the support of $x(n)$ is bounded and limited to the interval $[a, b]$.

Popular examples of finite-duration discrete-time signals are the unit delta pulse signal and the even-symmetric rectangular pulse signal. The *unit delta pulse* $\delta(n)$ (Fig. 1.11) is defined in the form

$$\delta(n) = \begin{cases} 1 & \text{for } n = 0, \\ 0 & \text{elsewhere.} \end{cases} \tag{1.7}$$

The discrete-time *rectangular pulse signal* $\Pi_N(n)$ is defined as

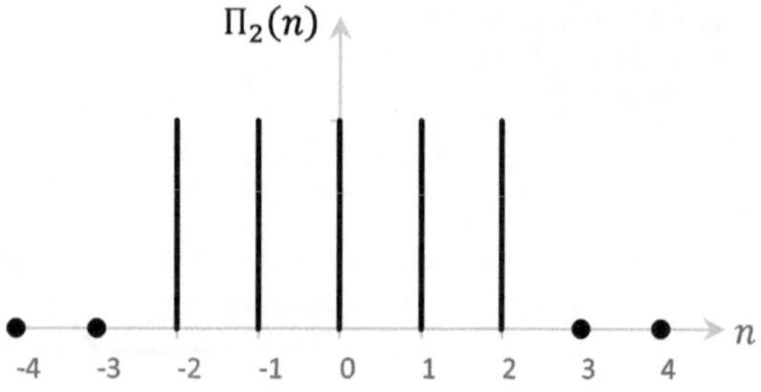

Fig. 1.12 Discrete-time rectangular pulse $\Pi_2(n)$

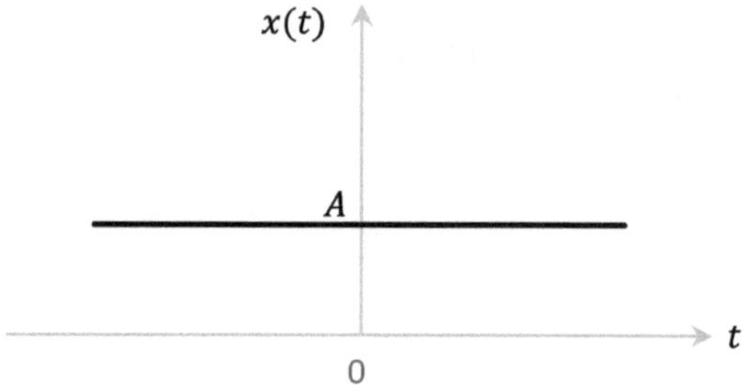

Fig. 1.13 Continuous-time constant signal

$$\Pi_N(n) = \begin{cases} 1 & \text{for } |n| \leq \mathrm{N}, \\ 0 & \text{elsewhere.} \end{cases} \tag{1.8}$$

It is an even-symmetric signal with exactly $2N + 1$ nonzero values (equal 1) in the interval $[-N, N]$, $n \in \mathbb{Z}$. The signal $\Pi_2(n)$ is displayed in Fig. 1.12.

1.1.3 Infinite-Duration Signals

In this part, we show examples of the most popular models of infinite-duration continuous-time and discrete-time signals.

The *continuous-time constant signal* (Fig. 1.13) is defined as

$$x(t) = A, \tag{1.9}$$

where $A \in \mathbb{R}$.

The *discrete-time constant signal* (Fig. 1.14) is the analog of (1.9) defined as

$$x(n) = A, \tag{1.10}$$

where $A \in \mathbb{R}$.

The *continuous-time unit step signal* (Fig. 1.15), denoted with $\mathbf{1}(t)$, $t \in \mathbb{R}$, is defined as

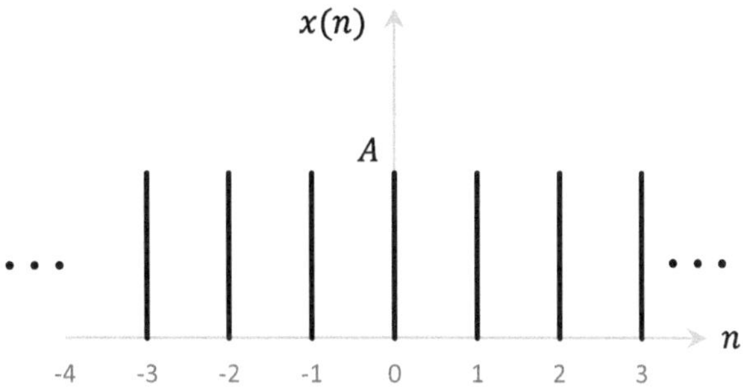

Fig. 1.14 Discrete-time constant signal

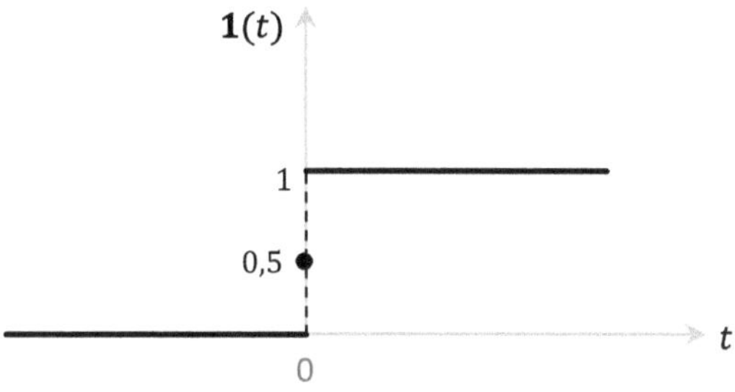

Fig. 1.15 Continuous-time unit step signal

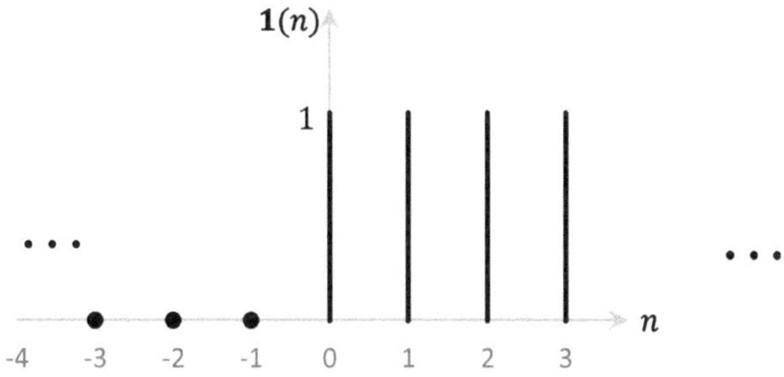

Fig. 1.16 Discrete-time unit step signal

$$\mathbf{1}(t) = \begin{cases} 1 & \text{for} \quad \tau \in (0, \infty), \\ 0,5 & \text{for} \quad \tau = 0, \\ 0 & \text{elsewhere.} \end{cases} \tag{1.11}$$

The *discrete-time unit step signal* (Fig. 1.16), denoted with $\mathbf{1}(n)$, $n \in \mathbb{Z}$, is the analogue of (1.11) defined as

$$\mathbf{1}(n) = \begin{cases} 1 & \text{for} \quad n \geq 0, \\ 0 & \text{elsewhere.} \end{cases} \tag{1.12}$$

Notice that differently than for $\mathbf{1}(t)$, here $\mathbf{1}(0) = 1$.

The unit step signals (1.11) and (1.12) are used in definitions of causal signals and also in analog and digital system theory (e.g., [5, 16, 20]).

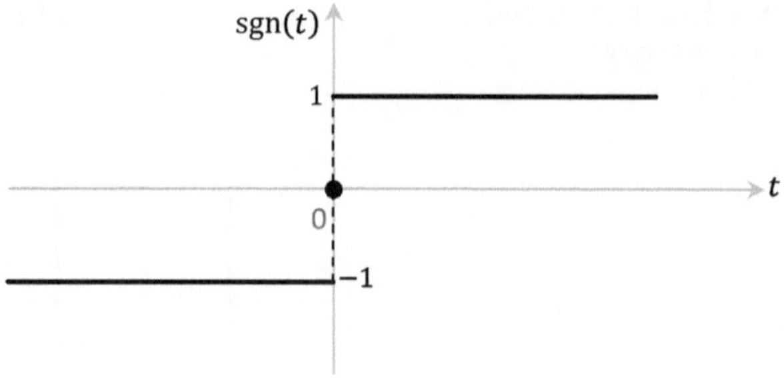

Fig. 1.17 Continuous-time signum signal

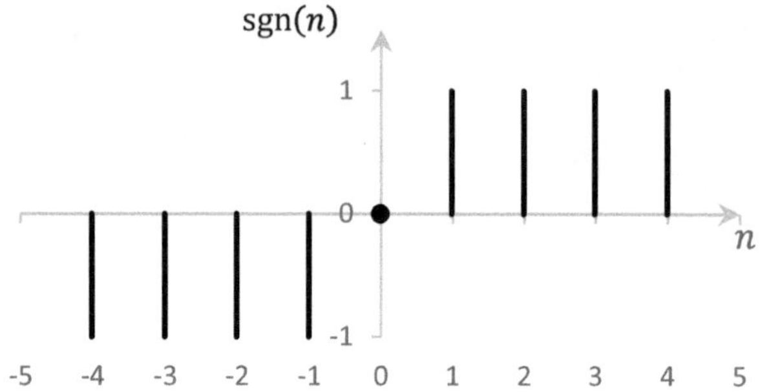

Fig. 1.18 Discrete-time signum signal

The *continuous-time signum signal* sgn(t), $t \in \mathbb{R}$ (Fig. 1.17), is defined as

$$\operatorname{sgn}(t) = \begin{cases} -1 & \text{for} \quad \tau \in (-\infty, 0), \\ 0 & \text{for} \quad \tau = 0, \\ 1 & \text{for} \quad \tau \in (0, \infty). \end{cases} \tag{1.13}$$

The *discrete-time signum signal* sgn(n), $n \in \mathbb{Z}$ (Fig. 1.18) is defined analogously to (1.13) in the form

$$\operatorname{sgn}(n) = \begin{cases} -1 & \text{for} \quad n < 0, \\ 0 & \text{for} \quad n = 0, \\ 1 & \text{for} \quad n > 0. \end{cases} \tag{1.14}$$

The *continuous-time ramp signal*
$\boldsymbol{r}(t)$, $t \in \mathbb{R}$ (Fig. 1.19) is defined as follows:

$$\boldsymbol{r}(t) = \begin{cases} t & \text{for} \quad t \geq 0, \\ 0 & \text{elsewhere}. \end{cases} \tag{1.15}$$

The *discrete-time ramp signal* $\boldsymbol{r}(n)$, $n \in \mathbb{Z}$ (Fig. 1.20) is defined as

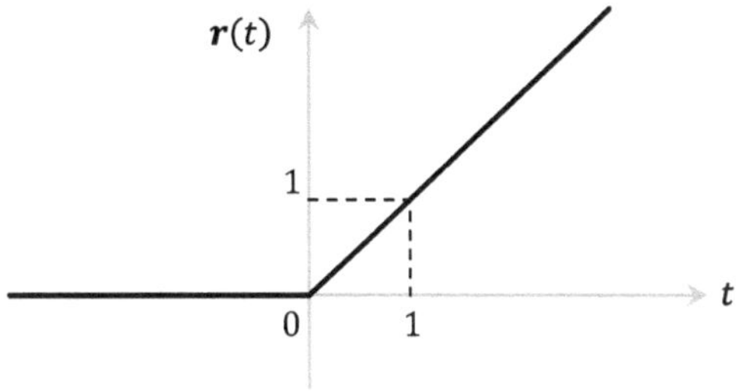

Fig. 1.19 Continuous-time ramp signal

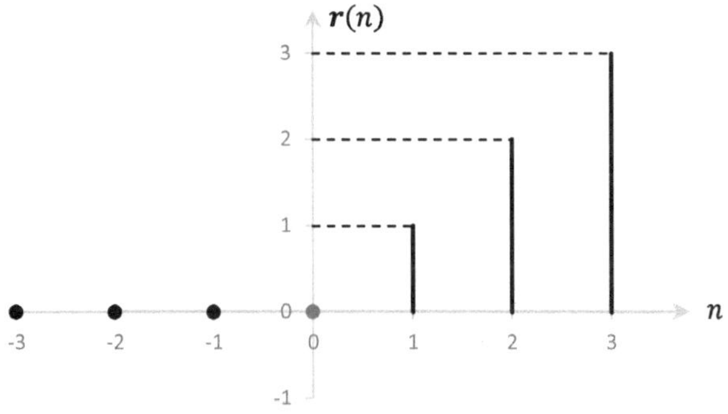

Fig. 1.20 Discrete-time ramp signal

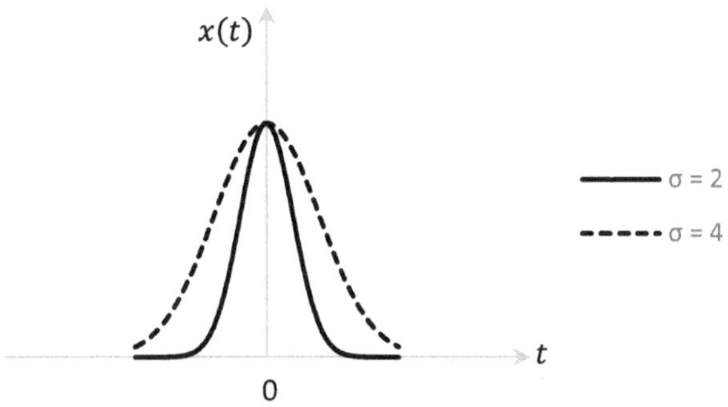

Fig. 1.21 Continuous-time Gaussian signal for two different values of σ

$$r(n) = \begin{cases} n & \text{for } n \geq 0, \\ 0 & \text{elsewhere.} \end{cases} \tag{1.16}$$

The *Gaussian* signal, widely used in mathematics, signal theory, filter theory, and mathematical physics (e.g., [7, 18, 19]), is defined by

$$x(t) = e^{-\frac{t^2}{2\sigma^2}}, \tag{1.17}$$

where σ is responsible for the slope of the Gaussian curve ("bell" shape curve). Let us illustrate it in Fig. 1.21 for two values of $\sigma = 2$ and $\sigma = 4$. Notice that values of the Gaussian signal approach zero if $t \to \pm\infty$.

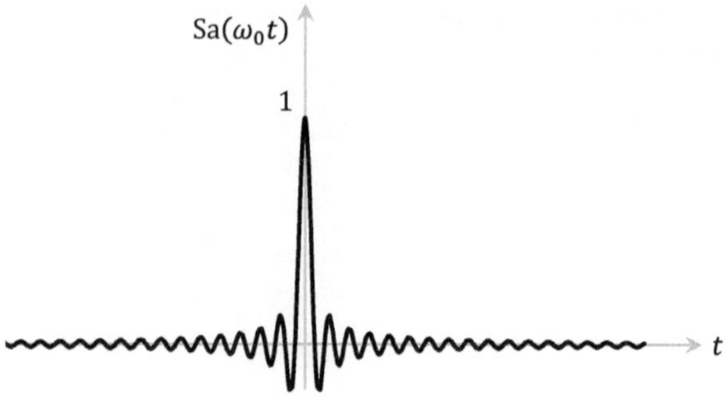

Fig. 1.22 Sampling signal Sa($\omega_0 t$)

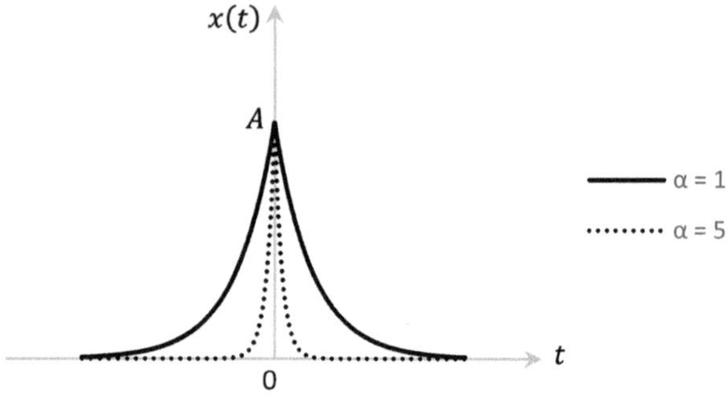

Fig. 1.23 Two-sided exponential signal

The next function widely applied in signal theory (e.g., [4, 15, 19]), especially in sampling, is the *sampling signal* Sa($\omega_0 t$), often denoted with sinc(·) and defined in the form

$$\mathrm{Sa}(\omega_0 t) = \begin{cases} \dfrac{\sin(\omega_0 t)}{\omega_0 t} & \text{for} \quad t \neq 0, \\ 1 & \text{for} \quad t = 0. \end{cases} \tag{1.18}$$

The graph of (1.18) is displayed in Fig. 1.22. We observe zero crossings for $t = k\pi$, $k \in \mathbb{Z} \neg \{0\}$, exactly the same as for sin($\omega_0 t$), The signal $\frac{1}{\omega_0 t}$ plays the role of an envelope (hyperbolic shape) responsible for attenuation of sine oscillations.

The *two-sided exponential signal* is defined as

$$x(t) = A \cdot e^{-\alpha|t|}, \tag{1.19}$$

where A is its maximum value for $t = 0$ and α is responsible for the slopes. Figure 1.23 shows the signal plot for $\alpha = 1$ and $\alpha = 5$. Let us notice that an exponential curve asymtotically approaches zero if $t \rightarrow \pm\infty$.

The *one-sided* (right-sided) *exponential signal,* often used in system theory (e.g., [2, 4, 15]), is defined as

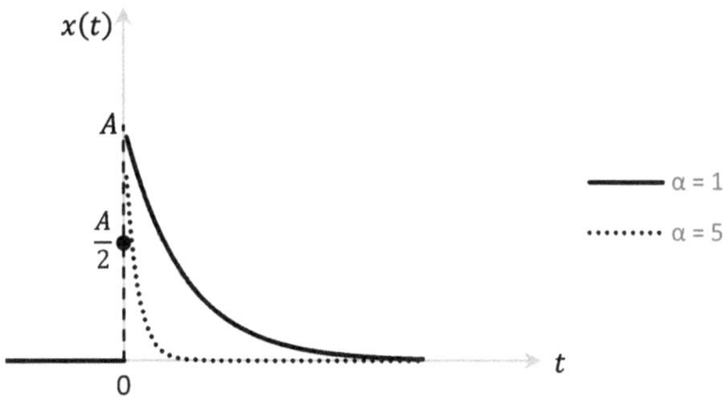

Fig. 1.24 Right-sided exponential signal

$$x(t) = A \cdot e^{-\alpha t} \cdot \mathbf{1}(t), \tag{1.20}$$

where parameters A and α play the same role as in (1.19). Figure 1.24 shows its plot for $\alpha = 1$ and $\alpha = 5$. We see that for $t \to +\infty$, $x(t)$ approaches 0.

Evidently, we can define discrete-time analogs of signals (1.17), (1.18), (1.19), and (1.20).

1.1.4 Distribution Signals

This part is dedicated to chosen distribution functions that play important role in diverse fields of signal theory. These are the Dirac delta distribution and the comb distribution.

The *Dirac delta distribution* $\delta(t)$, $t \in \mathbb{R}$ was introduced by Paul Dirac (1902–1984), the famous British theoretical physicist and Nobel laureate for the atom model theory. The Dirac delta is defined in the form

$$\delta(t) = \begin{cases} \infty & \text{for} \quad t = 0 \\ 0 & \text{elsewhere} \end{cases} \quad \text{and} \quad \int_{-\infty}^{\infty} \delta(t) dt = 1. \tag{1.21}$$

We can see that (1.21) is not an "ordinary" function since it does not assume a finite value for $t = 0$. It is represented by an arrow of length 1 fixed at $t = 0$ (see Fig. 1.25). In signal theory, it is used as a model of an infinitely short pulse signal. We will see in Chap. 2 that it also represents spectra of a specific class of signals. The Dirac delta has some interesting properties listed in Table 1.1.

The *comb distribution* $\delta_T(t)$, $t \in \mathbb{R}$ (Fig. 1.26) is a periodic signal defined as

$$\delta_T(t) = \sum\nolimits_{n \in \mathbb{Z}} \delta(t - nT), \tag{1.22}$$

where $\delta(\cdot)$ is the Dirac delta given by (1.21) and T is the fundamental period. It is used as a model of the ideal sampling signal (see Chap. 6).

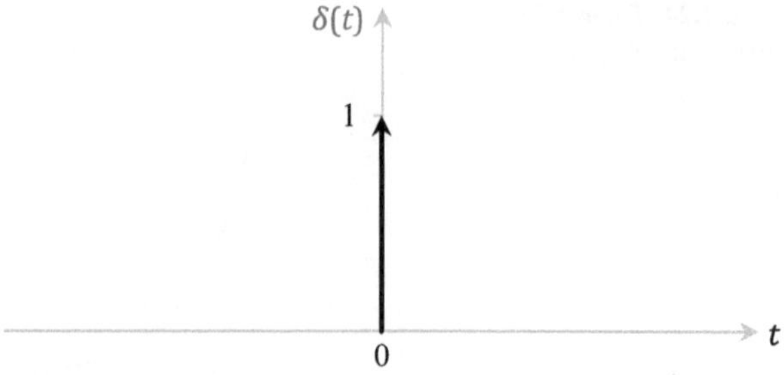

Fig. 1.25 Dirac delta distribution

Table 1.1 Properties of the Dirac delta distribution

No.	Property	Formula
1	Parity	$\delta(-t) = \delta(t)$
2	Multiplication by a constant $A \in \mathbb{R}$	$\int A\delta(t)dt = A$
3	Differentiation and integrability	$\frac{d}{dt}\mathbf{1}(t) = \delta(t)$ $\int_{-\infty}^{t}\delta(\tau)d\tau = \mathbf{1}(t)$
4	Sampling	$x(t)\delta(t - t_0) = x(t_0)\delta(t - t_0)$
5	Filtration	$\int x(t)\delta(t - t_0)dt = x(t_0)$
6	Time-scaling	$\delta(t/t_0) = \|t_0\|\delta(t)$
7	Convolution[a]	$x(t) * \delta(t - t_0) = x(t - t_0)$
8	Limit representations	$\delta(t) = \lim\limits_{\alpha \to 0} \frac{1}{\alpha}\Pi\left(\frac{t}{\alpha}\right)$ $\delta(t) = \lim\limits_{\alpha \to 0} \frac{1}{\alpha}\Lambda\left(\frac{t}{\alpha}\right)$ $\delta(t) = \lim\limits_{\alpha \to 0} \frac{1}{\alpha}\mathrm{Sa}\left(\pi\frac{t}{\alpha}\right)$ $\delta(t) = \lim\limits_{\alpha \to 0} \frac{1}{\alpha}\exp\left(-\frac{\pi t^2}{\alpha^2}\right)$

[a]The convolution of continuous-time signals is described in Sect. 1.4 of this chapter

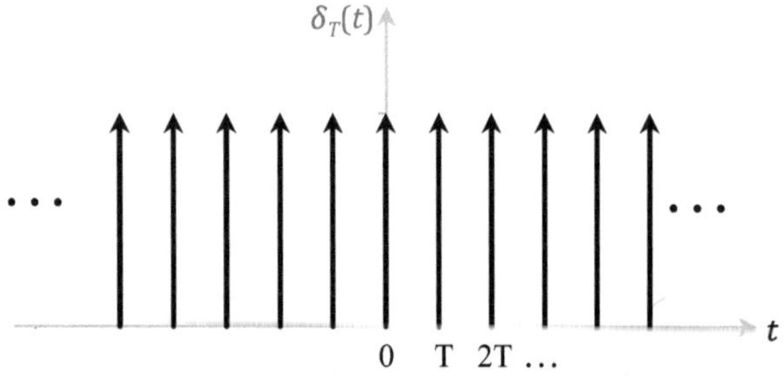

Fig. 1.26 Comb distribution

1.2 Parameters of Continuous- and Discrete-Time Signals

In time domain, signals are described by their basic parameters: average value, energy and average power. In Table 1.2 we collected general formulas of these parameters for continuous- and discrete-time signals.

Table 1.2 Parameters of continuous- and discrete-time signals

Parameter	Continuous-time $x(t)$, $t \in \mathbb{R}$	Discrete-time $x(n)$, $n \in \mathbb{Z}$
Average value	$\overline{x} = \lim\limits_{\tau \to +\infty} \frac{1}{2\tau} \int\limits_{-\tau}^{\tau} x(t)dt$	$\overline{x} = \lim\limits_{M \to +\infty} \frac{1}{2M+1} \sum_{n=-M}^{M} x(n)$
Energy	$E_x = \lim\limits_{\tau \to +\infty} \int\limits_{-\tau}^{\tau} \lvert x(t)\rvert^2 dt$	$E_x = \lim\limits_{M \to +\infty} \sum_{n=-M}^{M} \lvert x(n)\rvert^2$
Average power	$P_x = \lim\limits_{\tau \to +\infty} \frac{1}{2\tau} \int\limits_{-\tau}^{\tau} \lvert x(t)\rvert^2 dt$	$P_x = \lim\limits_{M \to +\infty} \frac{1}{2M+1} \sum_{n=-M}^{M} \lvert x(n)\rvert^2$

In signal theory, there are two important, from the practical point of view, signal classes: *energy* and *power signals*. Let us present their definitions and give some examples.

Definition 1.5 The signal $x(t)$ or $x(n)$ is called an *energy signal* if its energy E_x is finite, that is, $0 < E_x < \infty$.

Definition 1.6 The signal $x(t)$ or $x(n)$ is called a *power signal* if its average power P_x is finite, that is, $0 < P_x < \infty$.

Energy and power signals form two *separate* classes, since:

- If $0 < E_x < \infty$ then $P_x = 0$;
- If $0 < P_x < \infty$ then $E_x = \infty$.

Important The class of *energy signals* includes:

- Finite-duration signals,
- Signals for which $\lim\limits_{\tau \to +\infty} \int_{-\tau}^{\tau} \lvert x(t)\rvert^2 dt < \infty$ (resp. $\lim\limits_{M \to +\infty} \sum_{n=-M}^{M} \lvert x(n)\rvert^2 < \infty$) (e.g., Gaussian, two-sided exponential, right-sided exponential signals with $\alpha > 0$),
- Signals with attenuating oscillations (e.g., Sa(·)).

The class of *power signals* includes:

- Periodic signals,
- Signals for which $\lim\limits_{\tau \to +\infty} \frac{1}{2\tau} \int_{-\tau}^{\tau} \lvert x(t)\rvert^2 dt < \infty$ (resp. $\lim\limits_{M \to +\infty} \frac{1}{2M+1} \sum_{n=-M}^{M} \lvert x(n)\rvert^2 < \infty$) (e.g., constant, unit step, signum).

For *finite-duration* signals (Definitions 1.3 and 1.4), the *average value* $\overline{x}$ (for duration) is calculated as follows:

- For continuous-time signals:

$$\overline{x} = \frac{1}{b-a} \int_a^b x(t)dt, \tag{1.23}$$

- For discrete-time signals:

$$\overline{x} = \frac{1}{b-a+1}\sum_{n=a}^{b} x(n), \tag{1.24}$$

and the signal $x(\cdot)$ is zero outside the interval $[a, b]$.

For *finite-duration* signals, *energy* E_x can be calculated using following formulas:

- For continuous-time signals:

$$E_x = \int_a^b |x(t)|^2 dt, \tag{1.25}$$

- For discrete-time signals:

$$E_x = \sum_{n=a}^{b} |x(n)|^2, \tag{1.26}$$

and the signal $x(\cdot)$ is zero outside the interval $[a, b]$.

Problem 1.4 Calculate the average value $\overline{x}$ and energy E_x of: (a) the rectangular pulse $A\Pi\left(\frac{t}{\tau}\right)$, (b) the triangular pulse $A\Lambda\left(\frac{t}{\tau}\right)$.

Solution The rectangular pulse $A\Pi\left(\frac{t}{\tau}\right)$ is displayed in Fig. 1.9. Using (1.23) and (1.25), we immediately get

$$\overline{x} = \frac{1}{b-a}\int_a^b A dt = A,$$

$$E_x = \int_a^b A^2 dt = A^2(b-a).$$

The triangular pulse $A\Lambda\left(\frac{t}{\tau}\right)$ (Fig. 1.10) is defined by (1.6) as

$$x(t) = A\Lambda\left(\frac{t}{\tau}\right) = \begin{cases} \frac{A}{\tau}(t+\tau) & \text{for} \quad \tau \in [-\tau, 0), \\ -\frac{A}{\tau}(t-\tau) & \text{for} \quad \tau \in [0, \tau], \\ 0 & \text{elsewhere.} \end{cases}$$

Using (1.23), we calculate its average value:

$$\bar{x}=\frac{1}{2\tau}\int_{-\tau}^{\tau}x(t)dt=\frac{1}{2\tau}\left[\int_{-\tau}^{0}\frac{A}{\tau}(t+\tau)dt-\int_{0}^{\tau}\frac{A}{\tau}(t-\tau)dt\right]$$
$$=\frac{A}{2\tau^2}\left[\left(\frac{t^2}{2}+\tau t\right)\Big|_{-\tau}^{0}-\left(\frac{t^2}{2}-\tau t\right)\Big|_{0}^{\tau}\right]=\frac{A}{2\tau^2}\left(\frac{\tau^2}{2}+\frac{\tau^2}{2}\right)=\frac{A}{2}.$$

The energy of a triangular pulse is calculated using (1.25). We have

$$E_x=\int_{-\tau}^{0}\left[\frac{A}{\tau}(t+\tau)\right]^2dt+\int_{0}^{\tau}\left[-\frac{A}{\tau}(t-\tau)\right]^2dt=2\int_{0}^{\tau}\frac{A^2}{\tau^2}(t-\tau)^2dt$$
$$=2\frac{A^2}{\tau^2}\int_{0}^{\tau}(t^2-2t\tau+\tau^2)dt=2\frac{A^2}{\tau^2}\left(\frac{t^3}{3}-\tau t^2+\tau^2 t\right)\Big|_{0}^{\tau}=\frac{2}{3}A^2\tau.$$

For *periodic* signals (Definitions 1.1 and 1.2), the *average value* $\bar{x}$ (for period) is calculated as follows:

- For continuous-time signals:

$$\bar{x}=\frac{1}{T_0}\int_{\langle T_0\rangle}x(t)dt, \tag{1.27}$$

- For discrete-time signals:

$$\bar{x}=\frac{1}{N_0}\sum\nolimits_{\langle N_0\rangle}x(n), \tag{1.28}$$

where $\int_{\langle T_0\rangle} x(t)$ means integration over one period T_0. In practice, limits of the above integral can be arbitrarily chosen, for example, from $-\frac{T_0}{2}$ to $\frac{T_0}{2}$, from 0 to T_0, etc. Similarly, $\sum_{\langle N_0\rangle} x(n)$ means summation over one period N_0.

For *periodic* signals, the average power (for period) is given by

- For continuous-time signals:

$$P_x=\frac{1}{T_0}\int_{\langle T_0\rangle}|x(t)|^2dt, \tag{1.29}$$

- For discrete-time signals:

$$P_x = \frac{1}{N_0} \sum_{\langle N_0 \rangle} |x(n)|^2. \tag{1.30}$$

Problem 1.5 Calculate the average power P_x of the constant signal (1.9): $x(t) = A$ and of the harmonic signal (1.1): $x(t) = A\cos(\omega_0 t + \varphi_0)$.

Solution The average power of the constant signal $x(t) = A$ can be calculated using the formula from Table 1.2, that is,

$$P_x = \lim_{\tau \to +\infty} \frac{1}{2\tau} \int_{-\tau}^{\tau} |x(t)|^2 dt = \lim_{\tau \to +\infty} \frac{1}{2\tau} \int_{-\tau}^{\tau} A^2 dt = A^2.$$

The average power of the harmonic signal $x(t) = A\cos(\omega_0 t + \varphi_0)$ is calculated using (1.29) as follows:

$$\begin{aligned} P_x &= \frac{1}{T_0} \int_{\langle T_0 \rangle} |x(t)|^2 dt = \frac{1}{T_0} \int_0^{T_0} A^2 \cos^2(\omega_0 t + \varphi_0) dt \\ &= \frac{A^2}{T_0} \int_0^{T_0} \frac{1 + \cos[2(\omega_0 t + \varphi_0)]}{2} dt = \frac{A^2}{2T_0} \left(t\big|_0^{T_0} + \frac{\sin[2(\omega_0 t + \varphi_0)]}{2\omega_0} \bigg|_0^{T_0} \right) \\ &= \frac{A^2}{2} + \frac{A^2}{4\omega_0 T_0} \underbrace{[\sin(2\omega_0 T_0 + \varphi_0) - \sin(\varphi_0)]}_{=0} = \frac{A^2}{2}. \end{aligned}$$

Important The average power of a constant signal equals the square of its value, and the average power of a harmonic signal equals half of the square of its amplitude.

Problem 1.6 Given the discrete-time signal $x(n) = 2 + \cos\left(\frac{\pi}{6} n^2\right), n \in \mathbb{Z}$. Calculate its average power (for period).

Solution To verify that $x(n)$ is periodic, we can apply the periodicity condition from Definition 1.2:

$$x(n+N_0)=2+\cos\left[\frac{\pi}{6}(n+N_0)^2\right]=2+\cos\left[\frac{\pi}{6}\left(n^2+2nN_0+N_0^2\right)\right]$$
$$=2+\cos\left(\frac{\pi}{6}n^2+\frac{n\pi N_0}{3}+\frac{\pi N_0^2}{6}\right).$$

We only need to check when the above expression equals $x(n)=2+\cos\left(\frac{\pi}{6}n^2\right)$. It is possible if and only if N_0 is the smallest multiple of 6, that is $N_0=6$. Then for every $n\in\mathbb{Z}$, terms $\frac{n\pi N_0}{3}$ and $\frac{\pi N_0^2}{6}$ are multiples of 2π and $x(n)$ is periodic. The graph of $x(n)$ is displayed in Fig. 1.27.

The average power of $x(n)$ equals a sum of average powers: P_1 of the constant signal equal 2 and P_2 of $\cos\left(\frac{\pi}{6}n^2\right)$. Applying the appropriate formula from Table 1.2, we obtain:

$$P_1=\lim_{M\to+\infty}\frac{1}{2M+1}\sum_{n=-M}^{M}2^2=4\frac{2M+1}{2M+1}=4 \qquad \text{(i.e., the square of 2)}.$$

Then using (1.30), we get

$$P_2=\frac{1}{6}\sum_{n=0}^{5}\cos^2\left(\frac{\pi}{6}n^2\right)$$
$$=\frac{1}{6}\left[1+\cos^2\left(\frac{\pi}{6}\right)+\cos^2\left(\frac{2\pi}{3}\right)+\cos^2\left(\frac{3\pi}{2}\right)+\cos^2\left(\frac{8\pi}{3}\right)+\cos^2\left(\frac{25\pi}{6}\right)\right]$$
$$=\frac{1}{6}\left(1+\frac{3}{4}+\frac{1}{4}+0+\frac{1}{4}+\frac{3}{4}\right)=\frac{1}{2},$$

which exactly equals half of squared amplitude $=1$. We can see that results from Problem 1.5 concerning continuous-time signals can be easily adapted for discrete-time periodic signals. Finally, the average power of $x(n)=2+\cos\left(\frac{\pi}{6}n^2\right)$ is 4,5.

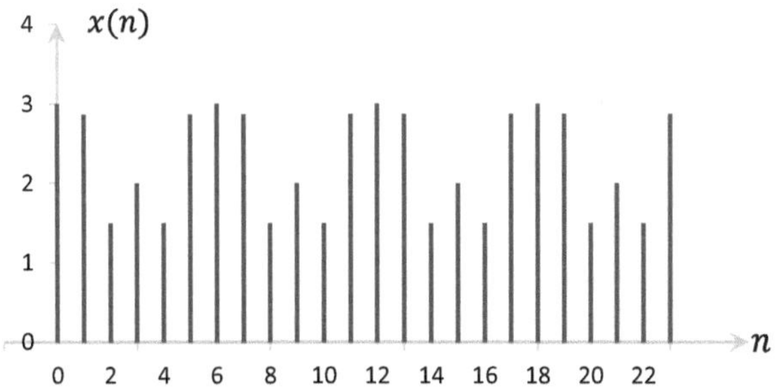

Fig. 1.27 Periodic signal $x(n)$ from Problem 1.6

1.3 Operations on Signals in Time Domain

In this section, we present basic mathematical operations performed on signals in continuous- and discrete-time domains (see Table 1.3). These are: simple operations on two or more signals as summation or multiplication; geometric operations (shift, time-reversal, scaling); derivation or integration and more advanced: convolution and cross-correlation of two signals.

Let us now introduce notions of even and odd components of a signal defined by means of basic operations of summation, time-reversal, and multiplication by a constant.

Definition 1.7 The *even component* $x_e(t)$ of a continuous-time signal $x(t)$, $t \in \mathbb{R}$ is defined as

$$x_e(t) = \frac{x(t) + x(-t)}{2}. \tag{1.31}$$

The *odd component* $x_o(t)$ of a continuous-time signal $x(t)$, $t \in \mathbb{R}$ is defined as

$$x_o(t) = \frac{x(t) - x(-t)}{2}. \tag{1.32}$$

Evidently we have $x(t) = x_e(t) + x_o(t)$. The components (1.31) and (1.32) are *orthogonal*, that is, $\int x_e(t) x_o(t) dt = 0$.

Problem 1.7 Given the signal $x(t) = \Pi\left(\frac{t-1/2}{2}\right), t \in \mathbb{R}$. Sketch plots of its even and odd components. Calculate the energies of both components $x_e(t)$ and $x_o(t)$ to prove that their sum equals the energy of $x(t)$. Check orthogonality of $x_e(t)$ and $x_o(t)$.

Solution The even component (1.31) is defined as a weighted (by 0,5) sum of $x(t)$ and $x(-t)$. We deal with three simple signal operations: time reversal, summation, and multiplication by a constant (see Table 1.3). The problem can easily be solved graphically. First, let us plot the graph of $x(t) = \Pi\left(\frac{t-1/2}{2}\right)$ (Fig. 1.28). Then the graph of $x(-t)$ is displayed in Fig. 1.29.

Applying (1.31), we can sketch the even component $x_e(t)$ as it is shown in Fig. 1.30. The energy E_{x_e} of $x_e(t)$ is calculated using (1.25) as follows:

Table 1.3 Operations on continuous- and discrete-time signals

Operation		Continuous-time, $t \in \mathbb{R}$	Discrete-time, $n \in \mathbb{Z}$	Remarks
Summation		$x_1(t) + x_2(t)$	$x_1(n) + x_2(n)$	
Multiplication		$x_1(t) \cdot x_2(t)$	$x_1(n) \cdot x_2(n)$	
	Multiplication by a constant $A \in \mathbb{R}$	$A \cdot x(t)$	$A \cdot x(n)$	If $A > 1$—*signal amplification*; If $A \in (0, 1)$—*signal attenuation*
	Multiplication by unit step	$x(t) \cdot \mathbf{1}(t)$	$x(n) \cdot \mathbf{1}(n)$	$x(\cdot)$ *causal*
	Multiplication by a window function	$x(t) \cdot w(t)$,	$x(n) \cdot w(n)$,	*Windowing* $w(\cdot)$—a window function
	Multiplication by a cosine/sine signal	$x(t) \cdot \cos(\omega_0 t)$ $x(t) \cdot \sin(\omega_0 t)$	$x(n) \cdot \cos(\omega_0 n)$ $x(n) \cdot \sin(\omega_0 n)$	*Modulation*
	Multiplication by "delta"	$x(t) \cdot \delta(t - t_0) = x(t_0) \cdot \delta(t - t_0)$, $\delta(t)$—Dirac delta distribution	$x(n) \cdot \delta(n - n_0) = x(n_0) \cdot \delta(n - n_0)$, $\delta(n)$—Unit delta pulse	
Shift		$x(t - t_0)$, $t_0 \in \mathbb{R}$	$x(n - n_0)$, $n_0 \in \mathbb{Z}$	If $t_0 > 0$ or $n_0 > 0$—*Right shift*; If $t_0 < 0$ or $n_0 < 0$—*Left shift*
Time-reversal		$x(-t)$	$x(-n)$	
Scaling		$x(a \cdot t)$, $a \in \mathbb{R}$	The signal can only be expanded, which is equivalent to adding a number of zeros between its every sample.	If $a > 1$—*Time compression*; If $a \in (0, 1)$—*Time expansion*
Derivation		$\frac{d^n x(t)}{dt^n}$	–	
Integration		$\int_{-\infty}^{t} x(\tau) d\tau$	–	
Convolution		$x_1(t) * x_2(t)$	$x_1(n) * x_2(n)$	See Sect. 1.4
Cross-correlation		$x_1(t) \odot x_2(t)$	$x_1(n) \odot x_2(n)$	See Sect. 1.4

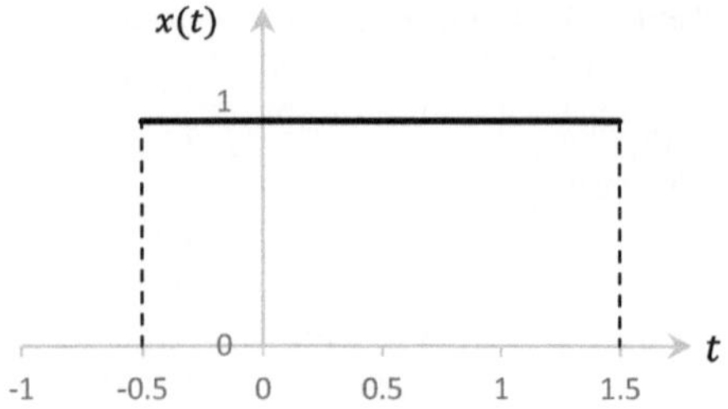

Fig. 1.28 Plot of $x(t)$ from Problem 1.7

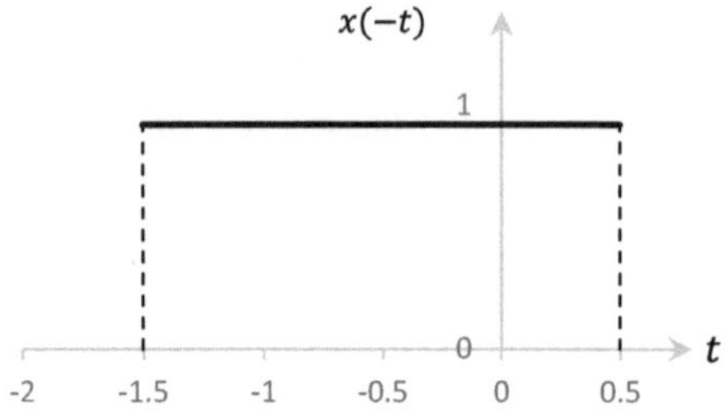

Fig. 1.29 Time-reversal of $x(t)$ from Fig. 1.28

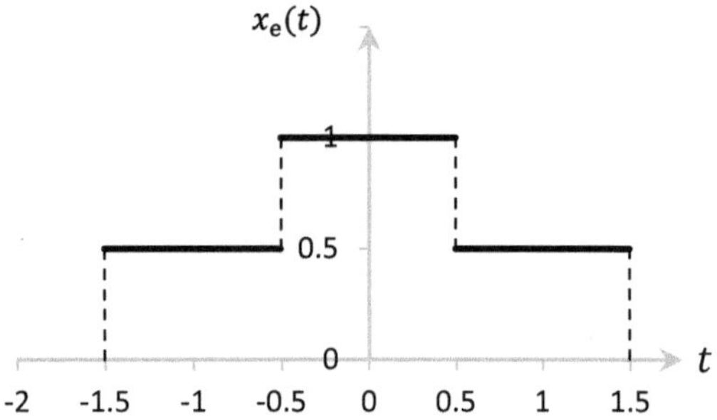

Fig. 1.30 Even component $x_e(t)$ of $x(t)$ from Problem 1.7

$$E_{x_e} = \int_{-1,5}^{1,5} x_e^2(t)dt = 2\left[\int_{0}^{0,5} dt + \int_{0,5}^{1,5} 0{,}25dt\right] = 1,5.$$

The odd component $x_o(t)$ is displayed in Fig. 1.31. Its energy

$$E_{x_0} = \int_{-1,5}^{1,5} x_o^2(t)dt = 2\int_{0,5}^{1,5} 0,25dt = 0,5.$$

The sum $E_{x_e} + E_{x_0} = 2$ equals the energy E_x of $x(t) = \Pi\left(\frac{t-1/2}{2}\right)$. Let us check the orthogonality of both components. We have

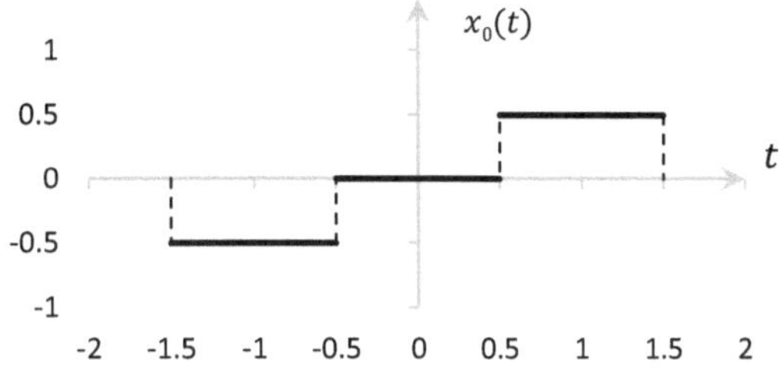

Fig. 1.31 Odd component $x_o(t)$ of $x(t)$ from Problem 1.7

$$\int_{-1,5}^{1,5} x_e(t)x_o(t)dt = \int_{-1,5}^{-0,5}\left[\frac{1}{2}\cdot\left(-\frac{1}{2}\right)\right]dt + \int_{-0,5}^{0,5}(1\cdot 0)dt + \int_{0,5}^{1,5}\left(\frac{1}{2}\cdot\frac{1}{2}\right)dt$$
$$= -\frac{1}{4}+\frac{1}{4}=0.$$

In analogy to (1.31) and (1.32) we can define even and odd components of discrete-time signals.

Definition 1.8 The *even component* $x_e(n)$ of a discrete-time signal $x(n)$, $n \in \mathbb{Z}$ is defined as

$$x_e(n) = \frac{x(n) + x(-n)}{2}. \tag{1.33}$$

The *odd component* $x_o(n)$ of a discrete-time signal $x(n)$, $n \in \mathbb{Z}$ is defined as

$$x_o(n) = \frac{x(n) - x(-n)}{2}. \tag{1.34}$$

Evidently we have $x(n) = x_e(n) + x_o(n)$. The components (1.33) and (1.34) are *orthogonal*, that is, $\sum_{n\in\mathbb{Z}} x_e(n)x_o(n) = 0$.

Problem 1.8 Given the discrete-time signal $x(n) = 3\cdot\delta(n) + 2\cdot\delta(n+1) + \delta(n+2)$, $n \in \mathbb{Z}$. Sketch plots of its even and odd components $x_e(n)$ and $x_o(n)$. Calculate the energies of both components to prove that their sum equals the energy of $x(n)$. Check orthogonality of $x_e(n)$ and $x_o(n)$.

Solution We deal with a finite-duration signal shown in Fig. 1.32. In analogy to the solution of Problem 1.7, we reverse $x(n)$ in time to get $x(-n)$ (Fig. 1.33), and then using (1.33) and (1.34), we can plot graphs of $x_e(n)$ and $x_o(n)$ (Figs. 1.34 and 1.35).

Using (1.26), we can calculate the energies of even and odd components as follows:

$$E_{x_e} = \frac{1}{4} + 1 + 9 + 1 + \frac{1}{4} = 11,5$$

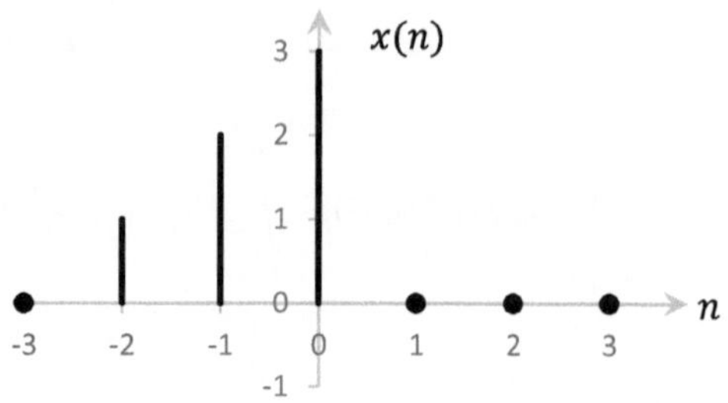

Fig. 1.32 Plot of $x(n)$ from Problem 1.8

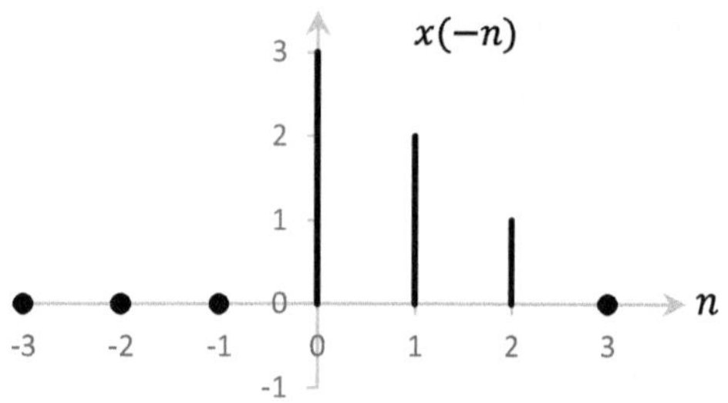

Fig. 1.33 Time-reversal of the signal from Fig. 1.32

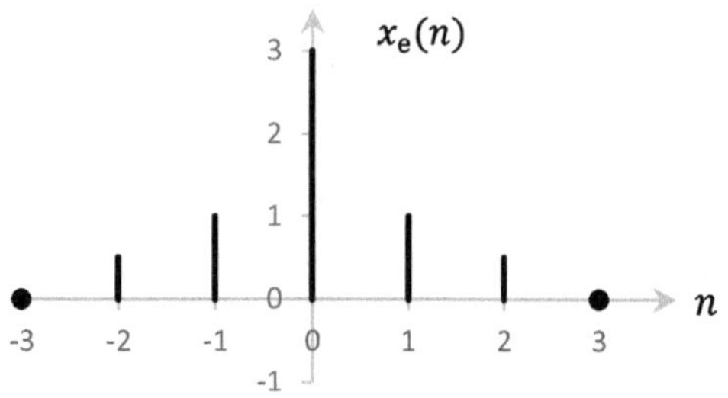

Fig. 1.34 Even component $x_e(n)$ of signal from Fig. 1.32

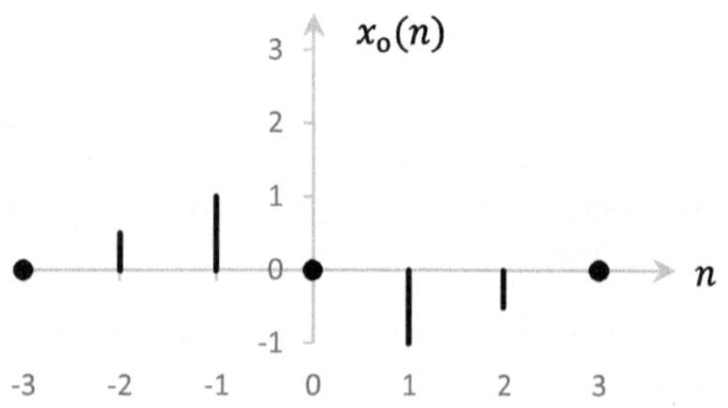

Fig. 1.35 Odd component $x_o(n)$ of signal from Fig. 1.32

$$E_{x_o} = \frac{1}{4} + 1 + 1 + \frac{1}{4} = 2,5.$$

The energy of $x(n)$ is $E_x = 1 + 4 + 9 = 14$ and equals $E_{x_e} + E_{x_o}$.
The orthogonality criterion for $x_e(n)$ and $x_o(n)$ is fulfilled because

$$\sum_n x_e(n)x_o(n) = \frac{1}{2} \cdot \frac{1}{2} + 1 \cdot 1 + 1 \cdot (-1) + \frac{1}{2} \cdot \left(-\frac{1}{2}\right) = 0.$$

Problem 1.9 Given the plot of a finite-duration signal $x(t)$ shown in Fig. 1.36. Express $x(t)$ in a general form

$$x(t) = y_1(t) \cdot \mathbf{1}(t - t_1) + y_2(t) \cdot \mathbf{1}(t - t_2) + y_3(t) \cdot \mathbf{1}(t - t_3),$$

where $y_i(t)$, $i = 1, 2, 3$ are linear functions and $\mathbf{1}(t - t_i)$, $i = 1, 2, 3$ are unit step signals shifted in time by t_i. Calculate the derivative $\frac{dx(t)}{dt}$ and plot its graph.

Solution The signal $x(t)$ can be written as

$$x(t) = \begin{cases} -t-1 & \text{for} \quad t \in [-1,0), \\ t-1 & \text{for} \quad t \in [0,2), \\ 0 & \text{elsewhere} \end{cases}$$

or equivalently as

$$x(t) = (-t-1) \cdot [\mathbf{1}(t+1) - \mathbf{1}(t)] + (t-1) \cdot [\mathbf{1}(t) - \mathbf{1}(t-2)].$$

In the above sum, we can notice that $[\mathbf{1}(t + 1) - \mathbf{1}(t)]$ is a rectangular window function that "cuts out" from an infinite-duration signal $(-t - 1)$ the fragment limited to $t \in [-1, 0]$. Of course, we can write

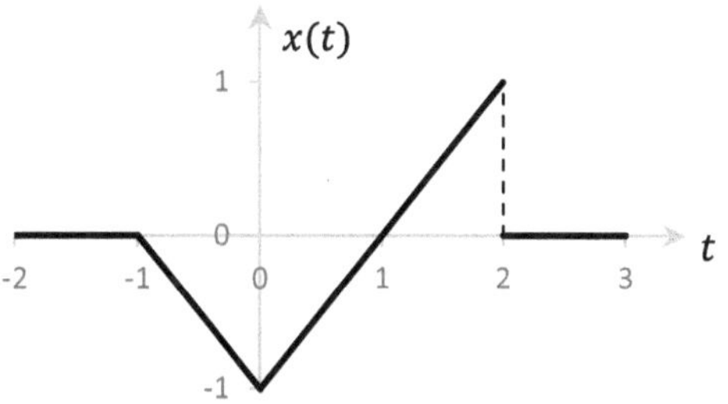

Fig. 1.36 Signal $x(t)$ from Problem 1.9

$$\mathbf{1}(t+1)-\mathbf{1}(t)=\Pi\left(t+\frac{1}{2}\right).$$

The same applies for $[\mathbf{1}(t)-\mathbf{1}(t-2)]=\Pi\left(\frac{t-1}{2}\right)$. Let us develop the above sum as follows:

$$\begin{aligned}x(t)&=(-t-1)\cdot\mathbf{1}(t+1)-(-t-1)\cdot\mathbf{1}(t)+(t-1)\cdot\mathbf{1}(t)-(t-1)\cdot\mathbf{1}(t-2)=\\&=(-t-1)\cdot\mathbf{1}(t+1)+(t+1)\cdot\mathbf{1}(t)+(t-1)\cdot\mathbf{1}(t)+(-t+1)\cdot\mathbf{1}(t-2)=\\&=(-t-1)\cdot\mathbf{1}(t+1)+2t\cdot\mathbf{1}(t)+(-t+1)\cdot\mathbf{1}(t-2).\end{aligned}$$

Comparing it with the general formula $x(t)=y_1(t)\cdot\mathbf{1}(t-t_1)+y_2(t)\cdot\mathbf{1}(t-t_2)+y_3(t)\cdot\mathbf{1}(t-t_3)$, we get

$$y_1(t)=(-t-1),t_1=-1,\ \ y_2(t)=2t,t_2=0 \text{ and } y_3(t)=(-t+1),t_3=2.$$

Having in mind that $\frac{d}{dt}\mathbf{1}(t)=\delta(t)$ (see Table 1.1), let us now calculate the derivative $\frac{dx(t)}{dt}$:

$$\begin{aligned}\frac{dx(t)}{dt}&=\frac{d}{dt}\{(-t-1)\cdot\mathbf{1}(t+1)+2t\cdot\mathbf{1}(t)+(-t+1)\cdot\mathbf{1}(t-2)\}\\&=\frac{d}{dt}\{(-t-1)\cdot\mathbf{1}(t+1)\}+\frac{d}{dt}\{2t\cdot 1(t)\}+\frac{d}{dt}\{(-t+1)\cdot\mathbf{1}(t-2)\}\\&=-\mathbf{1}(t+1)+(-t-1)\cdot\delta(t+1)+2\cdot\mathbf{1}(t)+2t\cdot\delta(t)-\mathbf{1}(t-2)\\&\quad+(-t+1)\cdot\delta(t-2).\end{aligned}$$

Now, let us apply the sampling property of the Dirac distribution: $x(t)\delta(t-t_0)=x(t_0)\delta(t-t_0)$. We have

$$\begin{aligned}\frac{dx(t)}{dt}&=-\mathbf{1}(t+1)+0\cdot\delta(t+1)+2\cdot\mathbf{1}(t)+0\cdot\delta(t)-\mathbf{1}(t-2)+(-1)\cdot\delta(t-2)\\&=-\mathbf{1}(t+1)+2\cdot\mathbf{1}(t)-\mathbf{1}(t-2)-\delta(t-2).\end{aligned}$$

To sketch the plot of the derivative $\frac{dx(t)}{dt}$ (see Fig. 1.37), we start from the left with the plot of $-\mathbf{1}(t+1)$ (left time shift of $\mathbf{1}(t)$ by 1 and multiplication by a constant $=-1$). Then at $t=0$, we add to $-\mathbf{1}(t+1)$ the signal $2\cdot\mathbf{1}(t)$ to obtain the constant signal $=1$ for $t\in(0,2)$. At $t=2$, we have to subtract $\mathbf{1}(t-2)$ and get 0. Since there is a discontinuity in $x(t)$ for $t=2$ (in other words, a "jump down" of the plot of $x(t)$), in the derivative $\frac{dx(t)}{dt}$, we have $-\delta(t-2)$ represented by the arrow of length -1 (marked in red) fixed at $t=2$.

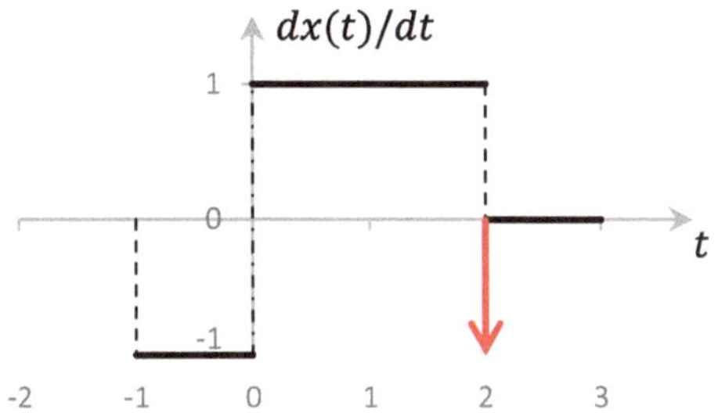

Fig. 1.37 Derivative of the signal $x(t)$ from Problem 1.9

1.4 Convolution and Correlation

This section is devoted to convolution and correlation of continuous- and discrete-time signals. A convolution is a basic operation used in analog and digital signal filtering (Chaps. 3, 4, 5, 6, 7, and 8), while auto- and cross-correlation functions are useful, for example, in telecommunications, radars, etc. [12, 13].

Definition 1.9 A *continuous-time convolution* of two signals $x_1(t)$ and $x_2(t)$, $t \in \mathbb{R}$ is defined by the integral

$$x_1(t) * x_2(t) = \int x_1(\tau)x_2(t-\tau)d\tau. \tag{1.35}$$

Since convolution is *commutative*, that is, $x_1(t) * x_2(t) = x_2(t) * x_1(t)$, we can use the equivalent definition as follows:

$$x_1(t) * x_2(t) = \int x_2(\tau)x_1(t-\tau)d\tau. \tag{1.36}$$

Definition 1.10 A *discrete-time convolution* of two signals $x_1(n)$ and $x_2(n)$, $n \in \mathbb{Z}$ is defined by a sum

$$x_1(n) * x_2(n) = \sum\nolimits_{k\in\mathbb{Z}} x_1(k)x_2(n-k). \tag{1.37}$$

Evidently, $x_1(n) * x_2(n) = x_2(n) * x_1(n)$, so (1.37) can be rewritten as

$$x_1(n) * x_2(n) = \sum\nolimits_{k\in\mathbb{Z}} x_2(k)x_1(n-k). \tag{1.38}$$

The properties of convolution are collected in Table 1.4.

The continuous- and discrete-time convolution can be calculated in three different ways: directly from formulas (1.35), (1.36), (1.37), and (1.38), by a geometrical method (see Problems 1.10 and 1.11), or using convolution theorems concerning Fourier-, Laplace-, and $\mathcal{Z}$-transformations presented in Chapters 2, 5, and 8 of this book.

Table 1.4 Properties of convolution

No.	Property	Formula
1	Commutativity	$x_1 * x_2 = x_2 * x_1$
2	Associativity	$x_1 * (x_2 * x_3) = (x_1 * x_2) * x_3$
3	Distributiveness over addition	$x_1 * (x_2 + x_3) = x_1 * x_2 + x_1 * x_3$
4	Convolution with Dirac distribution $\delta(t)$ (1.21)	$x(t) * \delta(t) = x(t)$
	Convolution with time-shifted Dirac distribution $\delta(t - t_0)$	$x(t) * \delta(t - t_0) = x(t - t_0)$
5	Convolution with unit delta pulse $\delta(n)$ (1.7)	$x(n) * \delta(n) = x(n)$
	Convolution with time-shifted unit pulse $\delta(n - n_0)$	$x(n) * \delta(n - n_0) = x(n - n_0)$

The geometrical method of calculation of a continuous-time convolution consists of three steps:

1. Since τ is the integral variable in (1.35) and (1.36), we should sketch plots of x_1 and x_2 as functions of τ.
2. Depending on which formula, (1.35) or (1.36), we choose, we sketch the plot of $x_2(-\tau)$ or $x_1(-\tau)$ (time-reversal operation, see Table 1.3).
3. We shift the plot of $x_2(-\tau)$ or $x_1(-\tau)$ in τ-domain by t and calculate the integral of $x_1(\tau)x_2(t - \tau)$ or $x_2(\tau)x_1(t - \tau)$, respectively. In other words, to calculate a convolution according to (1.35), we shift $x_2(-\tau)$ right and/or left while $x_1(\tau)$ is "fixed". The calculated area under the product $x_1(\tau)x_2(t - \tau)$ or $x_2(\tau)x_1(t - \tau)$ is a convolution for a given t

Let us explain this method solving Problem 1.10.

Problem 1.10 Using the geometrical method, compute the continuous-time convolution $y(t) = x_1(t) * x_2(t)$ of two signals: $x_1(t) = 2\Pi\left(\frac{t}{6}\right)$ and $x_2(t) = 3\Pi\left(\frac{t-3}{2}\right)$. Sketch the graph of $y(t)$.

Solution Let us sketch graphs of $x_1(\tau) = 2\Pi\left(\frac{\tau}{6}\right)$ and $x_2(\tau) = 3\Pi\left(\frac{\tau-3}{2}\right)$ as shown in Fig. 1.38. In this case, we can apply the formula (1.35), so we have to sketch the graph of $x_2(-\tau)$ (Fig. 1.38).

Then we come to the step 3 of the graphical procedure (Fig. 1.39). First, we shift $x_2(-\tau)$ *extreme* left by t in the τ-domain. Scenario No. 1 shows the situation when the "red pulse" "touches" the "black one" on the left, that is, $-2 + t = -3 \Leftrightarrow t = -1$). Evidently, the convolution integral is zero. Then we start to move the "red pulse" right till the moment, when it is "inside" the "black one" (Scenario No. 2), that is, $-4 + t = -3 \Leftrightarrow t = 1$). Then we have situation No. 3, when the "red one" is sliding inside the "black one" till the moment when $-2 + t = 3 \Leftrightarrow t = 5$. In this situation, the convolution integral is constant and equals 12. In the next scenario No. 4, both rectangles overlap to the right. Continuing shifting right, we come to situation

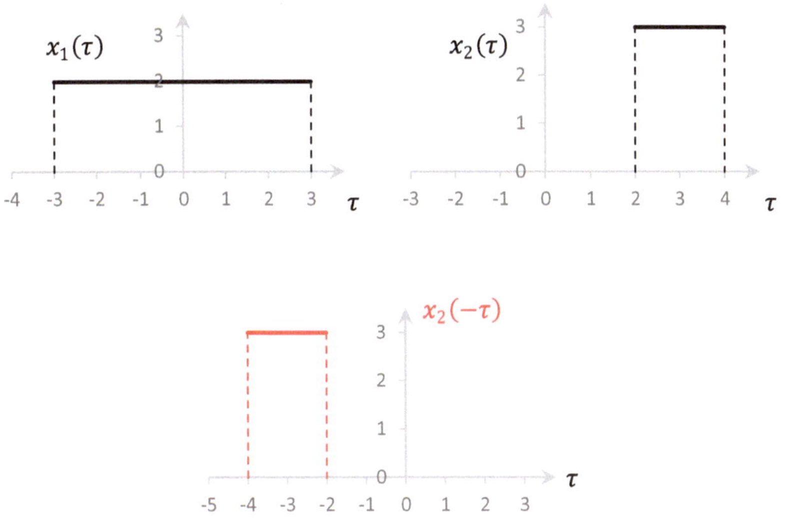

Fig. 1.38 Visualization of the first two steps of the graphical procedure for determining the convolution of two signals from Problem 1.10

No. 5, when the "red pulse" touches the "black one" on the right, that is, $-4 + t = 3 \Leftrightarrow t = 7$. The convolution integral equals zero once again.

Let us summarize the obtained results. The convolution $y(t)$ displayed in Fig. 1.40 is the following:

$$y(t) = \begin{cases} 6(t+1) & \text{for} \quad t \in (-1, 1] \\ 12 & \text{for} \quad t \in (1, 5] \\ 6(7-t) & \text{for} \quad t \in (5, 7] \\ 0 & \text{for} \quad \mathrm{t} \in (-\infty, -1] \cup (7, +\infty). \end{cases}$$

The graphical method can be applied for calculation of a convolution in the discrete-time domain. It looks similar to the procedure described above and also consists of three steps:

1. We plot graphs of x_1 and x_2 as functions of k, since k is the summation variable in (1.37) and (1.38);
2. Depending on which formula, (1.37) or (1.38), we choose, we plot the graph of $x_2(-k)$ or $x_1(-k)$ (time-reversal operation, see Table 1.3);
3. We shift $x_2(-k)$ or $x_1(-k)$ right and/or left by the step $n \in \mathbb{Z}$ and calculate the sum of products $x_1(k)x_2(n - k)$ or $x_2(k)x_1(n - k)$ for every k. In other words, to calculate a convolution according to (1.37), we shift $x_2(-k)$ right and/or left while $x_1(k)$ is "fixed". The calculated sum of products $x_1(k)x_2(n - k)$ equals the convolution for a given n.

Fig. 1.39 Visualization of five different situations representing the third step of the graphical procedure for determining the convolution of two signals from Problem 1.10

No.	Graphical representation and computation of $y(t)$
1	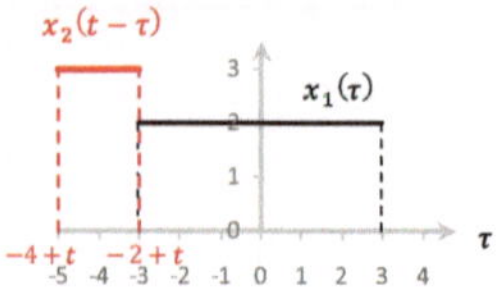 $t \in (-\infty, -1]$: $y(t) = 0$
2	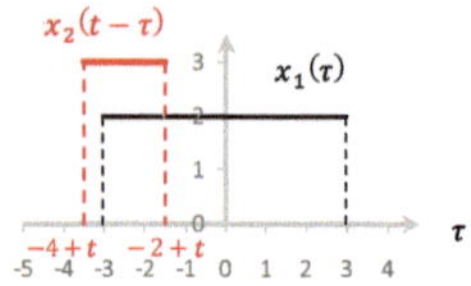 $t \in (-1, 1]$: $y(t) = \int_{-3}^{-2+t} 6\,d\tau = 6(1+t)$
3	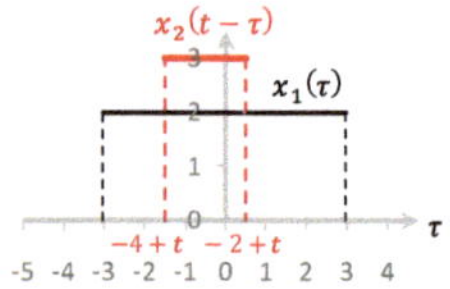 $t \in (1, 5]$: $y(t) = \int_{-4+t}^{-2+t} 6\,d\tau = 12$
4	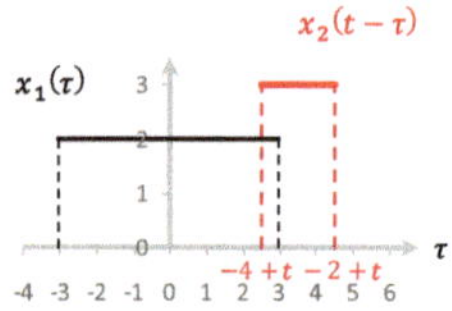 $t \in (5, 7]$: $y(t) = \int_{-4+t}^{3} 6\,d\tau = 6(7-t)$
5	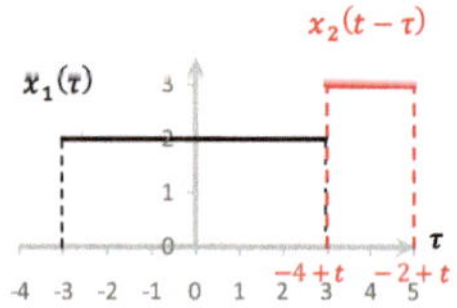 $t \in (7, +\infty)$: $y(t) = 0$

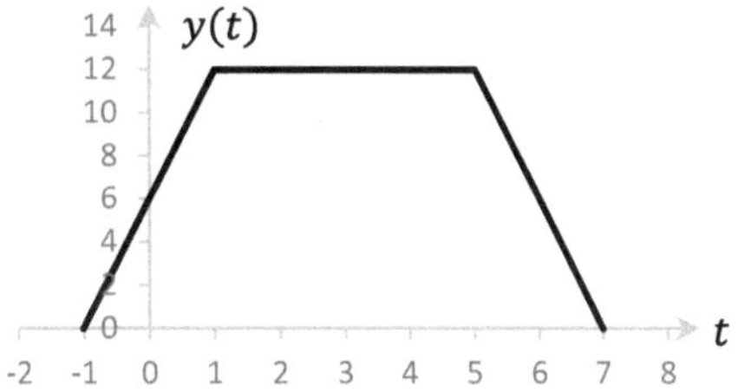

Fig. 1.40 Convolution $y(t)$ of two signals from Problem 1.10

Let us explain this method solving Problem 1.11.

Problem 1.11 Using the geometrical method calculate the discrete-time convolution $y(n) = x_1(n) * x_2(n)$ of two signals: $x_1(n) = \delta(n) + 2\delta(n-1) + 3\delta(n-2)$ and $x_2(n) = \delta(n+1) + \delta(n-1)$. Plot the graph of $y(n)$.

Solution Let us apply (1.38). We start with graphs of $x_1(k)$ and $x_2(k)$ as shown in Fig. 1.41. So, we have to reverse $x_1(k)$ in time to get $x_1(-k)$ (see Fig. 1.41).

The next step is shifting $x_1(-k)$ left by $1, 2, \ldots.$ to obtain $x_1(-1-k)$, $x_1(-2-k)$, etc. Let us notice that for $n \leq -2$, $x_2(k)x_1(n-k)$ is zero. Then, we have to shift $x_1(-k)$ right by $1, 2, \ldots.$ to obtain $x_1(1-k)$, $x_1(2-k)$, $x_1(3-k)$. For $n \geq 4$, $y(n) = 0$. All these steps are visualized in Fig. 1.42.

The graph of $y(n)$ is displayed in Fig. 1.43.

Important The convolution $y(n) = x_1(n) * x_2(n)$ of two signals from Problem 1.11, $x_1(n) = \delta(n) + 2\delta(n-1) + 3\delta(n-2)$ and $x_2(n) = \delta(n+1) + \delta(n-1)$, can be calculated using an alternative method.

Notice that $y(n) = x_1(n) * [\delta(n+1) + \delta(n-1)]$. Using distributiveness over addition for convolution, we can write

$$y(n) = x_1(n) * \delta(n+1) + x_1(n) * \delta(n-1).$$

Convolution with time-shifted unit pulse $\delta(n-n_0)$ (see Table 1.4) permits to write

$$y(n) = x_1(n+1) + x_1(n-1).$$

Since $x_1(n+1) = \delta(n+1) + 2\delta(n) + 3\delta(n-1)$ and $x_1(n-1) = \delta(n-1) + 2\delta(n-2) + 3\delta(n-3)$, we obtain

$$y(n) = \delta(n+1) + 2\delta(n) + 4\delta(n-1) + 2\delta(n-2) + 3\delta(n-3),$$

which is in accordance with the plot from Fig. 1.43.

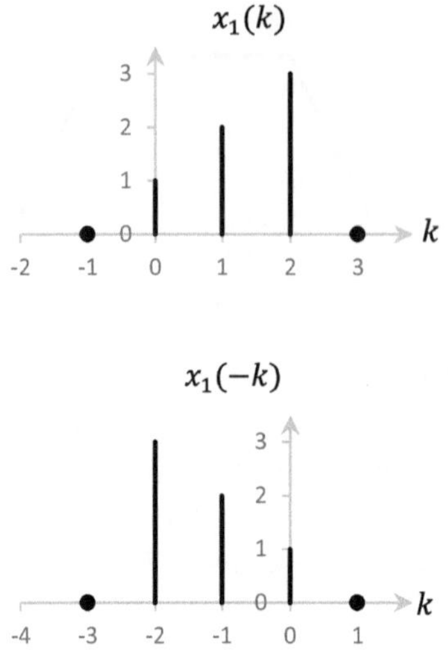

Fig. 1.41 Visualization of the first two steps of the graphical procedure for computing a discrete-time convolution of two signals from Problem 1.11

In the next Problem 1.12, we will analyze a special case of the convolution of a finite-duration signal $z(t)$ with the comb distribution $\delta_{T_0}(t) = \sum_{n \in \mathbb{Z}} \delta(t - nT_0)$ defined in (1.22). We have

$$z(t) * \delta_{T_0}(t) =$$
$$z(t) * \sum_{n \in \mathbb{Z}} \delta(t - nT_0) = \sum_{n \in \mathbb{Z}} z(t) * \delta(t - nT_0) = \sum_{n \in \mathbb{Z}} z(t - nT_0). \quad (1.39)$$

We can see that a result of the convolution of a finite-duration signal $z(t)$ with the comb distribution is a *periodic repetition* of $z(t)$ with period T_0. In other words, any periodic signal can be defined as a periodic repetition of an arbitrary chosen pulse signal under the condition that the time duration of $z(t)$ is smaller than T_0 (in some cases it can be equal T_0).

Problem 1.12 Determine the convolution $x(t) = z(t) * 2\sum_{n \in \mathbb{Z}} \delta(t - 4n)$, where $z(t) = \Pi(t)$. Sketch the graph of $x(t)$. Compute values of the following parameters of $x(t)$ and $z(t)$: average value, energy, average power.

Solution Applying (1.39) for $x(t)$, we get

$$x(t) = z(t) * 2\sum_{n \in \mathbb{Z}} \delta(t - 4n) = 2\sum_{n \in \mathbb{Z}} z(t - 4n) = 2\sum_{n \in \mathbb{Z}} \Pi(t - 4n).$$

Fig. 1.42 Visualization of five different situations representing the third step of the graphical procedure for computing a convolution of two signals from Problem 1.11

n	Graphical representation of $x_1(n-k)$ and computation of $y(n)$
-1	$x_1(-1-k)$ 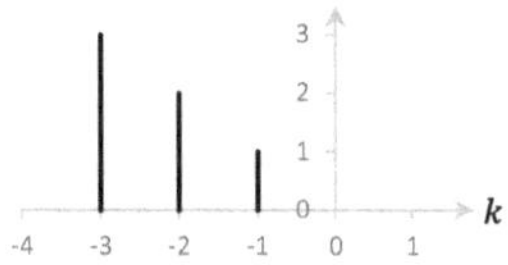 $y(-1)=\sum_{k\in\mathbb{Z}} x_2(k)x_1(-1-k)=1$
0	$x_1(-k)$ 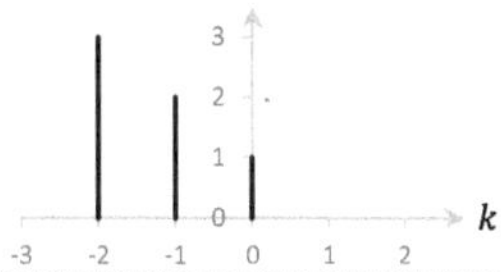 $y(0)=\sum_{k\in\mathbb{Z}} x_2(k)x_1(-k)=2$
1	$x_1(1-k)$ 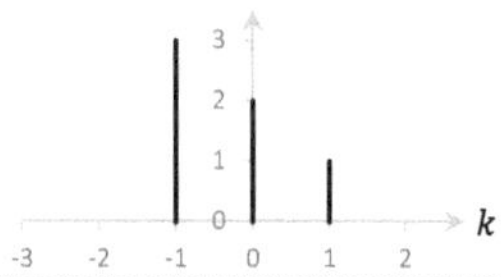 $y(1)=\sum_{k\in\mathbb{Z}} x_2(k)x_1(1-k)=4$
2	$x_1(2-k)$ 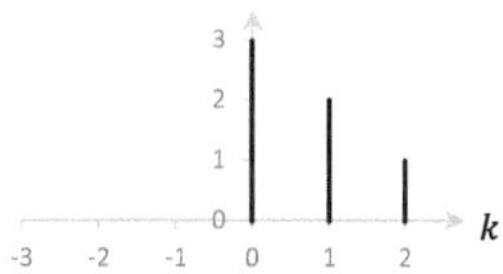 $y(2)=\sum_{k\in\mathbb{Z}} x_2(k)x_1(2-k)=2$
3	$x_1(3-k)$ 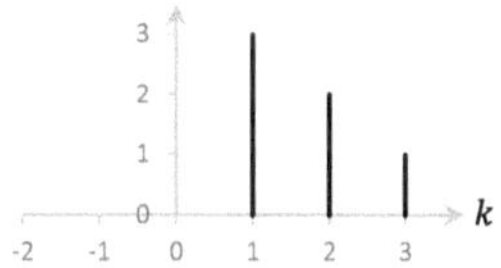 $y(3)=\sum_{k\in\mathbb{Z}} x_2(k)x_1(3-k)=3$

Fig. 1.43 Convolution $y(n)$ of two signals from Problem 1.11

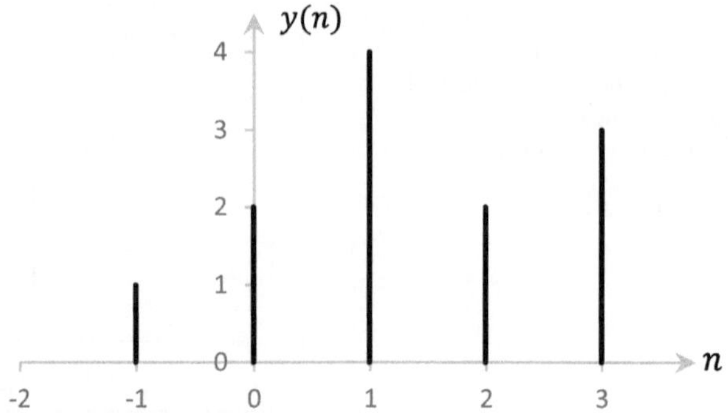

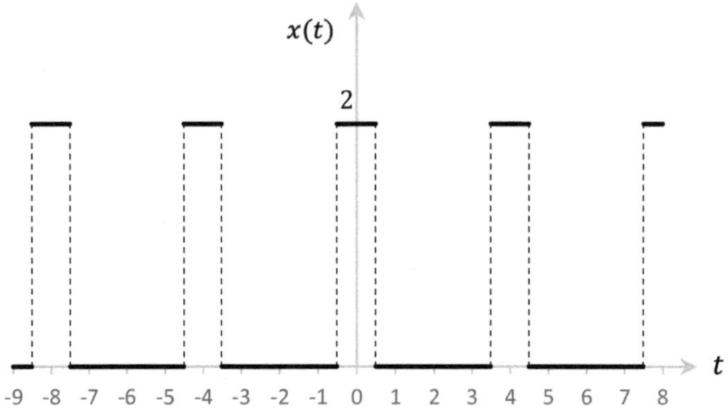

Fig. 1.44 Convolution of the rectangle pulse with the comb distribution from Problem 1.12

We can see that $x(t)$ is a unipolar rectangle pulse train (see Fig. 1.4), that is, the periodic repetition of the rectangle pulse $2 \cdot z(t) = 2\Pi(t)$ with period 4. Let us plot the graph of $x(t)$ as shown in Fig. 1.44.

Using (1.23) and (1.27), we can calculate the average values of signals $z(t)$ and $x(t)$:

$$\overline{z} = \int_{-1/2}^{1/2} dt - 1, \quad \overline{x} = \frac{1}{4}\int_{-1/2}^{1/2} 2dt = \frac{1}{2}.$$

Let us notice that $\overline{x} = \frac{2\overline{z}}{T_0}$, where the factor 2 equals "amplitude" of the comb distribution: $2\sum_{n\in\mathbb{Z}}\delta(t-4n)$. Energy E_z of $z(t)$ equals 1 (see Problem 1.4) and energy E_x of $x(t)$ is evidently ∞. Average power of $z(t)$ is 0, while average power P_x of $x(t)$, calculated using (1.29), is

$$P_x = \frac{1}{4} \int_{-1/2}^{1/2} 4dt = 1.$$

We can notice that $P_x = \frac{4E_z}{T_0}$, where 4 is the squared "amplitude" of the comb distribution: $2\sum_{n\in\mathbb{Z}}\delta(t-4n)$.

Let us now introduce *cross-correlation functions* of continuous- and discrete-time signals that express similarity between two signals (e.g., [11, 13, 15]). Let us notice that it is possible to define two different cross-correlation functions that usually are not equal (see Table 1.5 below).

Definition 1.11 The *cross-correlation function* $R_{12}(\tau)$ between two continuous-time signals $x_1(t)$ and $x_2(t)$, $t \in \mathbb{R}$ is defined as

$$R_{12}(\tau) = \lim_{T\to\infty}\left\{\frac{1}{2T}\int_{-T}^{T} x_1(t)x_2^*(t-\tau)dt\right\} \tag{1.40}$$

The *cross-correlation function* $R_{21}(\tau)$ between two continuous-time signals $x_2(t)$ and $x_1(t)$, $t \in \mathbb{R}$ is

$$R_{21}(\tau) = \lim_{T\to\infty}\left\{\frac{1}{2T}\int_{-T}^{T} x_2(t)x_1^*(t-\tau)dt\right\}. \tag{1.41}$$

Notice that if $x_1(t) \equiv x_2(t) = x(t)$, the formulas (1.40) and (1.41) define the *auto-correlation* function $R_x(\tau)$ of $x(t)$ as follows:

$$R_x(\tau) = \lim_{T\to\infty}\left\{\frac{1}{2T}\int_{-T}^{T} x(t)x^*(t-\tau)dt\right\}. \tag{1.42}$$

Definition 1.12 The *cross-correlation function* $R_{12}(k)$ between discrete-time signals $x_1(n)$ and $x_2(n)$, $n \in \mathbb{Z}$ is defined as

$$R_{12}(k) = \lim_{M\to+\infty}\left\{\frac{1}{2M+1}\sum\nolimits_{n=-M}^{M} x_1(n)x_2^*(n-k)\right\}. \tag{1.43}$$

The *cross-correlation function* $R_{21}(k)$ between discrete-time signals $x_2(n)$ and $x_1(n)$, $n \in \mathbb{Z}$ is

$$R_{21}(k) = \lim_{M\to+\infty}\left\{\frac{1}{2M+1}\sum\nolimits_{n=-M}^{M} x_2(n)x_1^*(n-k)\right\}. \tag{1.44}$$

If $x_1(n) \equiv x_2(n) = x(n)$, the formulas (1.43) and (1.44) define the *auto-correlation* function $R_x(k)$ of $x(n)$:

Table 1.5 Properties of cross-correlation functions $R_{12}(\cdot)$ and $R_{21}(\cdot)$

		Formula	
No.	Property	Continuous-time	Discrete-time
1	Complex conjugate symmetry	$R_{12}(\tau) = R_{21}^*(-\tau)$	$R_{12}(k) = R_{21}^*(-k)$
2	Relation to convolution	$R_{12}(\tau) = x_1(\tau) * x_2^*(-\tau)$ $R_{21}(\tau) = x_2(\tau) * x_1^*(-\tau)$	$R_{12}(k) = x_1(k) * x_2^*(-k)$ $R_{21}(k) = x_2(k) * x_1^*(-k)$
		If $x_2(t)$ is *Hermitian*, i.e., $x_2^*(t) = x_2(-t)$: $R_{12}(\tau) = x_1(\tau) * x_2(\tau)$ If $x_1(t)$ is *Hermitian*, i.e., $x_1^*(t) = x_1(-t)$: $R_{21}(\tau) = x_2(\tau) * x_1(\tau)$	*If* $x_2(n)$ is *Hermitian*, i.e., $x_2^*(n) = x_2(-n)$: $R_{12}(k) = x_1(k) * x_2(k)$ *If* $x_1(n)$ is *Hermitian*, i.e., $x_1^*(n) = x_1(-n)$: $R_{21}(k) = x_2(k) * x_1(k)$
		For real signals: $R_{12}(\tau) = x_1(\tau) * x_2(-\tau)$ $R_{21}(\tau) = x_2(\tau) * x_1(-\tau)$	*For real signals*: $R_{12}(k) = x_1(k) * x_2(-k)$ $R_{21}(k) = x_2(k) * x_1(-k)$
4	Commutativity	If *both* signals, $x_1(\cdot)$ and $x_2(\cdot)$, are *Hermitian*: $R_{12}(\cdot) = R_{21}(\cdot)$	
5	Zero value	If $R_{12}(0) = 0$, the signals $x_1(\cdot)$ and $x_2(\cdot)$ are *orthogonal.*	

$$R_x(k) = \lim_{M \to +\infty} \left\{ \frac{1}{2M+1} \sum\nolimits_{n=-M}^{M} x(n)x^*(n-k) \right\}. \tag{1.45}$$

Basic properties of cross-correlation functions $R_{12}(\cdot)$ and $R_{21}(\cdot)$ between two different continuous-time and discrete-time signals are collected in Table 1.5.

The formulas (1.40), (1.41), (1.42), (1.43), (1.44), and (1.45) can be simplified for *energy signals* (see Definition 1.5):

- For continuous-time energy signals:

$$R_{12}(\tau) = \int_{-\infty}^{\infty} x_1(t)x_2^*(t-\tau)dt, \tag{1.46}$$

$$R_{21}(\tau) = \int_{-\infty}^{\infty} x_2(t)x_1^*(t-\tau)dt, \tag{1.47}$$

$$R_x(\tau) = \int_{-\infty}^{\infty} x(t)x^*(t-\tau)dt. \tag{1.48}$$

- For discrete-time energy signals:

$$R_{12}(k) = \sum\nolimits_{n \in \mathbb{Z}} x_1(n)x_2^*(n-k), \tag{1.49}$$

$$R_{21}(k) = \sum\nolimits_{n \in \mathbb{Z}} x_2(n)x_1^*(n-k). \tag{1.50}$$

$$R_x(k) = \sum\nolimits_{n \in \mathbb{Z}} x(n)x^*(n-k). \tag{1.51}$$

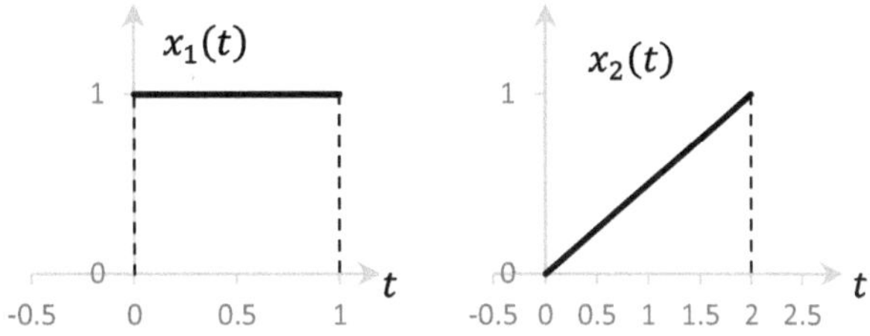

Fig. 1.45 Signals $x_1(t)$ and $x_2(t)$ from Problem 1.13

We can notice from (1.48) and (1.51) that $R_x(0)$ *is equal to the energy* E_x of the signal $x(\cdot)$.

Problem 1.13 Determine the cross-correlation functions $R_{12}(\tau)$ and $R_{21}(\tau)$ between two energy signals $x_1(t)$ and $x_2(t)$ shown in Fig. 1.45.

Solution Let us compute the function $R_{12}(\tau)$ given by (1.46). We can apply the graphical procedure basing on shifting the graph of $x_2(t) = \frac{1}{2}t$ (for $t \in [0, 2)$) left and right with a shift variable τ and calculating the integral of the product $x_1(t)x_2(t - \tau)$ (both signals are real-valued functions, so we can omit conjugation). The shifted signal is given by $x_2(t - \tau) = \frac{1}{2}(t - \tau)$ for $t \in [\tau, 2 + \tau)$.

Let us visualize all these calculations in Fig. 1.46. We easily can notice that for $\tau \in (-\infty, -2) \cup [1, +\infty)$, $R_{12}(\tau) = 0$. The cross-correlation function $R_{12}(\tau)$ (see Fig. 1.47) is

$$R_{12}(\tau) = \begin{cases} -\dfrac{1}{4}\tau^2 + 1 & \text{for} \quad \tau \in [-2, -1), \\ \dfrac{1}{2}\left(\dfrac{1}{2} - \tau\right) & \text{for } \tau \in [-1, 0), \\ \dfrac{1}{2}\left(\dfrac{1}{2} - \tau + \dfrac{\tau^2}{2}\right) & \text{for} \quad \tau \in [0, 1), \\ 0 & \text{for} \quad \tau \in (-\infty, -2) \cup [1, +\infty). \end{cases}$$

The cross-correlation function $R_{21}(\tau)$ (see Fig. 1.48) can be derived by applying complex conjugate property from Table 1.5, that is, $R_{21}(\tau) = R_{12}^*(-\tau) = R_{12}(-\tau)$ ($R_{12}(\tau)$ is a real-valued function). We get

$$R_{21}(\tau) = \begin{cases} -\dfrac{1}{4}\tau^2 + 1 & \text{for } \tau \in [1, 2), \\ \dfrac{1}{2}\left(\dfrac{1}{2} + \tau\right) & \text{for } \tau \in [0, 1), \\ \dfrac{1}{2}\left(\dfrac{1}{2} + \tau + \dfrac{\tau^2}{2}\right) & \text{for } \tau \in [-1, 0), \\ 0 & \text{for } \tau \in (-\infty, -1) \cup [2, +\infty). \end{cases}$$

<table>
<tr><th>No.</th><th>Graphical representation and computation of $R_{12}(\tau)$</th></tr>
<tr><td>1</td><td>
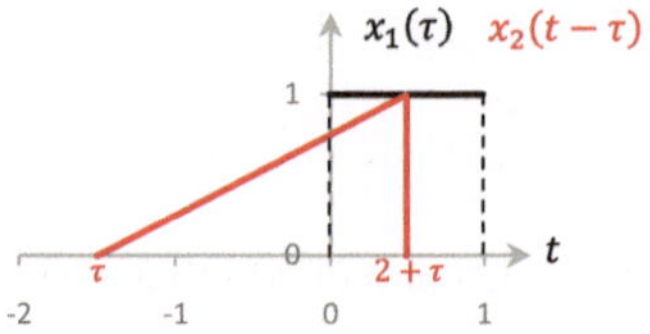

</td></tr>
<tr><td></td><td>
$\tau \in [-2,-1)$:

$$R_{12}(\tau)=\int_0^{2+\tau}\frac{1}{2}(t-\tau)\,dt=\frac{1}{2}\left(\frac{t^2}{2}-\tau t\right)\Bigg|_0^{2+\tau}=\frac{1}{2}\left(\frac{(2+\tau)^2}{2}-\tau(2+\tau)\right)=-\frac{1}{4}\tau^2+1$$
</td></tr>
<tr><td>2</td><td>
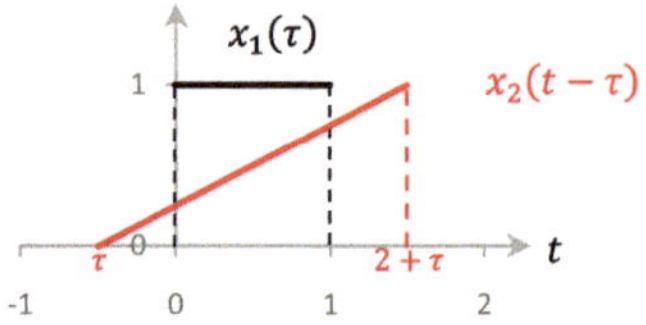

</td></tr>
<tr><td></td><td>
$\tau \in [-1,0)$:

$$R_{12}(\tau)=\int_0^{1}\frac{1}{2}(t-\tau)\,dt=\frac{1}{2}\left(\frac{t^2}{2}-\tau t\right)\Bigg|_0^{1}=\frac{1}{2}\left(\frac{1}{2}-\tau\right)$$
</td></tr>
<tr><td>3</td><td>
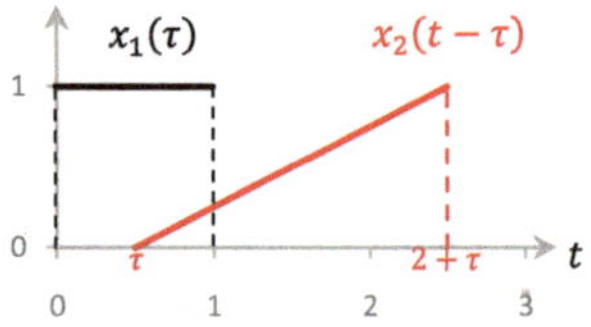

</td></tr>
<tr><td></td><td>
$\tau \in [0,1)$:

$$R_{12}(\tau)=\int_\tau^{1}\frac{1}{2}(t-\tau)\,dt=\frac{1}{2}\left(\frac{t^2}{2}-\tau t\right)\Bigg|_\tau^{1}=\frac{1}{2}\left(\frac{1}{2}-\tau+\frac{\tau^2}{2}\right)$$
</td></tr>
</table>

Fig. 1.46 Successive steps of the graphical method for computation of a cross-correlation function $R_{12}(\tau)$ between $x_1(t)$ and $x_2(t)$ from Fig. 1.45

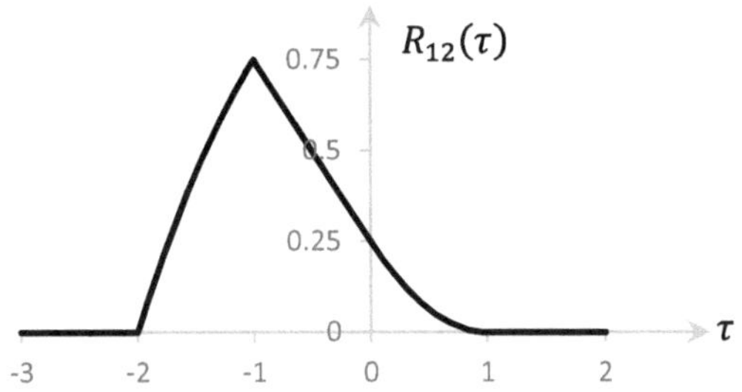

Fig. 1.47 Cross-correlation function $R_{12}(\tau)$ between $x_1(t)$ and $x_2(t)$ from Fig. 1.45

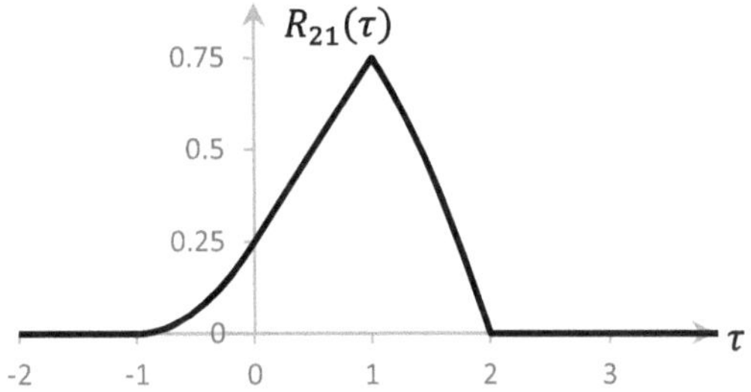

Fig. 1.48 Cross-correlation function $R_{21}(\tau)$ between signals $x_2(t)$ and $x_1(t)$ from Fig. 1.45

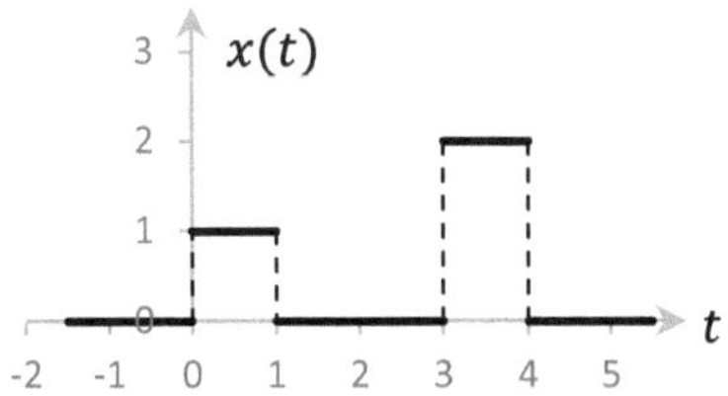

Fig. 1.49 Signal $x(t)$ from Problem 1.14

Problem 1.14 Determine the auto-correlation function $R_x(\tau)$ of the signal $x(t)$ shown in Fig. 1.49. Calculate $R_x(0)$ and compare it with the energy E_x of $x(t)$.

Solution Here also, it is comfortable to apply a graphical method. First, we have to shift right $x(t)$ with $\tau \geq 0$ (the shifted signal is marked in red in Fig. 1.50) and calculate the area under the product $x(t)x(t-\tau)$.There are four different situations shown in Fig. 1.50. For $\tau \geq 4$, $R_x(\tau) = 0$.

Let us summarize the obtained results for $\tau \geq 0$:

$$R_x(\tau) = \begin{cases} 5(1-\tau) & \text{for } \tau \in [0, 1), \\ 0 & \text{for } \tau \in [1, 2), \\ 2(\tau - 2) & \text{for } \tau \in [2, 3), \\ 2(4-\tau) & \text{for } \tau \in [3, 4), \\ 0 & \text{for } \tau \geq 4. \end{cases}$$

No.	Graphical representation and computation of $R_x(\tau)$ for $\tau \geq 0$
1	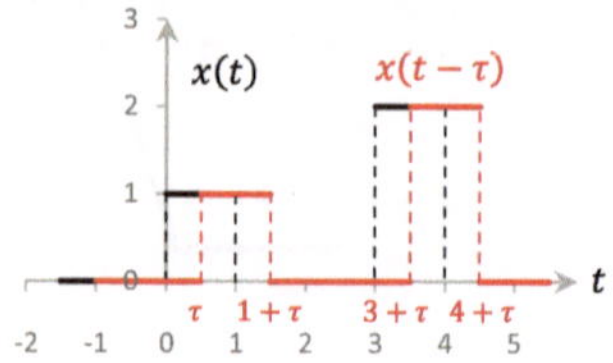
	$\tau \in [0,1)$: $$R_x(\tau) = \int_{\tau}^{1} dt + \int_{3+\tau}^{4} 4\,dt = 1 - \tau + 4(1-\tau) = 5(1-\tau)$$
2	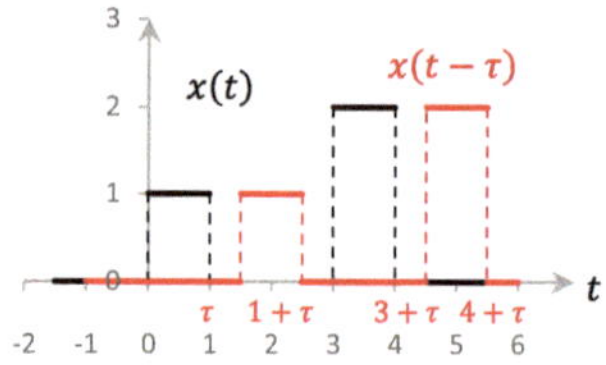
	$\tau \in [1,2)$: $R_x(\tau) = 0$
3	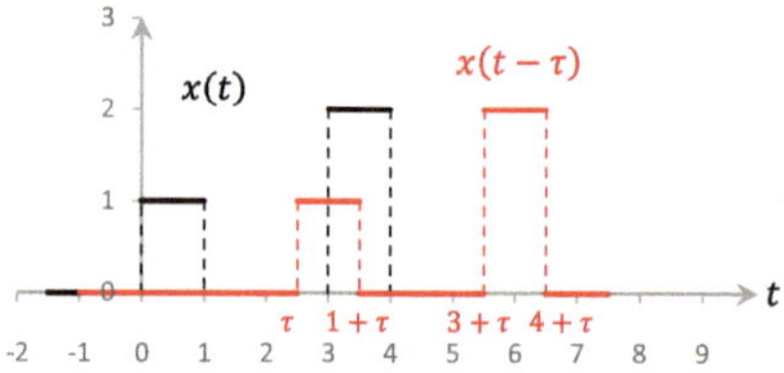
	$\tau \in [2,3)$: $R_x(\tau) = \int_3^{1+\tau} 2\,dt = 2(\tau - 2)$
4	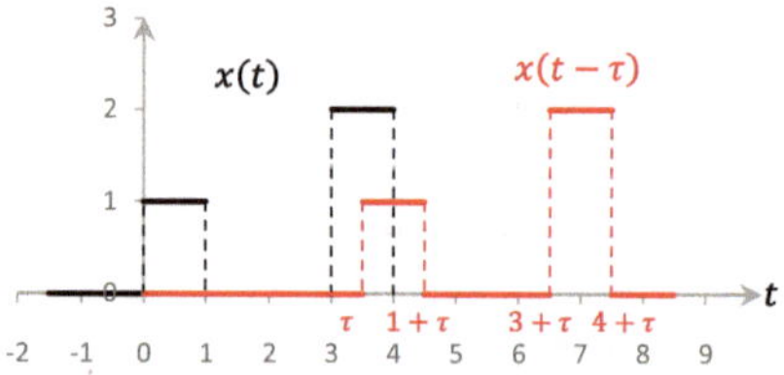
	$\tau \in [3,4)$: $R_x(\tau) = \int_{\tau}^{4} 2\,dt = 2(4 - \tau)$

Fig. 1.50 Illustration of the graphical procedure for computation of the auto-correlation function of the signal $x(t)$ from Fig. 1.48

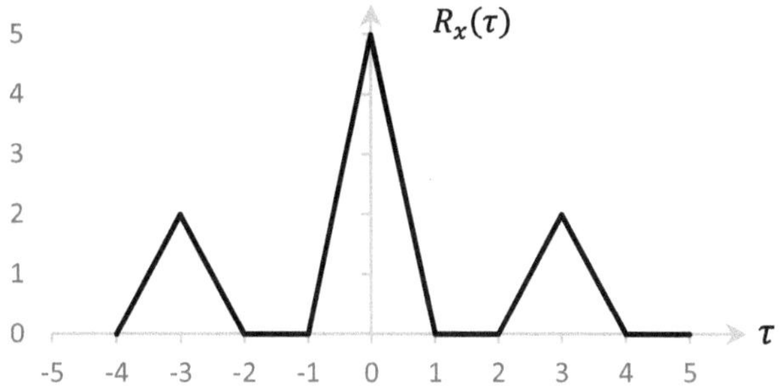

Fig. 1.51 Auto-correlation function of the signal $x(t)$ from Fig. 1.48

Since for real-valued signals, $R_x(\tau)$ exhibits even symmetry property, that is, $R_x(-\tau) = R_x(\tau)$, we do not need to consider the case $\tau < 0$ (that corresponds to left shift of $x(t)$). We can sketch the plot of $R_x(\tau)$ for $\tau \geq 0$ and then sketch the mirror image of it (w.r.t. the vertical axis $\tau = 0$). The autocorrelation function of $x(t)$ is illustrated in Fig. 1.51. Its maximum value $R_x(0) = 5$ equals the energy of $x(t)$ calculated from (1.25):

$$E_x = \int_0^1 dt + \int_3^4 4dt = 5.$$

Problem 1.15 Determine the cross-correlation functions $R_{12}(k)$ and $R_{21}(k)$ between two discrete-time signals from Problem 1.11: $x_1(n) = \delta(n) + 2\delta(n-1) + 3\delta(n-2)$ and $x_2(n) = \delta(n+1) + \delta(n-1)$.

Solution To compute $R_{12}(k)$, we can apply (1.49): $R_{12}(k) = \sum_{n \in \mathbb{Z}} x_1(n) x_2^*(n-k)$, in which we omit the symbol "$*$" since both signals are real-valued. Their plots are shown in Fig. 1.52.

We can see that for $k \leq -2$ and for $k \geq 4$, $R_{12}(k) = 0$. Let us calculate other values of $R_{12}(k)$ for $k \in \{-1, 0, 1, 2, 3\}$:

$$\begin{aligned} R_{12}(-1) &= \sum_{n \in \mathbb{Z}} x_1(n) x_2(n+1) \\ &= \sum_{n \in \mathbb{Z}} [\delta(n) + 2\delta(n-1) + 3\delta(n-2)] x_2(n+1) \\ &= \sum_{n \in \mathbb{Z}} [x_2(n+1) \cdot \delta(n) + 2x_2(n+1) \cdot \delta(n-1) + 3x_2(n+1) \cdot \delta(n-2)]. \end{aligned}$$

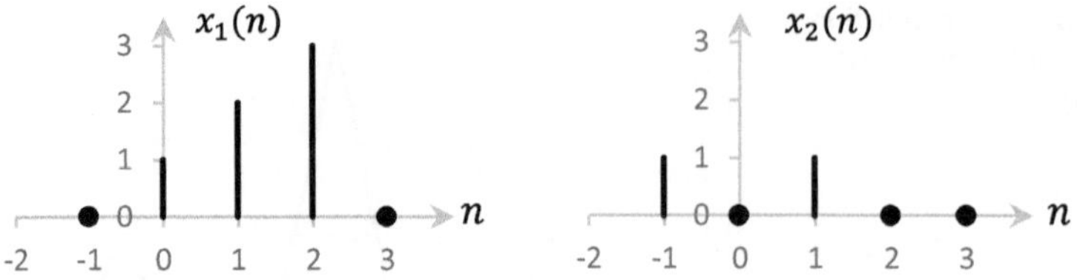

Fig. 1.52 Signals $x_1(n)$ and $x_2(n)$ from Problem 1.15

In the above expression, we deal with multiplication of $x_2(n + 1)$ by three unit delta pulses. Using the sampling property of the unit delta pulse from Table 1.3, we obtain

$$R_{12}(-1) = \sum_{n\in\mathbb{Z}} [x_2(1) \cdot \delta(n) + 2x_2(2) \cdot \delta(n-1) + 3x_2(3) \cdot \delta(n-2)] = 1.$$

Next,

$$\begin{aligned} R_{12}(0) &= \sum_{n\in\mathbb{Z}} x_1(n)x_2(n) = \sum_{n\in\mathbb{Z}} [x_2(n) \cdot \delta(n) + 2x_2(n) \cdot \delta(n-1) + 3x_2(n) \cdot \delta(n-2)] \\ &= \sum_{n\in\mathbb{Z}} [x_2(0) \cdot \delta(n) + 2x_2(1) \cdot \delta(n-1) + 3x_2(2) \cdot \delta(n-2)] \\ &= 2, \end{aligned}$$

$$\begin{aligned} R_{12}(1) &= \sum_{n\in\mathbb{Z}} x_1(n)x_2(n-1) \\ &= \sum_{n\in\mathbb{Z}} [x_2(n-1) \cdot \delta(n) + 2x_2(n-1) \cdot \delta(n-1) + 3x_2(n-1) \cdot \delta(n-2)] \\ &= \sum_{n\in\mathbb{Z}} [x_2(-1) \cdot \delta(n) + 2x_2(0) \cdot \delta(n-1) + 3x_2(1) \cdot \delta(n-2)] \\ &= 4, \end{aligned}$$

$$\begin{aligned} R_{12}(2) &= \sum_{n\in\mathbb{Z}} x_1(n)x_2(n-2) \\ &= \sum_{n\in\mathbb{Z}} [x_2(n-2) \cdot \delta(n) + 2x_2(n-2) \cdot \delta(n-1) + 3x_2(n-2) \cdot \delta(n-2)] \\ &= \sum_{n\in\mathbb{Z}} [x_2(-2) \cdot \delta(n) + 2x_2(-1) \cdot \delta(n-1) + 3x_2(0) \cdot \delta(n-2)] \\ &= 2, \end{aligned}$$

$$\begin{aligned} R_{12}(3) &= \sum_{n\in\mathbb{Z}} x_1(n)x_2(n-3) \\ &= \sum_{n\in\mathbb{Z}} [x_2(n-3)\cdot\delta(n) + 2x_2(n-3)\cdot\delta(n-1) + 3x_2(n-3)\cdot\delta(n-2)] \\ &= \sum_{n\in\mathbb{Z}} [x_2(-3)\cdot\delta(n) + 2x_2(-2)\cdot\delta(n-1) + 3x_2(-1)\cdot\delta(n-2)] \\ &= 3. \end{aligned}$$

In sum, we have

$$R_{12}(k) = \delta(k+1) + 2\delta(k) + 4\delta(k-1) + 2\delta(k-2) + 3\delta(k-3).$$

This result coincides with the convolution of the same signals calculated in Problem 1.11. We see that due to even parity of $x_2(n)$, the equality: $R_{12}(k) = x_1(k) * x_2(k)$ is satisfied (see Table 1.5). Since $R_{21}(k) = R_{12}(-k)$ (see Table 1.5), we have.

$$R_{21}(k) = 3\delta(k+3) + 2\delta(k+2) + 4\delta(k+1) + 2\delta(k) + \delta(k-1).$$

Correlation functions (1.40), (1.41), (1.42), (1.43), (1.44), and (1.45) *for periodic* continuous- and discrete-time signals are defined as follows:

- For continuous-time signals (both of the fundamental period T_0):

$$R_{12}(\tau) = \frac{1}{T_0}\int_{\langle T_0\rangle} x_1(t)x_2^*(t-\tau)dt, \tag{1.52}$$

$$R_{21}(\tau) = \frac{1}{T_0}\int_{\langle T_0\rangle} x_2(t)x_1^*(t-\tau)dt, \tag{1.53}$$

$$R_x(\tau) = \frac{1}{T_0}\int_{\langle T_0\rangle} x(t)x^*(t-\tau)dt, \tag{1.54}$$

- For discrete-time signals (both of the fundamental period N_0):

$$R_{12}(k) = \frac{1}{N_0}\sum_{\langle N_0\rangle} x_1(n)x_2^*(n-k), \tag{1.55}$$

$$R_{21}(k) = \frac{1}{N_0}\sum_{\langle N_0\rangle} x_2(n)x_1^*(n-k). \tag{1.56}$$

$$R_x(k) = \frac{1}{N_0}\sum_{\langle N_0\rangle} x(n)x^*(n-k). \tag{1.57}$$

We can notice from (1.54) and (1.57) that $R_x(0)$ *equals the average power* P_x (for period) of the signal $x(\cdot)$.

Problem 1.16 Given two periodic signals:

$$x_1(t) = 2\sin(\pi t) + \cos(2\pi t), x_2(t) = \frac{1}{2}\cos(\pi t), \quad t \in \mathbb{R}.$$

Determine the cross-correlation functions $R_{12}(\tau)$ oraz $R_{21}(\tau)$.

Solution Both signals are periodic with period $T_0 = 2$ [s]. Using (1.52), we obtain

$$R_{12}(\tau) = \frac{1}{2}\int_0^2 [2\sin(\pi t) + \cos(2\pi t)] \cdot \frac{1}{2}\cos(\pi t - \pi\tau)dt$$
$$= \frac{1}{2}\int_0^2 \sin(\pi t) \cdot \cos(\pi t - \pi\tau)dt + \frac{1}{4}\int_0^2 \cos(2\pi t) \cdot \cos(\pi t - \pi\tau)dt.$$

Using two trigonometric identities: $\sin\alpha\cos\beta = \frac{1}{2}[\sin(\alpha - \beta) + \sin(\alpha + \beta)]$ and $\cos\alpha\cos\beta = \frac{1}{2}[\cos(\alpha - \beta) + \cos(\alpha + \beta)]$, we get

$$R_{12}(\tau) = \frac{1}{4}\int_0^2 [\sin(\pi\tau) + \sin(2\pi t - \pi\tau)]dt + \frac{1}{8}\int_0^2 [\cos(\pi t + \pi\tau) + \cos(3\pi t - \pi\tau)]dt$$
$$= \frac{1}{4}\int_0^2 \sin(\pi\tau)dt + \frac{1}{4}\int_0^2 \sin(2\pi t - \pi\tau)dt + \frac{1}{8}\int_0^2 \cos(\pi t + \pi\tau)dt$$
$$+ \frac{1}{8}\int_0^2 \cos(3\pi t - \pi\tau)dt.$$

All integrals, except the first one, are zero (the reader can easily check it), and finally,

$$R_{12}(\tau) = \frac{1}{2}\sin(\pi\tau).$$

Since $R_{21}(\tau) = R_{12}^*(-\tau)$ (see Table 1.5), we have $R_{21}(\tau) = -\frac{1}{2}\sin(\pi\tau)$.

Problem 1.17 Determine the auto-correlation function $R_x(\tau)$ of the harmonic signal $x(t) = A\cos(\omega_0 t + \varphi_0)$ with fundamental period T_0. Calculate $R_x(0)$ and compare it with the average power P_x of $x(t)$.

Solution We can apply the formula (1.54), $\omega_0 = 2\pi/T_0$:

$$
\begin{aligned}
R_x(\tau) &= \frac{1}{T_0}\int_0^{T_0} A\cos(\omega_0 t + \varphi_0)A\cos[\omega_0(t-\tau)+\varphi_0]dt \\
&= \frac{A^2}{2T_0}\int_0^{T_0}[\cos(2\omega_0 t - \omega_0\tau + 2\varphi_0) + \cos(\omega_0\tau)]dt \\
&= \frac{A^2}{2T_0}\left[\int_0^{T_0}\cos(2\omega_0 t - \omega_0\tau + 2\varphi_0)dt + \int_0^{T_0}\cos(\omega_0\tau)dt\right] \\
&= \frac{A^2}{2T_0}\frac{1}{2\omega_0}\sin(2\omega_0 t - \omega_0\tau + 2\varphi_0)|_0^{T_0} + \frac{A^2}{2T_0}T_0\cos(\omega_0\tau) \\
&= \frac{A^2}{8\pi}[\sin(2\omega_0 T_0 - \omega_0\tau + 2\varphi_0) - \sin(-\omega_0\tau + 2\varphi_0)] + \frac{A^2}{2}\cos(\omega_0\tau) \\
&= \frac{A^2}{8\pi}[\sin(4\pi - \omega_0\tau + 2\varphi_0) - \sin(-\omega_0\tau + 2\varphi_0)] + \frac{A^2}{2}\cos(\omega_0\tau).
\end{aligned}
$$

The first term is zero, then finally the auto-correlation function is

$$R_x(\tau) = \frac{A^2}{2}\cos(\omega_0\tau).$$

We can see that $R_x(\tau)$ of the harmonic signal $x(t) = A\cos(\omega_0 t + \varphi_0)$ is a cosine function with the same fundamental angular frequency ω_0. Moreover, it does not keep information about the initial phase φ_0. The value $R_x(0) = \frac{A^2}{2}$ equals the average power (per period) P_x of $x(t)$ (see Problem 1.5).

Problem 1.18 Determine the auto-correlation function $R_x(k)$ of the discrete-time periodic signal $x(n) = \sin\left(\frac{\pi}{2}n\right), n \in \mathbb{Z}$.

Solution First, let us determine the fundamental period of $x(n) = \sin\left(\frac{\pi}{2}n\right)$. Using (1.4), we easily obtain $N_0 = \frac{2\pi k}{\frac{\pi}{2}} = 4k$, so the smallest $k \in \mathbb{N}$, such that N_0 is an integer, is $k = 1$. We have $N_0 = 4$. Using (1.57), we get

$$R_x(k) = \frac{1}{4}\sum_{n=0}^{3} \sin\left(\frac{\pi}{2}n\right)\sin\left(\frac{\pi}{2}(n-k)\right)$$
$$= \frac{1}{4}\sin\left(\frac{\pi}{2}\right)\sin\left(\frac{\pi}{2}-\frac{\pi}{2}k\right)+\frac{1}{4}\sin\left(\frac{3\pi}{2}\right)\sin\left(\frac{3\pi}{2}-\frac{\pi}{2}k\right)=\frac{1}{2}\cos\left(\frac{\pi}{2}k\right).$$

Notice that the auto-correlation function $R_x(k)$ is a cosine with the same fundamental angular frequency. Its value $R_x(0)=\frac{1}{2}$ equals the average power (for period) of $x(n)$, that is, the squared amplitude = 1 divided by 2.

1.5 Orthogonalization and Approximation

In Problems 1.7 and 1.8, we determined even and odd components of a given signal and proved that they are orthogonal. In this section, we treat this problem larger in the context of a signal approximation with a finite sum of orthogonal harmonic signals and Walsh functions [22].

Definition 1.13 The set of complex-valued continuous-time signals $\{\phi_n(t)\}$, $n = 1, 2, \ldots$ is said to be *orthogonal* over the interval $[a, b]$ if it satisfies the following condition:

$$\int_a^b \phi_n(t)\phi_m^*(t)dt = \begin{cases} 0 & \text{for} \quad n \neq m \\ K_n & \text{for} \quad n = m. \qquad (K_n = \text{const}) \end{cases} \tag{1.58}$$

If $K_n = 1$, the set of complex-valued continuous-time signals $\{\phi_n(t)\}$, $n = 1, 2, \ldots$ is said to be *orthonormal.*

Definition 1.14 The *generalized Fourier series* is the expansion of a continuous-time signal $x(t)$, $t \in \mathbb{R}$ into a series of orthogonal functions:

$$x(t) = \sum\nolimits_{n=0}^{\infty} c_n\phi_n(t), \tag{1.59}$$

where coefficients c_n are calculated as

$$c_n = \frac{1}{K_n}\int_a^b x(t)\phi_n^*(t)dt. \tag{1.60}$$

If we take the partial sum of (1.59), that is,

$$x_N(t) = \sum\nolimits_{n=1}^{N} c_n\phi_n(t), \tag{1.61}$$

we can deal with *approximation* of a signal $x(t)$ with a finite sum of N orthogonal functions $\phi_n(t)$, $n = 1, 2, \ldots, N$, weighted by c_n. The *mean-square error* of this approximation

$$\varepsilon_N = \int_a^b |x(t) - x_N(t)|^2 dt \tag{1.62}$$

should be minimal, which means that $\lim\limits_{N\to\infty} \varepsilon_N = 0$. The mean-square error (1.62) can equivalently be expressed as

$$\varepsilon_N = \int_a^b |x(t)|^2 dt - \sum\nolimits_{n=1}^{N} K_n |c_n|^2. \tag{1.63}$$

Problem 1.19 Given the signal $x(t) = \Pi(t) - \Pi(t-1)$. Find its expansion into the generalized Fourier series with orthogonal functions $\phi_n(t) = \cos(n\pi t)$, $n = 1, 2, \ldots, N$ over the interval $\left[-\frac{1}{2}, \frac{3}{2}\right]$. Sketch the partial sum $x_9(t)$ and calculate the error of this approximation.

Solution First, let us show that functions $\phi_n(t) = \cos(n\pi t)$ are orthonormal over $\left[-\frac{1}{2}, \frac{3}{2}\right]$. For $n = m$:

$$\int_{-1/2}^{3/2} \cos^2(n\pi t) dt = \int_{-1/2}^{3/2} \frac{1 + \cos(2n\pi t)}{2} dt = 1 + \frac{\sin(2n\pi t)}{4n\pi}\Big|_{-1/2}^{3/2} = 1,$$

then $K_n = 1$. For $n \neq m$, we have

$$\begin{aligned}
&\int_{-1/2}^{3/2} \cos(n\pi t)\cos(m\pi t) dt = \frac{1}{2}\int_{-\frac{1}{2}}^{\frac{3}{2}} \cos((n+m)\pi t) dt + \frac{1}{2}\int_{-\frac{1}{2}}^{\frac{3}{2}} \cos((n-m)\pi t) dt \\
&= \frac{1}{2(n+m)\pi} \sin((n+m)\pi t)\Big|_{-\frac{1}{2}}^{\frac{3}{2}} + \frac{1}{2(n-m)\pi} \sin((n-m)\pi t)\Big|_{-\frac{1}{2}}^{\frac{3}{2}} \\
&= \frac{1}{2(n+m)\pi}\left[\sin\left(\frac{3}{2}(n+m)\pi\right) + \sin\left(\frac{1}{2}(n+m)\pi\right)\right] \\
&+ \frac{1}{2(n-m)\pi}\left[\sin\left(\frac{3}{2}(n-m)\pi\right) + \sin\left(\frac{1}{2}(n-m)\pi\right)\right] = 0.
\end{aligned}$$

Using (1.60), we can calculate coefficients c_n:

$$c_n = \int_{-1/2}^{3/2} x(t)\cos(n\pi t)dt = \int_{-1/2}^{1/2} \cos(n\pi t)dt - \int_{1/2}^{3/2} \cos(n\pi t)dt$$
$$= \tfrac{1}{n\pi}\sin(n\pi t)\Big|_{-1/2}^{1/2} - \tfrac{1}{n\pi}\sin(n\pi t)\Big|_{1/2}^{3/2}$$
$$= \frac{1}{n\pi}\left[\sin\left(n\frac{\pi}{2}\right) + \sin\left(n\frac{\pi}{2}\right) - \sin\left(n\frac{3\pi}{2}\right) + \sin\left(n\frac{3\pi}{2}\right)\right].$$

We can see that $c_1 = \frac{4}{\pi}, c_3 = -\frac{4}{3\pi}, c_5 = \frac{4}{5\pi}, c_7 = -\frac{4}{7\pi}, c_9 = \frac{4}{9\pi}$ and for even subscripts n: $c_n = 0$. Finally, the partial sum of 9 terms of the generalized Fourier series (1.61) is

$$x_9(t) = \sum_{n=1}^{9} c_n \cos(n\pi t)$$
$$= \frac{4}{\pi}\cos(\pi t) - \frac{4}{3\pi}\cos(3\pi t) + \frac{4}{5\pi}\cos(5\pi t) - \frac{4}{7\pi}\cos(7\pi t) + \frac{4}{9\pi}\cos(9\pi t).$$

It is displayed in red in Fig. 1.53. The approximation error calculated using (1.63) is

$$\varepsilon_9 = \int_{-1/2}^{3/2} |x(t)|^2 dt - \sum_{n=1}^{9} |c_n|^2$$
$$= 2 - \left[\left(\tfrac{4}{\pi}\right)^2 + \left(-\tfrac{4}{3\pi}\right)^2 + \left(\tfrac{4}{5\pi}\right)^2 + \left(-\tfrac{4}{7\pi}\right)^2 + \left(\tfrac{4}{9\pi}\right)^2\right] \cong 0,08079.$$

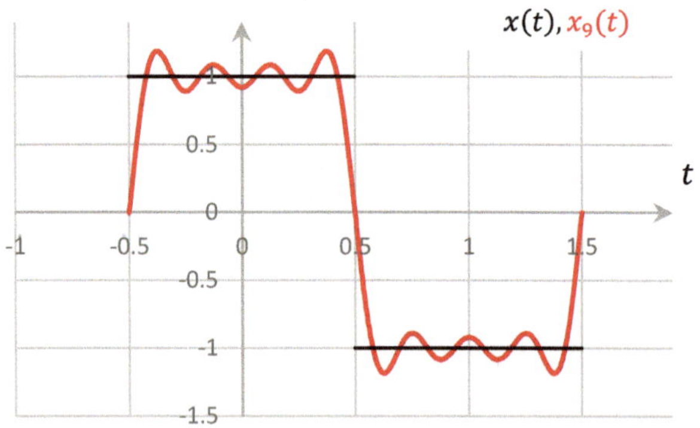

Fig. 1.53 Approximation of the signal $x(t)$ from Problem 1.19 with the partial sum (1.61)

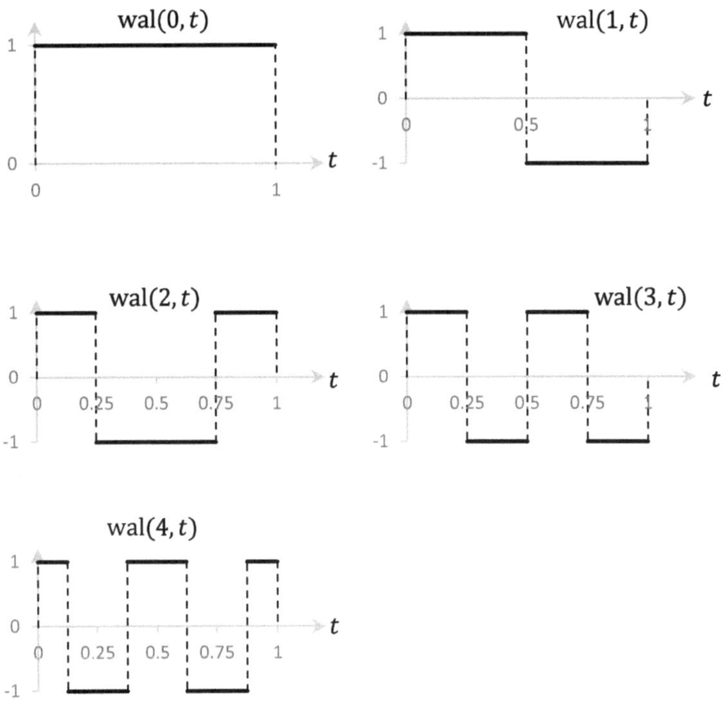

Fig. 1.54 Walsh orthogonal functions wal(n, t), $n = 0, 1, 2, 3, 4$

Problem 1.20 Given the signal $x(t) = \cos(\pi t)$. Approximate $x(t)$ in the interval $t \in [0, 1]$ with the partial sum of five ($N = 5$) orthonormal Walsh functions wal(n, t), $n = 0, 1, 2, 3, 4$ displayed in Fig. 1.54.

Solution The formula (1.60) gets a form

$$c_n = \int_0^1 x(t)\text{wal}(n, t)dt.$$

For $n = 0$, we have

$$c_0 = \int_0^1 x(t)\text{wal}(0, t)dt = \int_0^1 \cos(\pi t)\text{wal}(0, t)dt = \int_0^1 \cos(\pi t)dt$$
$$= \frac{\sin(\pi t)}{\pi}\bigg|_0^1 = 0.$$

For $n = 1$:

$$c_1 = \int_0^1 x(t)\mathrm{wal}(1,t)dt = \int_0^{0,5} \cos(\pi t)dt - \int_{0,5}^1 \cos(\pi t)dt$$
$$= \frac{1}{\pi}\left(\sin(\pi t)|_0^{0,5} - \sin(\pi t)|_{0,5}^1 \right) = \frac{2}{\pi} \cong 0,6366.$$

For $n = 2$:

$$c_2 = \int_0^1 x(t)\mathrm{wal}(2,t)dt = \int_0^{0,25} \cos(\pi t)dt - \int_{0,25}^{0,75} \cos(\pi t)dt + \int_{0,75}^1 \cos(\pi t)dt$$
$$= \frac{1}{\pi}\left(\sin(\pi t)|_0^{0,25} - \sin(\pi t)|_{0,25}^{0,75} + \sin(\pi t)|_{0,75}^1 \right) = 0.$$

For $n = 3$:

$$c_3 = \int_0^1 x(t)\mathrm{wal}(3,t)dt$$
$$= \int_0^{0,25} \cos(\pi t)dt - \int_{0,25}^{0,5} \cos(\pi t)dt + \int_{0,5}^{0,75} \cos(\pi t)dt - \int_{0,75}^1 \cos(\pi t)dt$$
$$= \frac{1}{\pi}\left(\sin(\pi t)|_0^{0,25} - \sin(\pi t)|_{0,25}^{0,5} + \sin(\pi t)|_{0,5}^{0,75} - \sin(\pi t)|_{0,75}^1 \right)$$
$$= \frac{2\sqrt{2}-2}{\pi} \cong 0,2637.$$

For $n = 4$:

$$c_4 = \int_0^1 x(t)\mathrm{wal}(4,t)dt$$
$$= \int_0^{0,125} \cos(\pi t)dt - \int_{0,125}^{0,375} \cos(\pi t)dt + \int_{0,375}^{0,625} \cos(\pi t)dt$$
$$- \int_{0,625}^{0,875} \cos(\pi t)dt + \int_{0,875}^1 \cos(\pi t)dt$$
$$= \frac{1}{\pi}\left(\sin(\pi t)|_0^{0,125} - \sin(\pi t)|_{0,125}^{0,375} + \sin(\pi t)|_{0,375}^{0,625} - \sin(\pi t)|_{0,625}^{0,875} + \sin(\pi t)|_{0,875}^1 \right) = 0.$$

The approximation $x_5(t)$ (see Fig. 1.55) with the partial weighted sum of the first five Walsh functions is

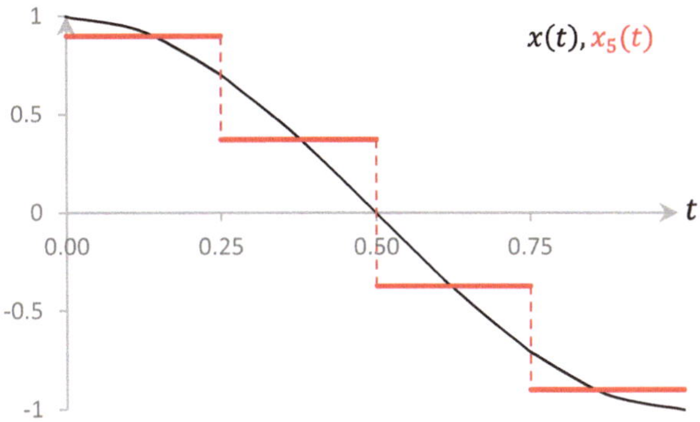

Fig. 1.55 Approximation of $x(t) = \cos(\pi t)$ in the interval $t \in [0, 1]$ with a partial weighted sum of Walsh orthonormal functions wal(n, t), $n = 0, 1, 2, 3, 4$

$$x_5(t) = \frac{2}{\pi}\,\mathrm{wal}(1,t) + \frac{2\sqrt{2}-2}{\pi}\mathrm{wal}(3,t) = 0,6366\,\mathrm{wal}(1,t) + 0,2637\mathrm{wal}(3,t).$$

Equivalently, for $t \in (0, 1)$, we can write

$$x_5(t) = \begin{cases} \frac{2}{\pi} + \frac{2\sqrt{2}-2}{\pi} = \frac{2\sqrt{2}}{\pi} \cong 0,9 & \text{for} \quad t \in (0, 1/4) \\ \frac{2}{\pi} - \frac{2\sqrt{2}-2}{\pi} = \frac{4-2\sqrt{2}}{\pi} \cong 0,3729 & \text{for} \quad t \in (1/4, 1/2) \\ -\frac{2}{\pi} + \frac{2\sqrt{2}-2}{\pi} = \frac{-4+2\sqrt{2}}{\pi} \cong -0,3729 & \text{for} \quad t \in (1/2, 3/4) \\ -\frac{2}{\pi} - \frac{2\sqrt{2}-2}{\pi} = -\frac{2\sqrt{2}}{\pi} \cong -0,9 & \text{for} \quad t \in (3/4, 1). \end{cases}$$

1.5.1 The Gram-Schmidt Orthogonalization Procedure

Given the set of M energy signals $\{s_i(t), i = 1, 2, \ldots, M\}$. It is possible to transform it into the set of N orthonormal signals $\{\phi_i(t), i = 1, 2, \ldots, N\}$, $N \leq M$ in the interval $[a, b]$. The procedure known as the *Gram-Schmidt orthogonalization* consists of a few steps:

Construction of $\phi_1(t)$.

$$\phi_1(t) = \frac{s_1(t)}{\sqrt{E_1}}, \qquad E_1 = \int_a^b s_1^2(t)dt. \tag{1.63}$$

Construction of $\phi_2(t)$.

$$\phi_2(t) = \frac{\theta_2(t)}{\sqrt{E_2}}, \qquad E_2 = \int_a^b \theta_2^2(t)dt, \tag{1.64}$$

where $\theta_2(t) = s_2(t) - (s_2, \phi_1)\phi_1(t)$ and $(s_2, \phi_1) = \int_a^b s_2(t)\phi_1(t)dt$ is the *inner product* of signals $s_2(t)$ and $\phi_1(t)$.

Construction of other $\phi_i(t)$.

Given determined signals $\phi_1(t), \ldots, \phi_{L-1}(t)$, we want to calculate $\phi_L(t)$:

$$\phi_L(t) = \frac{\theta_L(t)}{\sqrt{E_L}}, \qquad E_L = \int_a^b \theta_L^2(t)dt, \tag{1.65}$$

where $\theta_L(t) = s_L(t) - \sum_{j=1}^{L-1} (s_L, \phi_j)\phi_j(t)$ and $(s_L, \phi_j) = \int_a^b s_L(t)\phi_j(t)dt$ is inner product of signals $s_L(t)$ and $\phi_j(t)$, $j = 1, \ldots, L-1$. If $\theta_L(t) = 0$, the orthognalization process stops.

Important The functions $\phi_i(t)$, $i = 1, 2, \ldots, L$ are orthogonal over the interval $[a, b]$ since, for $l \neq i$, we have

$$\begin{aligned} \int_a^b \phi_l(t)\phi_i(t)dt &= \frac{1}{\sqrt{E_l}} \int_a^b \theta_l(t)\phi_i(t)dt \\ &= \frac{1}{\sqrt{E_l}} \left[\int_a^b s_l(t)\phi_i(t)dt - \sum_{j=1}^{L-1} (s_l, \phi_j) \int_a^b \phi_j(t)\phi_i(t)dt \right] \\ &= \frac{1}{\sqrt{E_l}} [(s_l, \phi_i) - (s_l, \phi_i)] = 0. \end{aligned}$$

Problem 1.21 Apply the Gram-Schmidt orthogonalization procedure for the set of four signals $s_i(t)$, $i = 1, 2, 3, 4$ shown in Fig. 1.56 to transform it into a set $\phi_i(t)$ of orthonormal functions. Present $s_i(t)$ in the form of linear combinations of $\phi_i(t)$.

Solution Applying (1.63), we get

$$E_1 = \int_a^b s_1^2(t)dt = 4, \quad \phi_1(t) = \frac{s_1(t)}{2}.$$

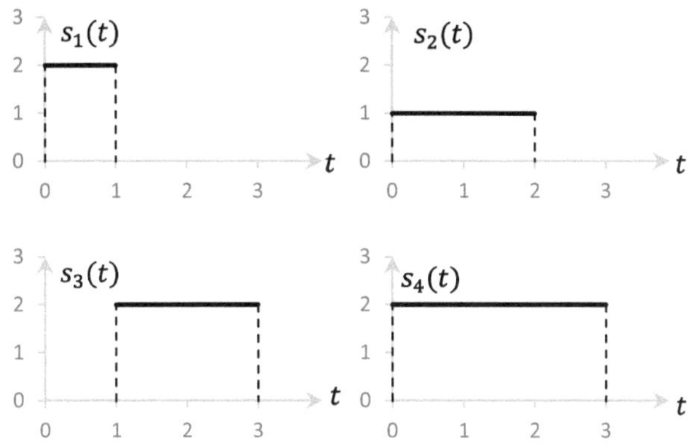

Fig. 1.56 Signals $s_i(t)$, $i = 1, 2, 3, 4$ from Problem 1.21

Then we calculate an inner product: $(s_2, \phi_1) = \int_0^2 s_2(t)\phi_1(t)dt = 1$. We get

$$\theta_2(t) = s_2(t) - (s_2, \phi_1)\phi_1(t) = s_2(t) - \phi_1(t) = s_2(t) - \frac{1}{2}s_1(t).$$

Using (1.64), we have

$$E_2 = \int_1^2 \theta_2^2(t)dt = 1, \qquad \phi_2(t) = \theta_2(t) = s_2(t) - \frac{1}{2}s_1(t)$$

Now, we come to the next step defined by (1.65).

$$\theta_3(t) = s_3(t) - \sum_{j=1}^{2} (s_3, \phi_j)\phi_j(t),$$

where $(s_3, \phi_1) = \int_1^3 s_3(t)\phi_1(t)dt = 0$ and $(s_3, \phi_2) = \int_1^2 s_3(t)\phi_2(t)dt = 2$. We have

$$\theta_3(t) = s_3(t) - 2\phi_2(t) = s_3(t) - 2s_2(t) + s_1(t).$$

The energy of $\theta_3(t)$ is $E_3 = \int_2^3 \theta_3^2(t)dt = 4$, so

$$\phi_3(t) = \frac{1}{2}\theta_3(t) = \frac{1}{2}s_3(t) - s_2(t) + \frac{1}{2}s_1(t).$$

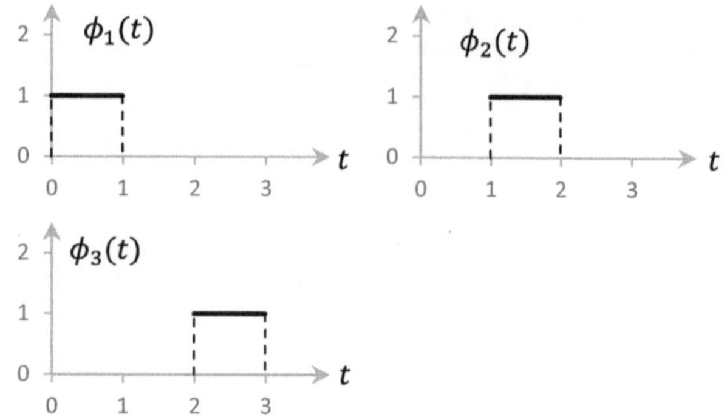

Fig. 1.57 Orthogonal signals $\phi_i(t)$, $i = 1, 2, 3$ from Problem 1.21

The plots of $\phi_1(t)$, $\phi_2(t)$ and $\phi_3(t)$ are displayed in Fig. 1.57. The Gram-Schmidt procedure stops since

$$\theta_4(t) = s_4(t) - \sum_{j=1}^{3} (s_4, \phi_j)\phi_j(t) = 0.$$

It is easy to check that $(s_4, \phi_j) = 2$ for $j = 1, 2, 3$ and $\theta_4(t) = s_4(t) - 2\phi_1(t) - 2\phi_2(t) - 2\phi_3(t) = 0$. In this case, we constructed the set of 3 orthogonal signals $\phi_i(t)$ ($i = 1, 2, 3$) from the set of 4 signals $s_i(t)$ ($i = 1, 2, 3, 4$). We can notice that $s_4(t)$ equals $s_1(t) + s_3(t)$ (in other words, we say that $s_4(t)$ is not a linearly independent—in analogy to vector operations).

Let us now express signals $s_i(t)$ in terms of $\phi_i(t)$.

We have $\phi_1(t) = \frac{s_1(t)}{2}$, so $s_1(t) = 2\phi_1(t)$. Then $\phi_2(t) = s_2(t) - \frac{1}{2}s_1(t)$ which is equivalent to $s_2(t) = \phi_2(t) + \frac{1}{2}s_1(t) = \phi_1(t) + \phi_2(t)$. Finally, $\phi_3(t) = \frac{1}{2}s_3(t) - s_2(t) + \frac{1}{2}s_1(t)$, then $s_3(t) = 2\phi_2(t) + 2\,\phi_3(t)$.

Exercises

1.1 Check periodicity of the following signals: (a) $x(t) = \cos(2\pi t) - 2\sin(2,1\pi t)$, $t \in \mathbb{R}$; (b) the even component of $x(t) = \sin(3\pi t) \cdot \mathbf{1}(t)$, $t \in \mathbb{R}$; (c) $x(n) - \cos\left(\frac{\pi}{4}n\right)\cos\left(\frac{\pi}{2}n\right), n \in \mathbb{Z}$. In case a signal is periodic, determine its fundamental period. For every signal (a), (b), (c), sketch the graphs and calculate average powers.

1.2 Sketch the graph of the periodic signal:

$$x(t) = \begin{cases} 2 & \text{for } t \in [2n, 2n+1), \\ 1 + \cos(2\pi t) & \text{for } t \in [2n+1, 2n+2), \quad n \in \mathbb{Z}. \end{cases}$$

Calculate its average value and average power for period.

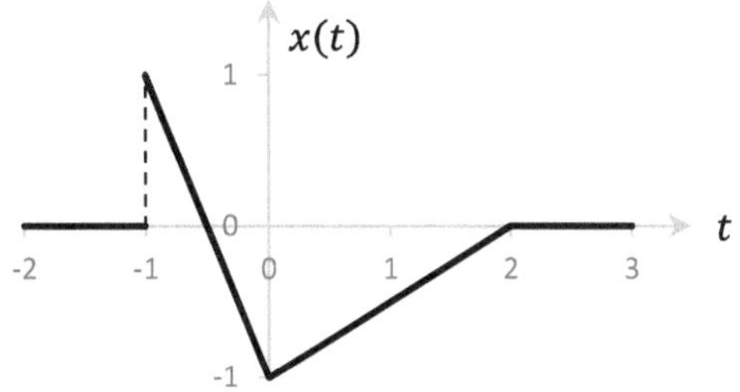

Fig. 1.58 Signal $x(t)$ from Exercise 1.5

1.3 Determine and sketch graphs of even and odd components of the following signals: (a) $x(t) = \left(1 - \frac{t}{2}\right) \cdot (\mathbf{1}(t) - \mathbf{1}(t-2))$; (b) $x(t) = e^{-2t} \cdot \mathbf{1}(t)$; (c) $x(n) = 3\delta(-n+1) + 3\delta(n) + 3\delta(n-1) + 2\delta(n-2) + \delta(n-3)$. In each case, prove that both components are orthogonal and that a sum of their energies equals the energy of $x(\cdot)$.

1.4 Determine and sketch graphs of even and odd components of the following signals: (a) $x(t) = 2 \cdot \mathbf{1}(t)$; (b) $x(t) = (1 - e^{-2t}) \cdot \mathbf{1}(t)$; (c) $x(n) = \mathbf{1}(n-2)$. In each case, prove that both components are orthogonal and that a sum of their average powers equals the average power of $x(\cdot)$.

1.5 Given a plot of the finite-duration signal $x(t)$ shown in Fig. 1.58. Express $x(t)$ in a general form

$$x(t) = y_1(t) \cdot \mathbf{1}(t - t_1) + y_2(t) \cdot \mathbf{1}(t - t_2) + y_3(t) \cdot \mathbf{1}(t - t_3),$$

where $y_i(t)$, $i = 1, 2, 3$ are linear functions and $\mathbf{1}(t - t_i)$, $i = 1, 2, 3$ are unit step signals shifted in time by t_i. Calculate the derivative $\frac{dx(t)}{dt}$ and plot its graph.

1.6 Using the geometrical method, calculate the convolution $y(t) = x_1(t) * x_2(t)$ of the following signals: $x_1(t) = \mathbf{1}(t) - 2 \cdot \mathbf{1}(t-1) + \mathbf{1}(t-2)$, $x_2(t) = 2 \cdot [\mathbf{1}(t) - \mathbf{1}(t-1)]$. Plot the graph of $y(t)$.

1.7 Using the geometrical method, calculate the convolution $y(n) = x_1(n) * x_2(n)$ of the following signals

$$x_1(n) = \begin{cases} 1 & \text{for } n \in [0,4] \\ 0 & \text{elsewhere} \end{cases}, \quad x_2(n) = \begin{cases} \left(\frac{1}{2}\right)^n & \text{for } n \in [0,3] \\ 0 & \text{elsewhere} \end{cases}.$$

Plot the graph of $y(n)$.

1.8 Compute and sketch the auto-convolution of the signal shown in Fig. 1.49. Compare the computed convolution with the auto-correlation function calculated in Problem 1.14.

1.9 Compute the convolution $x(t) = z(t) * 0,5\sum_{n\in\mathbb{Z}}\delta(t-4n)$ where $z(t) = 2\Lambda(t)$. Sketch the graph of $x(t)$. Calculate values of the following parameters of $x(t)$ and $z(t)$: average value, energy, average power.

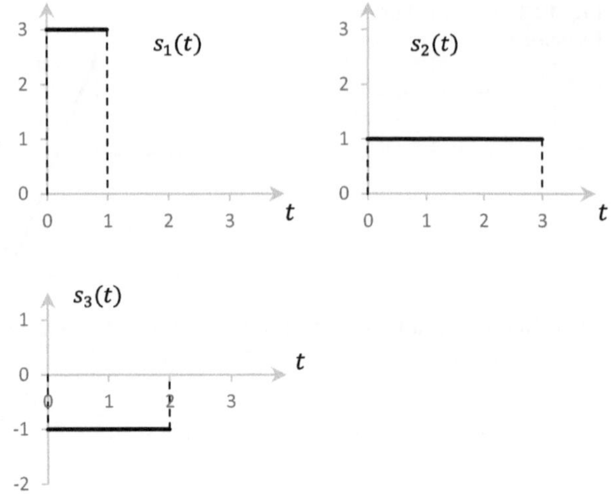

Fig. 1.59 Signals $s_i(t)$, $i = 1, 2, 3$ from Exercise 1.13

1.10 Compute and sketch the auto-correlation function $R_x(\tau)$ of the signal $x(t) = \Pi(t + 0,5) - \Pi(t - 0,5)$. Calculate $R_x(0)$ and compare it with the signal energy E_x.

1.11 Compute cross-correlation functions $R_{12}(\tau)$ and $R_{21}(\tau)$ between two energy signals $x_1(t)$ and $x_2(t)$ from Exercise 1.6.

1.12 Compute cross-correlation functions $R_{12}(k)$ and $R_{21}(k)$ between two energy signals $x_1(n)$ and $x_2(n)$ from Exercise 1.7.

1.13 Given the signal $x(t) = \Lambda(2t - 1)$. Approximate $x(t)$ with the partial sum of five ($N = 5$) orthonormal Walsh functions wal(n, t), $n = 0, 1, 2, 3, 4$ in the interval $t \in [0, 1]$.

1.14 Apply the Gram-Schmidt orthogonalization procedure for the set of signals $s_i(t)$, $i = 1, 2, 3$ shown in Fig. 1.59

Chapter 2
Frequency Analysis of Continuous-Time Signals

This chapter is devoted to the frequency representation of continuous-time signals. We start with the trigonometric and exponential Fourier series of a continuous-time periodic signal. The reader will get familiar with notions of amplitude-, phase- and power density spectra of periodic signals and will learn how to calculate the average power of periodic signals using Parseval's equality. Then we introduce the Fourier transform (FT) of energy signals, present its properties, and discuss some practical problems. Amplitude-, phase- and energy density spectra are also defined. We show how to use the Plancherel theorem for the calculation of the signal energy and the Wiener-Khintchine theorem for the calculation of the auto-correlation function of energy signals. At last, using a limit approach, the Fourier transform of power signals is defined. For periodic signals, we give relations between the exponential Fourier series and the Fourier transform.

2.1 Trigonometric and Exponential Fourier Series

Let us consider a continuous-time periodic signal $x(t)$, $t \in \mathbb{R}$ (see Definition 1.1) with fundamental period T_0 (in sec), fundamental frequency $f_0 = 1/T_0$ (in Hz) and fundamental angular frequency $\omega_0 = 2\pi f_0$ (in rad/sec). It can be represented as an infinite sum of harmonic components known as the Fourier series. Here, below, we give the definition of the trigonometric Fourier series, and then, using Euler's formula, we come to the notion of the exponential Fourier series.

K. M. Snopek, *Hands-on Signals and Systems Theory*,
https://doi.org/10.1007/978-3-031-56081-1_2

2.1.1 Trigonometric (Cosine) Fourier Series

Definition 2.1 The *trigonometric (cosine) Fourier series* of the signal $x(t)$ is defined as the sum

$$x(t) = A_0 + \sum_{n=1}^{\infty} A_n \cos(n\omega_0 t + \varphi_n) = A_0 + \sum_{n=1}^{\infty} A_n \cos(2\pi n f_0 t + \varphi_n), \quad (2.1)$$

where A_0 is the amplitude of the d.c. term, A_n is the amplitude of the nth harmonic with the fundamental angular frequency $n\omega_0 = 2\pi n f_0$, and φ_n is its initial phase angle.

Definition 2.2 The set $\{A_0, A_1, A_2, \ldots\}$ is called the *one-sided amplitude spectrum*, the set $\{\varphi_1, \varphi_2, \ldots\}$—the *one-sided phase spectrum*, and $\{A_0^2, A_1^2/2, A_2^2/2, \ldots\}$

—the *one-sided power spectrum* of a periodic signal $x(t)$, $t \in \mathbb{R}$.

All spectra are represented as functions of $n \in N \cup \{0\}$ (the harmonic index) in the form of peaks (discrete representation). It should be noted that the index n corresponds to the frequency nf_0.

Theorem 2.1 (*Parseval's equality*) The average power (for period) of a periodic signal $x(t)$, $t \in \mathbb{R}$ is

$$P_x = A_0^2 + \sum_{n=1}^{\infty} \frac{A_n^2}{2}, \quad (2.2)$$

where A_0^2 is the average power of the d.c. term and $\frac{A_n^2}{2}$—the average power of the nth harmonic. Parseval's equality is a basis for the definition of the total harmonic distorsion (THD) coefficient, given by

$$\text{THD} = \sqrt{\frac{\sum_{n=2}^{\infty} A_n^2}{\sum_{n=1}^{\infty} A_n^2}}, \quad (2.3)$$

expressing the average power contained in higher harmonics ($n \geq 2$) w.r.t. the average power in all harmonics ($n \geq 1$), usually in percents.

Problem 2.1 Given the signal $x(t) = 2 - 2\cos\left(\frac{1}{3}\pi t\right) + \sin\left(\frac{1}{2}\pi t\right)$. Determine and plot its amplitude- and phase spectra (one-sided), and calculate the percentage of the average power in the frequency band $[0, 2f_0]$.

Solution First, let us calculate the fundamental frequency of $x(t)$. We can apply the method proposed in Problem 1.1. The fundamental frequency of the component $\cos\left(\frac{1}{3}\pi t\right)$ is $f_1 = \frac{1}{6}$ [Hz], so $T_1 = 6$ [s]. For $\sin\left(\frac{1}{2}\pi t\right)$, we have $f_2 = \frac{1}{4}$ [Hz], then

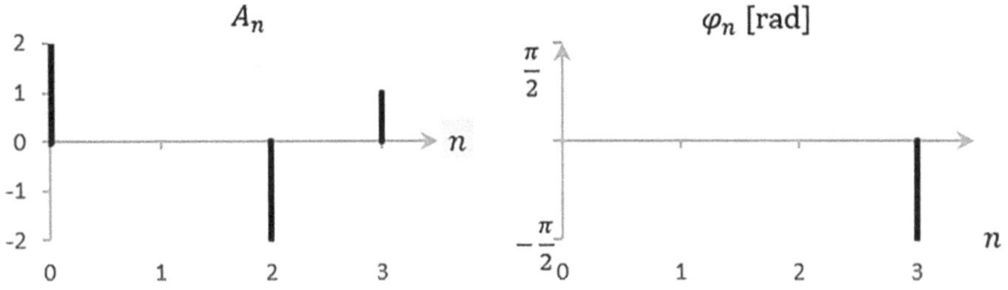

Fig. 2.1 Amplitude spectrum (left) and phase spectrum (right) of $x(t)$ from Problem 2.1

$T_2 = 4$ [s]. The fundamental period of $x(t)$, $T_0 = \text{LCM}(6, 4) = 12$, the fundamental frequency $f_0 = \frac{1}{12}$ [Hz]. We have, $f_1 = 2f_0$ and $f_2 = 3f_0$. The signal $x(t)$ is a sum of the constant equal 2 and of its 2nd- and 3rd harmonics.

To obtain the trigonometric Fourier series (2.1) of $x(t) = 2 - 2\cos\left(\frac{1}{3}\pi t\right) + \sin\left(\frac{1}{2}\pi t\right)$, we can apply a known trigonometric equality: $\sin(x) = \cos\left(x - \frac{\pi}{2}\right)$ and get

$$x(t) = 2 - 2\cos\left(\frac{1}{3}\pi t\right) + \cos\left(\frac{1}{2}\pi t - \frac{\pi}{2}\right).$$

The amplitude spectrum is $\{A_0, A_1, A_2, A_3\} = \{2, 0, -2, 1\}$ and the phase spectrum is $\{\varphi_1, \varphi_2, \varphi_3\} = \{0, 0, -\frac{\pi}{2}\}$. Their plots are displayed in Fig. 2.1.

Important The equivalent frequency representation of $x(t)$ can be the amplitude spectrum $\{2, 0, 2, 1\}$ and the phase spectrum $\{0, \pi, \frac{\pi}{2}\}$ since $-2\cos\left(\frac{1}{3}\pi t\right) = 2\cos\left(\frac{1}{3}\pi t + \pi\right)$. Evidently, the angle $+\pi$ can be replaced with $-\pi$.

The average power in the frequency band $[0, 2f_0]$ can be calculated using Parseval's equality (2.2):

$$P_{[0,2f_0]} = A_0^2 + \sum_{n=1}^{2} \frac{A_n^2}{2} = 5$$

while the average power of $x(t)$ is

$$P_x = A_0^2 + \sum_{n=1}^{3} \frac{A_n^2}{2} = 5,5.$$

The calculated power percentage is $10/11 \cdot 100\ \% \approx 91\ \%$.

If we develop the formula (2.1):

$$x(t) = A_0 + \sum_{n=1}^{\infty} A_n \cos(n\omega_0 t + \varphi_n),$$

we can get an equivalent trigonometric Fourier series of $x(t)$ in the form of a sum of a constant term and of sine and cosine harmonic components:

$$x(t) = A_0 + \sum_{n=1}^{\infty} (A_n \cos \varphi_n) \cos(n\omega_0 t) + \sum_{n=1}^{\infty} (-A_n \sin \varphi_n) \sin(n\omega_0 t).$$

Usually, the coefficients (A_n cos φ_n) are denoted with a_n and ($-A_n$ sin φ_n) as b_n. We have

$$\begin{gathered} x(t) = A_0 + \sum\nolimits_{n=1}^{\infty} a_n \cos(n\omega_0 t) + \sum\nolimits_{n=1}^{\infty} b_n \sin(n\omega_0 t), \\ a_n = A_n \cos \varphi_n, b_n = -A_n \sin \varphi_n \end{gathered} \tag{2.4a}$$

or equivalently,

$$x(t) = \frac{a_0}{2} + \sum\nolimits_{n=1}^{\infty} a_n \cos(n\omega_0 t) + \sum\nolimits_{n=1}^{\infty} b_n \sin(n\omega_0 t). \tag{2.4b}$$

The coefficients a_n ($n = 0, 1, 2\ldots$) and b_n ($n = 1, 2\ldots$) of (2.4a and 2.4b) are defined as

$$a_n = \frac{2}{T_0} \int_{\langle T_0 \rangle} x(t) \cos(n\omega_0 t) dt, \quad b_n = \frac{2}{T_0} \int_{\langle T_0 \rangle} x(t) \sin(n\omega_0 t) dt. \tag{2.5}$$

Important For periodic signals with even parity, we have $b_n = 0$ for all n. If a signal is odd, all $a_n = 0$. Moreover, if a signal is antisymmetric, that is, $x(t) = -x(t + T_0/2)$, in its Fourier series (2.4a and 2.4b), there are only odd-indexed components and $A_0 = 0$.

There are the following relations between coefficients (2.1) and (2.5):

$$A_0 = \frac{a_0}{2}, \quad A_n = \sqrt{a_n^2 + b_n^2}. \tag{2.6}$$

In Problem 2.2, we show how to exploit relations (2.6) to obtain the trigonometric Fourier series (2.1) of a specific periodic signal.

Problem 2.2 Given the periodic signal $x(t) = 2 - \frac{1}{2}\cos(\pi t) + \frac{\sqrt{3}}{2}\sin(\pi t)$ $+\frac{1}{2}\sin(3\pi t)$. Determine its one-sided amplitude- and phase spectra, and calculate the THD coefficient.

Solution Let us notice that $x(t)$ is a sum of the constant equal 2, of the 1st- and of the 3rd harmonics:

$$x(t) = A_0 + A_1\cos(2\pi f_0 t + \varphi_1) + A_3\cos(2\pi 3 f_0 t + \varphi_3).$$

The fundamental frequency $f_0 = 0.5$ Hz. Comparing $x(t)$ with (2.4a), we immediately get $A_0 = 2$, $a_1 = -\frac{1}{2}$, $b_1 = \frac{\sqrt{3}}{2}$ and $b_3 = \frac{1}{2}$. Using (2.6), we can obtain the amplitude of the 1st harmonic: $A_1 = \sqrt{a_1^2 + b_1^2} = \pm 1$. Assuming that $A_1 = 1$, we can calculate the initial phase of the 1st harmonic using (2.4a) as follows:

$$\begin{cases} a_1 = A_1\cos\varphi_1 \Leftrightarrow -\dfrac{1}{2} = \cos\varphi_1 \\ b_1 = -A_1\sin\varphi_1 \Leftrightarrow \dfrac{\sqrt{3}}{2} = -\sin\varphi_1 \end{cases}$$

The above equations are true for $\varphi_1 = \frac{4}{3}\pi$. If we assume that $A_1 = -1$, the reader can easily check that $\varphi_1 = \frac{\pi}{3}$. Having all the above considerations in mind, we can write $x(t)$ as

$$x(t) = 2 + \cos\left(2\pi t + \frac{4}{3}\pi\right) + \frac{1}{2}\cos\left(3\pi t - \frac{\pi}{2}\right)$$

or equivalently as

$$x(t) = 2 - \cos\left(2\pi t + \frac{\pi}{3}\right) + \frac{1}{2}\cos\left(3\pi t - \frac{\pi}{2}\right).$$

It is now easy to plot one-sided amplitude- and phase spectra (see Fig. 2.2) analogously as in Problem 2.1.

Let us now calculate THD using (2.3):

$$\text{THD} = \sqrt{\frac{A_3^2}{A_1^2 + A_3^2}} = \sqrt{\frac{0,25}{1,25}} \cong 0,4472 = 44,72\%.$$

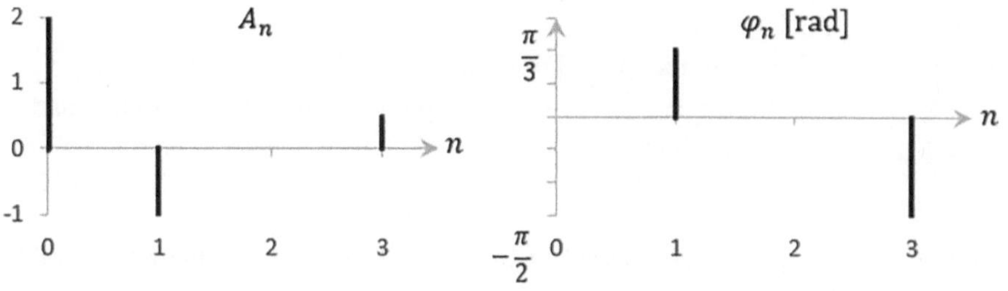

Fig. 2.2 Amplitude spectrum (left) and phase spectrum (right) of $x(t)$ from Problem 2.2

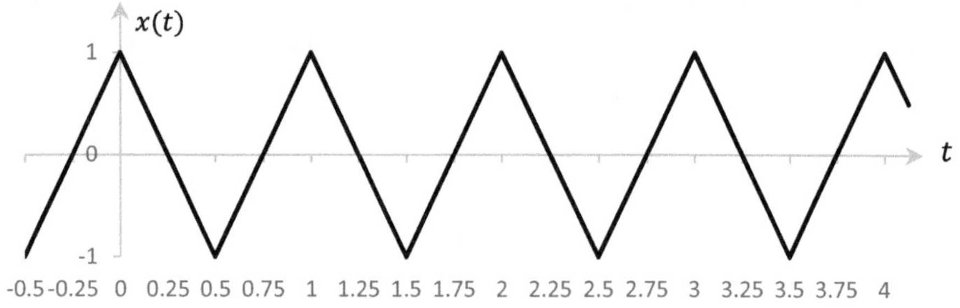

Fig. 2.3 Triangular waveform $x(t)$ from Problem 2.3

In Problems 2.1 and 2.2, we dealt with signals in the form of a finite sum of a few harmonics. In Problems 2.3 and 2.4, we develop the formula for the infinite Fourier series of a triangular bipolar waveform.

Problem 2.3 Given a triangular even-symmetric bipolar waveform with the duty factor $d = \frac{1}{2}$, the amplitude $A = 1$ and the fundamental period $T_0 = 1$ [s] displayed in Fig. 2.3. Calculate the coefficients of the Fourier series (2.4a).

Solution It can be noticed that $x(t)$ is an even-symmetric function defined in the interval $t \in [0, 1)$ by

$$x(t) = \begin{cases} -4\left(t - \frac{1}{4}\right) & \text{for } t \in \left[0, \frac{1}{2}\right), \\ 4\left(t - \frac{3}{4}\right) & \text{for } t \in \left[\frac{1}{2}, 1\right). \end{cases}$$

We can see that the average value of $x(t)$ is $A_0 = 0$ (the signal is antisymmetric, as shown in Fig. 2.3). Since $T_0 = 1$, then $\omega_0 = 2\pi$ [rad/s]. Using (2.5), we can calculate the coefficients a_n of the trigonometric Fourier series (2.4a):

$$a_n = 2\left[-\int_0^{\frac{1}{2}} 4\left(t-\frac{1}{4}\right)\cos(2\pi nt)dt + \int_{\frac{1}{2}}^{1} 4\left(t-\frac{3}{4}\right)\cos(2\pi nt)dt\right]$$
$$= 2\left[-4\int_0^{\frac{1}{2}} t\cos(2\pi nt)dt + \int_0^{\frac{1}{2}} \cos(2\pi nt)dt + 4\int_{\frac{1}{2}}^{1} t\cos(2\pi nt)dt - 3\int_{\frac{1}{2}}^{1} \cos(2\pi nt)dt\right].$$

The second and the fourth integrals in the above expression are zero (the reader can easily check this). Continuing, we get

$$a_n = 2\left\{-4\left[\frac{t}{2\pi n}\sin(2\pi nt)\Big|_0^{\frac{1}{2}} - \frac{1}{2\pi n}\int_0^{\frac{1}{2}} \sin(2\pi nt)dt\right]\right.$$
$$\left. +4\left[\frac{t}{2\pi n}\sin(2\pi nt)\Big|_{\frac{1}{2}}^{1} - \frac{1}{2\pi n}\int_{\frac{1}{2}}^{1} \sin(2\pi nt)dt\right]\right\}$$
$$= \frac{4}{\pi n}\left\{\left[\frac{-1}{2\pi n}\cos(2\pi nt)\right]\Big|_0^{\frac{1}{2}} - \left[\frac{-1}{2\pi n}\cos(2\pi nt)\right]\Big|_{\frac{1}{2}}^{1}\right\}$$
$$= \frac{2}{\pi^2 n^2}[-\cos(\pi n) + 1 + \cos(2\pi n) - \cos(\pi n)] = \frac{4}{\pi^2 n^2}[1-(-1)^n].$$

The Fourier series of an even- and antisymmetric bipolar triangle waveform is

$$x(t) = \sum_{n=1}^{\infty} \frac{4}{\pi^2 n^2}[1-(-1)^n]\cos(2\pi nt)$$
$$= \frac{8}{\pi^2}\left[\cos(2\pi t) + \frac{1}{9}\cos(6\pi t) + \frac{1}{25}\cos(10\pi t) + \frac{1}{49}\cos(14\pi t) + \ldots\right].$$

Problem 2.4 Using the results of Problem 2.3, determine the coefficients of the Fourier series of a periodic waveform *y*(*t*) shown in Fig. 2.4.

Solution We can notice that $y(t) = \frac{1}{2}x\left(t-\frac{1}{4}\right)$ where *x*(*t*) is represented by the Fourier series developed in Problem 2.3. We have

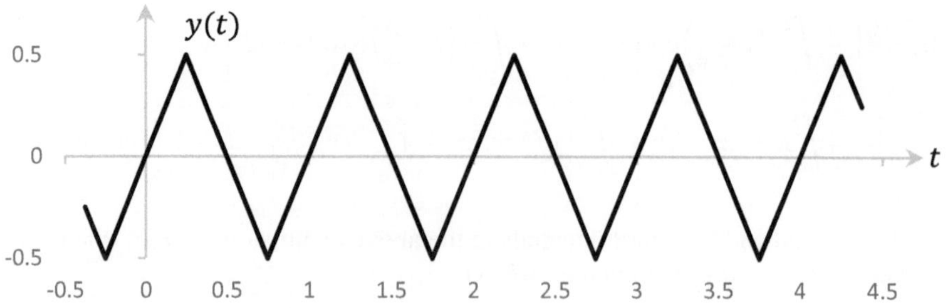

Fig. 2.4 Triangular waveform $y(t)$ from Problem 2.4

$$y(t) = \frac{1}{2}x\left(t - \frac{1}{4}\right) = \sum_{n=1}^{\infty} \frac{2}{\pi^2 n^2}[1 - (-1)^n] \cos\left(2\pi n\left(t - \frac{1}{4}\right)\right)$$
$$= \sum_{n=1}^{\infty} \frac{2}{\pi^2 n^2}[1 - (-1)^n] \cos\left(2\pi nt - \frac{\pi}{2}n\right)$$
$$= \frac{4}{\pi^2}\left[\cos\left(2\pi t - \frac{\pi}{2}\right) + \frac{1}{9}\cos\left(6\pi t - \frac{3\pi}{2}\right) + \frac{1}{25}\cos\left(10\pi t - \frac{5\pi}{2}\right) + \frac{1}{49}\cos\left(14\pi t - \frac{7\pi}{2}\right) + \ldots\right]$$
$$= \frac{4}{\pi^2}\left[\sin(2\pi t) - \frac{1}{9}\sin(6\pi t) + \frac{1}{25}\sin(10\pi t) - \frac{1}{49}\sin(14\pi t) + \ldots\right].$$

We have shown that an odd- and antisymmetric periodic signal can be developed into a sine Fourier series of odd-indexed harmonics.

2.1.2 *Exponential Fourier Series*

In this subsection, we will apply Euler's formulas

$$\cos(\omega t) = \frac{1}{2}\left(e^{j\omega t} + e^{-j\omega t}\right), \tag{2.7a}$$

$$\sin(\omega t) = \frac{1}{2j}\left(e^{j\omega t} - e^{-j\omega t}\right), \tag{2.7b}$$

to obtain the equivalent form of the (2.4a) known as the *exponential Fourier series.* We have

$$x(t) = A_0 + \sum_{n=1}^{\infty} a_n \cos(n\omega_0 t) + \sum_{n=1}^{\infty} b_n \sin(n\omega_0 t)$$
$$= A_0 + \sum_{n=1}^{\infty} a_n \frac{1}{2}\left(e^{jn\omega_0 t} + e^{-jn\omega_0 t}\right) + \sum_{n=1}^{\infty} b_n \frac{1}{2j}\left(e^{jn\omega_0 t} - e^{-jn\omega_0 t}\right).$$

Let us reorganize the above expression as follows:

$$x(t) = A_0 + \sum_{n=1}^{\infty} \frac{1}{2}(a_n - jb_n)e^{jn\omega_0 t} + \sum_{n=1}^{\infty} \frac{1}{2}(a_n + jb_n)e^{-jn\omega_0 t}.$$

Introducing the new notation:

$$c_n = \frac{1}{2}(a_n - jb_n), \;\; c_n^* = \frac{1}{2}(a_n + jb_n) \text{ and } c_0 = A_0,$$

we can rewrite $x(t)$ to get the definition of the exponential Fourier series.

Definition 2.3 The *exponential Fourier series* of a periodic signal $x(t)$, $t \in \mathbb{R}$ is defined as

$$x(t) = c_0 + \sum\nolimits_{n=1}^{\infty} c_n e^{jn\omega_0 t} + \sum\nolimits_{n=1}^{\infty} c_n^* e^{-jn\omega_0 t} \tag{2.8}$$

or equivalently as

$$x(t) = \sum_{n \in \mathbb{Z}} c_n e^{jn\omega_0 t} \tag{2.9}$$

where the coefficients c_n are defined by

$$c_n = \frac{1}{T_0} \int_{\langle T_0 \rangle} x(t) e^{-jn\omega_0 t} dt. \tag{2.10}$$

There are some relations between coefficients (2.10) and (2.5):

$$c_0 = \frac{a_0}{2}, \;\; a_n = 2 \cdot \mathrm{Re}\{c_n\}, b_n = -2 \cdot \mathrm{Im}\{c_n\} \;\text{ for }\; n > 1 \tag{2.11}$$

Definition 2.4 The set $\{|c_n|, n \in \mathbb{Z}\}$ is called the *two-sided amplitude spectrum*, $\{\arg c_n, n \in \mathbb{Z}\}$—the *two-sided phase spectrum*, and $\{|c_n|^2, n \in \mathbb{Z}\}$—the *two-sided power spectrum* of a periodic signal $x(t)$, $t \in \mathbb{R}$.

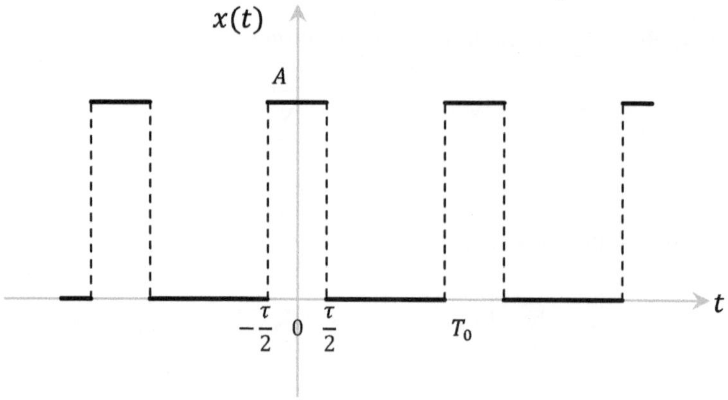

Fig. 2.5 Unipolar rectangular waveform $x(t)$ from Problem 2.3

All spectra are represented as functions of $n \in \mathbb{Z}$ in the form of peaks (discrete representation). Evidently, the index $n \geq 0$ corresponds to frequencies nf_0. For $n < 0$, we deal with "negative frequencies"—a pure mathematical concept. The exponential Fourier series (2.9) is a basis for the definition of the complex Fourier transform described in Sect. 2.2 of this chapter.

Theorem 2.2 (*Parseval's equality*) The average power (per period) of a periodic signal $x(t)$, $t \in \mathbb{R}$ is calculated as

$$P_x = \sum\nolimits_{n \in \mathbb{Z}} |c_n|^2. \tag{2.12}$$

Problem 2.5 Given the even-symmetric unipolar rectangular waveform with fundamental period T_0, amplitude A, and pulse duration τ displayed in Fig. 2.5. Calculate the coefficients of the exponential Fourier series (2.10).

Solution We can calculate the exponential Fourier series coefficients using (2.10). We have ($\omega_0 = 2\pi/T_0$)

$$c_n = \frac{1}{T_0} \int\limits_{-\frac{\tau}{2}}^{\frac{\tau}{2}} Ae^{-jn\omega_0 t} dt = \frac{A}{T_0} \frac{e^{-jn\omega_0 t}}{-jn\omega_0} \Bigg|_{-\frac{\tau}{2}}^{\frac{\tau}{2}} = \frac{A}{2\pi jn} \left(e^{jn\omega_0 \frac{\tau}{2}} - e^{-jn\omega_0 \frac{\tau}{2}} \right).$$

In the above expression, we can recognize Euler's formula for sinus (2.7b). Further,

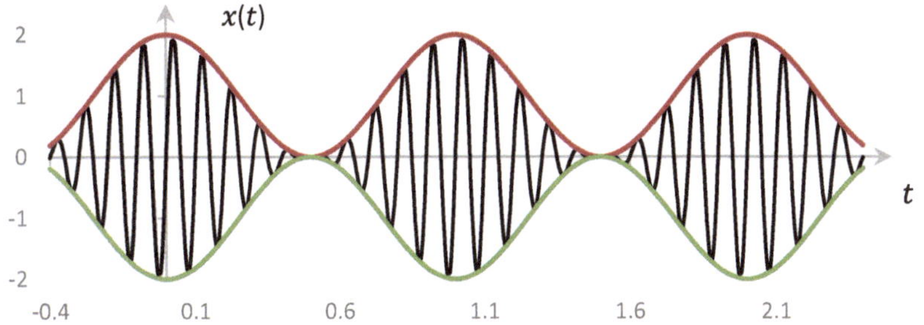

Fig. 2.6 Amplitude-modulated signal $x(t)$ from Problem 2.6

$$c_n = \frac{A}{2\pi jn} 2j \sin\left(n\omega_0 \frac{\tau}{2}\right) = \frac{A}{\pi n} \sin\left(n\omega_0 \frac{\tau}{2}\right) = \frac{A}{\pi n} \sin\left(n\pi \frac{\tau}{T_0}\right).$$

Using (1.18), we can express the exponential Fourier series coefficients as $c_n = \frac{A\tau}{T_0} \mathrm{Sa}\left(n\pi \frac{\tau}{T_0}\right)$.

Problem 2.6 Given the periodic signal

$$x(t) = [1 + \cos(2\pi t)] \sin(20\pi t).$$

Find its exponential Fourier series and sketch its two-sided amplitude- and phase spectra. Calculate the percentage of power included in the sidebands.

Remark: The signal $x(t)$ is the *amplitude-modulated (AM) signal*, $\sin(20\pi t)$ is called a *carrier*, and $\cos(2\pi t)$ is a *modulating signal.*

Solution First, let us sketch three periods of $x(t)$ (see Fig. 2.6). We see that the amplitude of the high-frequency carrier is changing periodically (in other words, it is *modulated*). The upper envelope (marked in red) is $1 + \cos(2\pi t)$, and the lower envelope (in green) is $-[1 + \cos(2\pi t)]$. The fundamental period of $x(t)$ equals 1 [s].

Now, let us apply Euler's formulas (2.7a and 2.7b) for $x(t)$:

$$x(t) = \left[1 + \frac{1}{2}\left(e^{j2\pi t} + e^{-j2\pi t}\right)\right] \frac{1}{2j}\left(e^{j20\pi t} - e^{-j20\pi t}\right)$$

$$= \frac{1}{2j} e^{j20\pi t} - \frac{1}{2j} e^{-j20\pi t} + \frac{1}{4j} e^{j22\pi t} - \frac{1}{4j} e^{-j18\pi t} + \frac{1}{4j} e^{j18\pi t} - \frac{1}{4j} e^{-j22\pi t}$$

$$= \frac{1}{4} je^{-j22\pi t} + \frac{1}{2} je^{-j20\pi t} + \frac{1}{4} je^{-j18\pi t} - \frac{1}{4} je^{j18\pi t} - \frac{1}{2} je^{j20\pi t} - \frac{1}{4} je^{j22\pi t}.$$

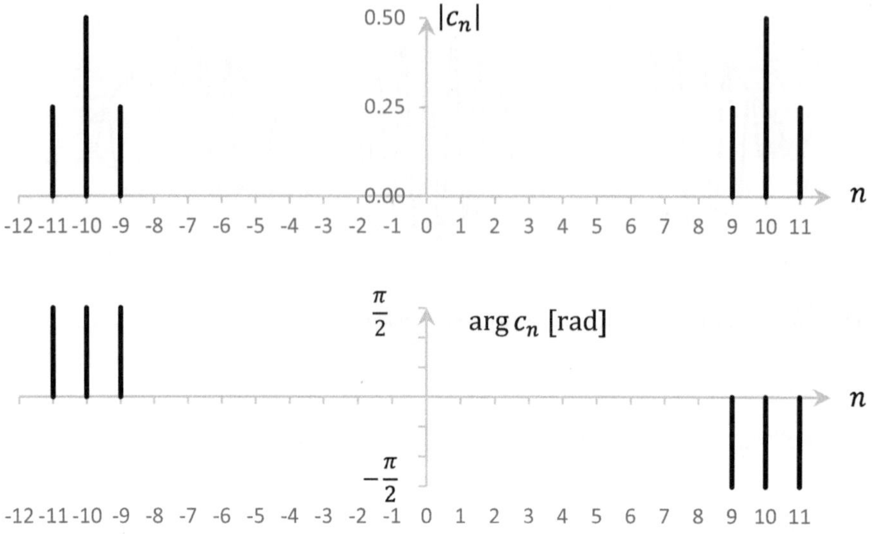

Fig. 2.7 Two-sided amplitude- (up) and phase (down) spectra of $x(t)$ from Problem 2.6

We can see that $x(t)$ is a sum of its 9th-, 10th -, and 11th harmonics. Then we have: $c_{-11} = c_{-9} = \frac{1}{4}j$, $c_{-10} = \frac{1}{2}j$, $c_{10} = -\frac{1}{2}j$, $c_9 = c_{11} = -\frac{1}{4}j$.

The two-sided amplitude spectrum: $|c_{-11}| = |c_{-9}| = |c_9| = |c_{11}| = \frac{1}{4}$, $|c_{-10}| = |c_{10}| = \frac{1}{2}$. The two-sided phase spectrum: $\arg c_{-11} = \arg c_{-10} = \arg c_{-9} = \frac{\pi}{2}$, $\arg c_9 = \arg c_{10} = \arg c_{11} = -\frac{\pi}{2}$. Both spectra are displayed in Fig. 2.7. In the amplitude spectrum, we can observe two peaks of amplitude $\frac{1}{2}$ at ± 10 [Hz] (for $n = \pm 10$) and two sideband peaks at ± 9 [Hz] and ± 11 [Hz] (resp. $n = \pm 9$ and $n = \pm 11$).

Let us calculate the power P_{sideband} of the signal $x(t)$ in the sideband. According to (2.11), we have

$$P_{\text{sideband}} = |c_{-11}|^2 + |c_{-9}|^2 + |c_9|^2 + |c_{11}|^2 = 4 \cdot \frac{1}{16} = \frac{1}{4},$$

while the total power P_x (for period) is

$$P_x = P_{\text{sideband}} + |c_{-10}|^2 + |c_{10}|^2 = \frac{1}{4} + 2 \cdot \frac{1}{4} = \frac{3}{4}.$$

The power percentage in the sideband is

$$\frac{P_{\text{sideband}}}{P_x} \cdot 100\% = 33,3(3)\%.$$

2.1.3 Fourier Synthesis

If we come back to Sect. 1.5 (Orthogonalization and approximation) from Chap. 1, we can notice analogies between the formula (1.59) of the generalized Fourier series and expressions (2.1) and (2.9) defining trigonometric and exponential Fourier series of a periodic continuous-time signal $x(t)$. Now, let us define the partial sum of N harmonics of (2.1):

$$x_N(t) = A_0 + \sum_{n=1}^{N} A_n \cos(n\omega_0 t + \varphi_n) \tag{2.13}$$

or, since $A_0 = \frac{a_0}{2}$ as in (2.6),

$$x_N(t) = \frac{a_0}{2} + \sum_{n=1}^{N} a_n \cos(n\omega_0 t) + \sum_{n=1}^{N} b_n \sin(n\omega_0 t). \tag{2.14}$$

Both the above formulas can be used for the approximation of a periodic signal $x(t)$ with a finite sum of N successive harmonics and a constant $A_0 = \frac{a_0}{2}$. Such a procedure is called the *Fourier synthesis*. The mean-square error of this approximation is defined similarly as in (1.62) and equals the sum of average powers of all nth order harmonics for $n = N + 1, N + 2, \ldots$, that is,

$$\varepsilon_N = \sum_{n=N+1}^{\infty} \frac{A_n^2}{2} \tag{2.15a}$$

or equivalently, using (2.6), as

$$\varepsilon_N = \sum_{n=N+1}^{\infty} \frac{a_n^2 + b_n^2}{2} \tag{2.15b}$$

Parseval's equality (2.11 and 2.12) permits also to define ε_N in the following form:

$$\varepsilon_N = 2\sum_{n=N+1}^{\infty} |c_n|^2. \tag{2.16}$$

Let us now apply the solution of Problem 2.5 to approximate the unipolar even-symmetric rectangular waveform with a finite sum of N harmonics and to calculate the error ε_N of such approximation.

Problem 2.7 Given the even-symmetric unipolar rectangular waveform $x(t)$ with fundamental period $T_0 = 4$ [s], amplitude $A = 1$, and pulse duration $\tau = 1$ [s] displayed in Fig. 2.5. Using (2.13), approximate $x(t)$ with the sum $x_N(t)$ of a constant A_0 and N harmonics, $N = 1, 3, 5, 7, 9$. In each case, sketch plots of $x_N(t)$ and calculate the approximation error ε_N.

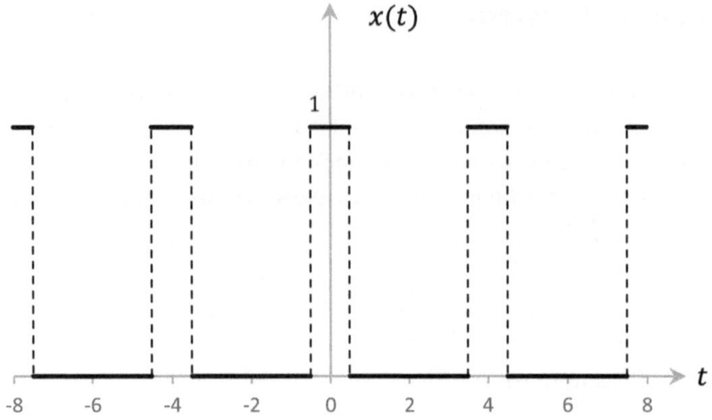

Fig. 2.8 Unipolar rectangular waveform $x(t)$ from Problem 2.7

Solution Let us sketch the plot of $x(t)$ (see Fig. 2.8) and calculate its average power P_x using (1.29). The average power will serve for the calculation of the approximation error ε_N. We have

$$P_x = \frac{1}{4} \int_{-\frac{1}{2}}^{\frac{1}{2}} 1^2 dt = \frac{1}{4}.$$

The coefficients c_n of the exponential Fourier series of $x(t)$ calculated in Problem 2.5 are $c_n = \frac{A\tau}{T_0} \mathrm{Sa}\left(n\pi \frac{\tau}{T_0}\right)$. For $T_0 = 4$, $A = 1$, $\tau = 1$, we get $c_n = \frac{1}{4} \mathrm{Sa}\left(n\frac{\pi}{4}\right)$ and $\omega_0 = \frac{\pi}{2}$ [rad/s]. For $n = 0, 1, \ldots, 9$, the successive coefficients c_n are:

$$\boldsymbol{c_0} = \frac{1}{4}\mathrm{Sa}(0) = \frac{1}{4}; \boldsymbol{c_1} = \frac{1}{4}\mathrm{Sa}\left(\frac{\pi}{4}\right) = \frac{1}{4}\frac{\sin\left(\frac{\pi}{4}\right)}{\frac{\pi}{4}} = \frac{\sqrt{2}}{2\pi};$$

$$\boldsymbol{c_2} = \frac{1}{4}\mathrm{Sa}\left(\frac{\pi}{2}\right) = \frac{1}{4}\frac{\sin\left(\frac{\pi}{2}\right)}{\frac{\pi}{2}} = \frac{1}{2\pi}; \boldsymbol{c_3} = \frac{1}{4}\mathrm{Sa}\left(\frac{3\pi}{4}\right) = \frac{1}{4}\frac{\sin\left(\frac{3\pi}{4}\right)}{\frac{3\pi}{4}} = \frac{\sqrt{2}}{6\pi};$$

$$\boldsymbol{c_4} = \frac{1}{4}\mathrm{Sa}(\pi) = 0; \boldsymbol{c_5} = \frac{1}{4}\mathrm{Sa}\left(\frac{5\pi}{4}\right) = \frac{1}{4}\frac{\sin\left(\frac{5\pi}{4}\right)}{\frac{5\pi}{4}} = -\frac{\sqrt{2}}{10\pi};$$

$$\boldsymbol{c_6} = \frac{1}{4}\mathrm{Sa}\left(\frac{3\pi}{2}\right) = \frac{1}{4}\frac{\sin\left(\frac{3\pi}{2}\right)}{\frac{3\pi}{2}} = -\frac{1}{6\pi}; \boldsymbol{c_7} = \frac{1}{4}\mathrm{Sa}\left(\frac{7\pi}{4}\right) = \frac{1}{4}\frac{\sin\left(\frac{7\pi}{4}\right)}{\frac{7\pi}{4}} = -\frac{\sqrt{2}}{14\pi};$$

$$\boldsymbol{c}_8 = \frac{1}{4}\mathrm{Sa}(2\pi) = 0; \boldsymbol{c}_9 = \frac{1}{4}\mathrm{Sa}\left(\frac{9\pi}{4}\right) = \frac{1}{4}\frac{\sin\left(\frac{9\pi}{4}\right)}{\frac{9\pi}{4}} = \frac{\sqrt{2}}{18\pi}.$$

Applying (2.11 and 2.12): $a_n = 2 \cdot \mathrm{Re}\,\{c_n\}$, $b_n = -2 \cdot \mathrm{Im}\,\{c_n\}$, we can obtain

$$a_0 = \frac{1}{2}; a_1 = \frac{\sqrt{2}}{\pi};\ a_2 = \frac{1}{\pi};\ a_3 = \frac{\sqrt{2}}{3\pi};\ a_4 = 0;\ a_5 = -\frac{\sqrt{2}}{5\pi};\ a_6 = -\frac{1}{3\pi};$$

$$a_7 = -\frac{\sqrt{2}}{7\pi};\ a_8 = 0; a_9 = \frac{\sqrt{2}}{9\pi}$$

and all b_n are zero.

Then using (2.14) for $N = 1, 3, 5, 7, 9$, we can get the corresponding partial Fourier sums $x_N(t)$:

$$x_1(t) = \frac{1}{4} + \frac{\sqrt{2}}{\pi}\cos\left(\frac{\pi t}{2}\right);$$

$$x_3(t) = \frac{1}{4} + \frac{\sqrt{2}}{\pi}\cos\left(\frac{\pi t}{2}\right) + \frac{1}{\pi}\cos(\pi t) + \frac{\sqrt{2}}{3\pi}\cos\left(\frac{3\pi t}{2}\right);$$

$$x_5(t) = \frac{1}{4} + \frac{\sqrt{2}}{\pi}\cos\left(\frac{\pi t}{2}\right) + \frac{1}{\pi}\cos(\pi t) + \frac{\sqrt{2}}{3\pi}\cos\left(\frac{3\pi t}{2}\right) - \frac{\sqrt{2}}{5\pi}\cos\left(\frac{5\pi t}{2}\right);$$

$$x_7(t) = \frac{1}{4} + \frac{\sqrt{2}}{\pi}\cos\left(\frac{\pi t}{2}\right) + \frac{1}{\pi}\cos(\pi t) + \frac{\sqrt{2}}{3\pi}\cos\left(\frac{3\pi t}{2}\right) - \frac{\sqrt{2}}{5\pi}\cos\left(\frac{5\pi t}{2}\right)$$
$$- \frac{1}{3\pi}\cos(3\pi t) - \frac{\sqrt{2}}{7\pi}\cos\left(\frac{7\pi t}{2}\right);$$

$$x_9(t) = \frac{1}{4} + \frac{\sqrt{2}}{\pi}\cos\left(\frac{\pi t}{2}\right) + \frac{1}{\pi}\cos(\pi t) + \frac{\sqrt{2}}{3\pi}\cos\left(\frac{3\pi t}{2}\right) - \frac{\sqrt{2}}{5\pi}\cos\left(\frac{5\pi t}{2}\right)$$
$$- \frac{1}{3\pi}\cos(3\pi t) - \frac{\sqrt{2}}{7\pi}\cos\left(\frac{7\pi t}{2}\right) + \frac{\sqrt{2}}{9\pi}\cos\left(\frac{9\pi t}{2}\right).$$

All partial sums $x_N(t)$ for $N = 1, 3, 5, 7, 9$ are shown in Fig. 2.9. We can notice *Gibb's phenomenon* characterized by the presence of oscillations of higher amplitude near a jump discontinuity in a signal.

The corresponding approximation errors ε_N are calculated using (2.15) from the formula:

$$\varepsilon_N = P_x - \left(\frac{a_0^2}{4} + \sum_{n=1}^{N}\frac{a_n^2}{2}\right),$$

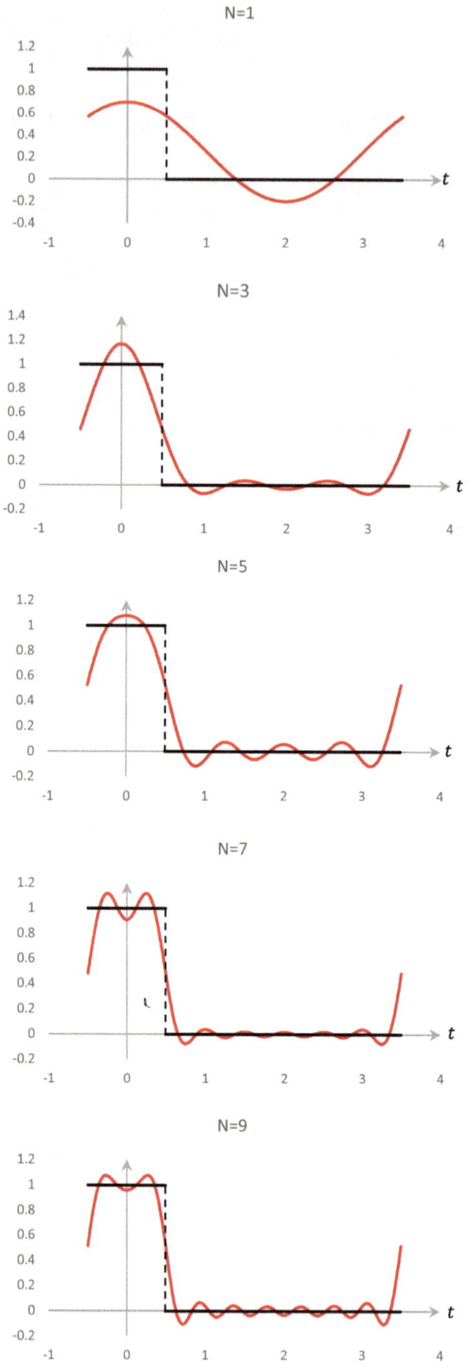

Fig. 2.9 Approximation of $x(t)$ from Problem 2.7 in the interval $t \in \left[-\frac{1}{2}, \frac{7}{2}\right]$ (in black) by the Fourier partial sums $x_N(t)$ for $N = 1, 3, 5, 7, 9$ (in red)

where $\frac{a_0^2}{4}$ is the average power of a constant term $\frac{a_0}{2}$ (compare with Problem 1.5, Chap. 1) and $\sum_{n=1}^{N} \frac{a_n^2}{2}$ is the sum of average powers of all harmonics for $n = 1, 2, \ldots,$ N. We have

$$\varepsilon_1 = \frac{1}{4} - \frac{1}{16} - \frac{1}{\pi^2} \approx 0,0861;$$

$$\varepsilon_3 = \frac{1}{4} - \frac{1}{16} - \left(\frac{1}{\pi^2} + \frac{1}{2\pi^2} + \frac{1}{9\pi^2}\right) \approx 0,0243;$$

$$\varepsilon_5 = \frac{1}{4} - \frac{1}{16} - \left(\frac{1}{\pi^2} + \frac{1}{2\pi^2} + \frac{1}{9\pi^2} + \frac{1}{25\pi^2}\right) \approx 0,0202;$$

$$\varepsilon_7 = \frac{1}{4} - \frac{1}{16} - \left(\frac{1}{\pi^2} + \frac{1}{2\pi^2} + \frac{1}{9\pi^2} + \frac{1}{25\pi^2} + \frac{1}{18\pi^2} + \frac{1}{49\pi^2}\right) \approx 0,0125;$$

$$\varepsilon_9 = \frac{1}{4} - \frac{1}{16} - \left(\frac{1}{\pi^2} + \frac{1}{2\pi^2} + \frac{1}{9\pi^2} + \frac{1}{25\pi^2} + \frac{1}{18\pi^2} + \frac{1}{49\pi^2} + \frac{1}{81\pi^2}\right) \approx 0,0113.$$

In Problem 2.8, we will demonstrate the equivalence of time- and frequency representations of a periodic signal $x(t)$, $t \in \mathbb{R}$. Starting from its amplitude- and phase spectra, we will show how to compute a given signal $x(t)$.

Problem 2.8 Fig. 2.10 shows two-sided amplitude- and phase spectra of a periodic signal $x(t)$ with the angular fundamental frequency $\omega_0 = 20\pi$ [rad/s]. Determine $x(t)$.

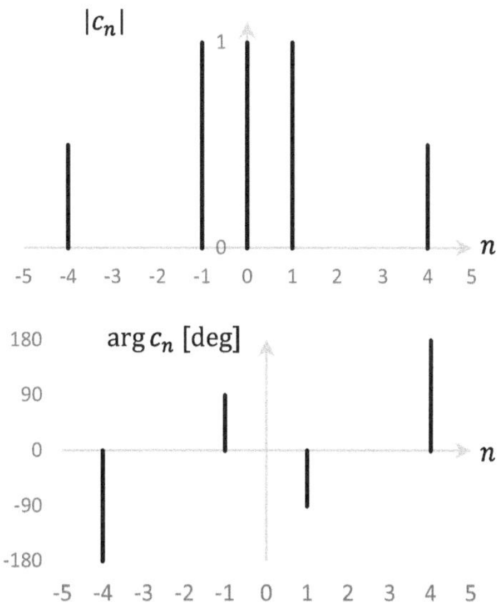

Fig. 2.10 Two-sided amplitude- and phase spectra of $x(t)$ from Problem 2.8

Solution Basing on the values and "position" of peaks of spectra shown in Fig. 2.10, we immediately get coefficients c_n of the exponential Fourier series:

$$c_0 = 1;\ \ c_1 = 1 \cdot e^{-\frac{j\pi}{2}} = -j;\ \ c_{-1} = 1 \cdot e^{\frac{j\pi}{2}} = j;$$

$$c_4 = \frac{1}{2} \cdot e^{j\pi} = -\frac{1}{2};\ \ c_{-4} = \frac{1}{2} \cdot e^{-j\pi} = -\frac{1}{2}.$$

Using (2.11a and 2.11b), we can obtain: $a_1 = 0$; $b_1 = 2$; $a_4 = -1$; $b_4 = 0$, and finally, using (2.4b), we get

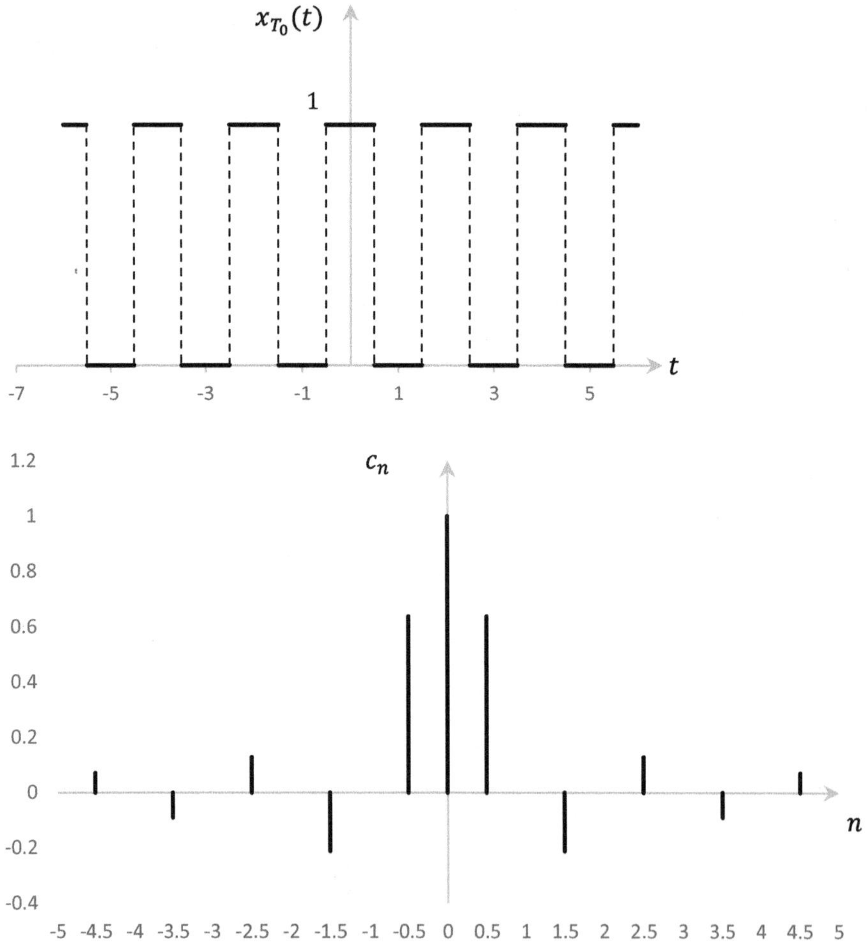

Fig. 2.11 Unipolar rectangular waveform $x_{T_0}(t)$ with fundamental period $T_0 = 2$ and its spectrum

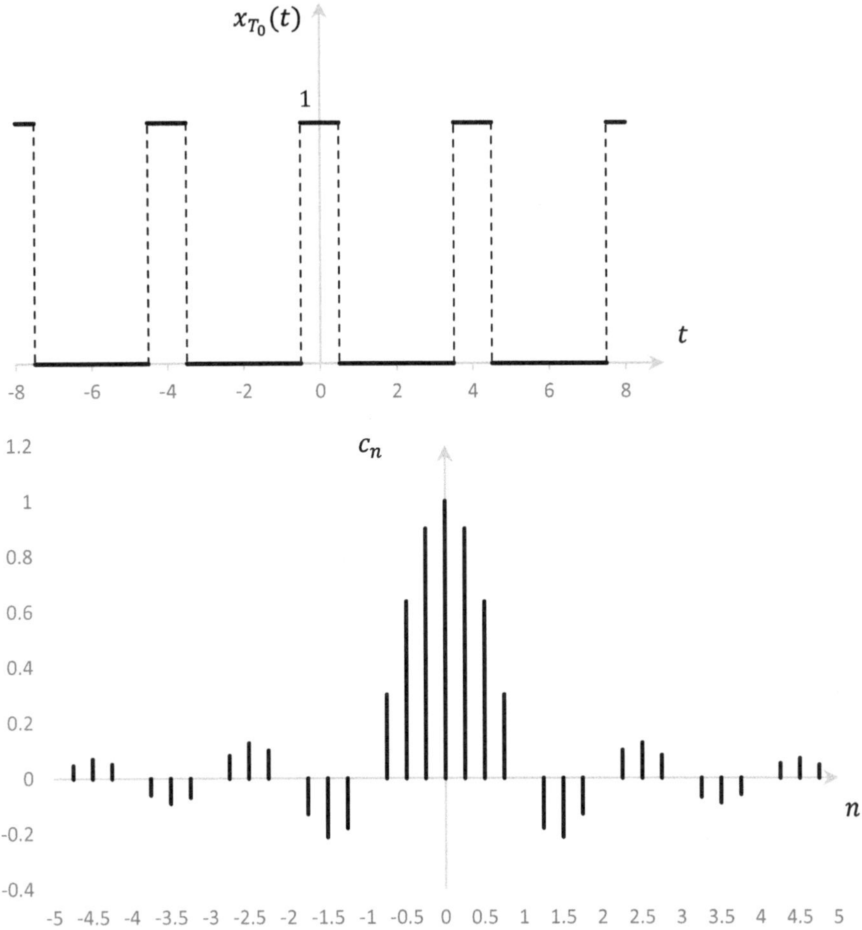

Fig. 2.12 Unipolar rectangular waveform $x_{T_0}(t)$ with fundamental period $T_0 = 4$ and its spectrum

$$x(t) = 1 + 2\sin(\omega_0 t) - \frac{1}{2}\cos(4\omega_0 t) = 1 + 2\sin(20\pi t) - \frac{1}{2}\cos(80\pi t).$$

2.2 Complex Fourier Transform

Let $x_{T_0}(t)$ be a periodic signal with fundamental period T_0. Imagine that T_0 is going up step by step to finally reach ∞. Figures 2.11, 2.12, and 2.13 show results of such operation in time- and frequency domains by the example of a unipolar rectangular waveform with fundamental period $T_0 = 2$, 4, and 8. We can see that when T_0 is

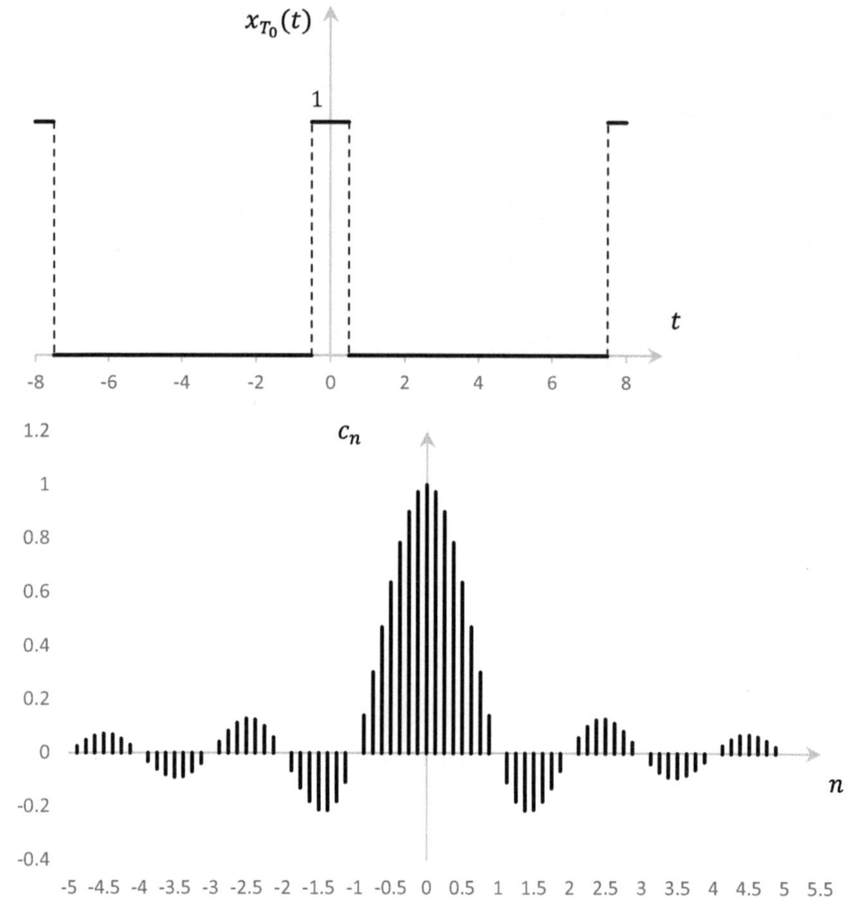

Fig. 2.13 Unipolar rectangular waveform $x_{T_0}(t)$ with fundamental period $T_0 = 8$ and its spectrum

increasing, rectangular pulses are "growing apart" while spectrum peaks (given by coefficients c_n of the exponential Fourier series) at frequencies $\pm nf_0 = \pm \frac{n}{T_0}$ are "bringing closer" (see Fig. 2.14). In the limit $T_0 \to \infty$, we can get a finite-duration signal $x(t)$ with a *continuous-frequency* spectrum $X(f)$.

2.2.1 *Fourier Transform* in the Ordinary Sense

Definition 2.5 The *complex Fourier transform* $X(\omega)$ (*spectrum*) of an *energy signal* $x(t)$ is defined as

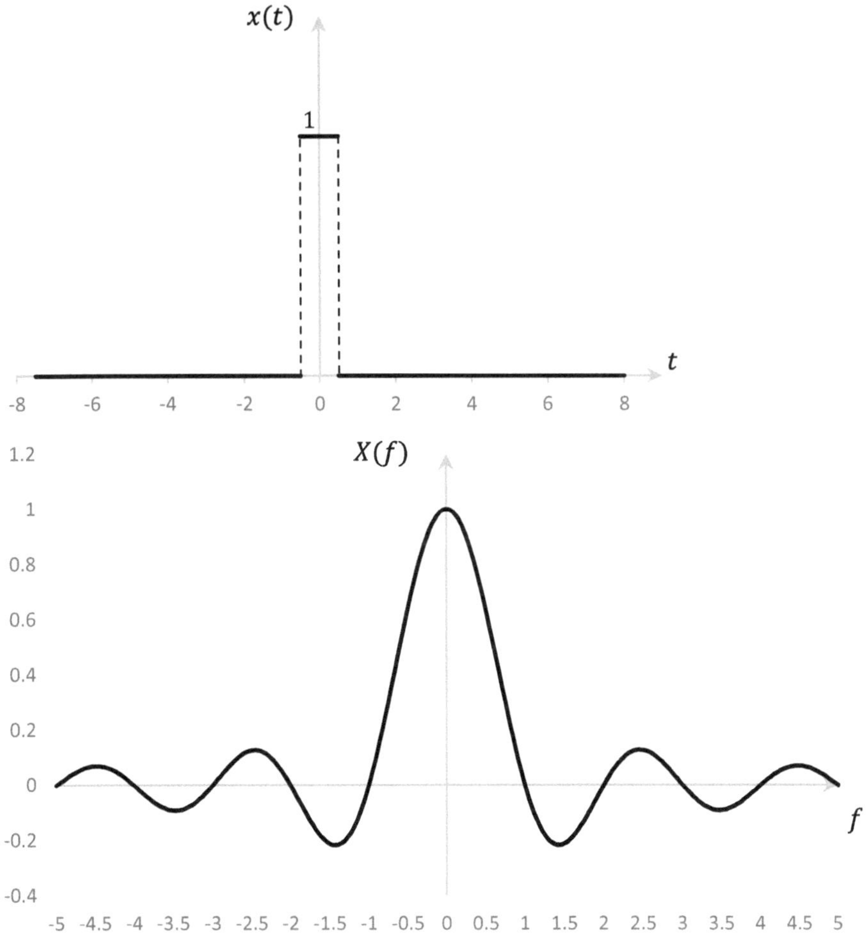

Fig. 2.14 Rectangular pulse $x(t)$ and its continuous-frequency spectrum

$$X(\omega) = \int x(t)e^{-j\omega t}dt. \tag{2.17}$$

Since $\omega = 2\pi f$, the complex FT of $x(t)$ can be expressed as

$$X(f) = \int x(t)e^{-j2\pi ft}dt. \tag{2.18}$$

The transformation defined by (2.17)–(2.18) connects two domains: time with frequency and permits to study signal properties *equivalently* in both domains.

Table 2.1 Fourier transforms of chosen energy and power signals

Signal $x(t)$, $t \in \mathbb{R}$	$X(\omega)$, $\omega \in \mathbb{R}$	$X(f), f \in \mathbb{R}$
$e^{-\alpha t} \cdot \mathbf{1}(t)$ $\alpha > 0$	$\frac{1}{\alpha + j\omega}$	$\frac{1}{\alpha + j2\pi f}$
$t^n e^{-\alpha t} \cdot \mathbf{1}(t)$ $\alpha > 0, n = 1, 2, \ldots$	$\frac{n!}{(\alpha + j\omega)^{n+1}}$	$\frac{n!}{(\alpha + j2\pi f)^{n+1}}$
$e^{-\alpha\|t\|}$ $\alpha > 0$	$\frac{2\alpha}{\alpha^2 + \omega^2}$	$\frac{2\alpha}{\alpha^2 + 4\pi^2 f^2}$
$\frac{1}{t^2 + \alpha^2}$ $\alpha > 0$	$\frac{\pi}{\alpha} e^{-\alpha\|\omega\|}$	$\frac{\pi}{\alpha} e^{-2\pi\alpha\|f\|}$
$\Pi\left(\frac{t}{\tau}\right)$	$\tau \mathrm{Sa}\left(\omega \frac{\tau}{2}\right)$	$\tau \mathrm{Sa}(\pi f \tau)$
$\mathrm{Sa}(\omega_0 t)$ $= \mathrm{Sa}(2\pi f_0 t)$	$\frac{\pi}{\omega_0} \Pi\left(\frac{\omega}{2\omega_0}\right)$	$\frac{1}{2f_0} \Pi\left(\frac{f}{2f_0}\right)$
$\Lambda\left(\frac{t}{\tau}\right)$	$\tau \mathrm{Sa}^2\left(\omega \frac{\tau}{2}\right)$	$\tau \mathrm{Sa}^2(\pi f \tau)$
$\mathrm{Sa}^2(\omega_0 t)$ $= \mathrm{Sa}^2(2\pi f_0 t)$	$\frac{\pi}{\omega_0} \Lambda\left(\frac{\omega}{2\omega_0}\right)$	$\frac{1}{2f_0} \Lambda\left(\frac{f}{2f_0}\right)$
$e^{-\frac{t^2}{2\alpha^2}}$ $\alpha > 0$	$\sqrt{2\pi}\alpha e^{-\frac{\alpha^2\omega^2}{2}}$	$\sqrt{2\pi}\alpha e^{-2\alpha^2\pi^2 f^2}$
$\delta(t)$	1	1
1	$2\pi\delta(\omega)$	$\delta(f)$
$\delta(t - t_0)$	$e^{-j\omega t_0}$	$e^{-j2\pi f t_0}$
$e^{j\omega_0 t}$ $= e^{j2\pi f_0 t}$	$2\pi\delta(\omega - \omega_0)$	$\delta(f - f_0)$
$\mathrm{sgn}(t)$	$\frac{2}{j\omega}$	$\frac{1}{j\pi f}$
$\frac{1}{\pi t}$	$-j\ \mathrm{sgn}\ (\omega)$	$-j\ \mathrm{sgn}\ (f)$
$\mathbf{1}(t)$	$\pi\delta(\omega) + \frac{1}{j\omega}$	$\frac{1}{2}\delta(f) + \frac{1}{j2\pi f}$
$\cos(\omega_0 t)$ $= \cos\ (2\pi f_0 t)$	$\pi\ \delta(\omega + \omega_0) + \pi\delta(\omega - \omega_0)$	$\frac{1}{2}\ \delta(f + f_0) + \frac{1}{2}\delta(f - f_0)$
$\sin(\omega_0 t)$ $= \sin\ (2\pi f_0 t)$	$j\pi\ [\delta(\omega + \omega_0) - \delta(\omega - \omega_0)]$	$-\frac{1}{2j}\ [\delta(f + f_0) - \delta(f - f_0)]$

Definition 2.6 The *inverse Fourier transform* $x(t)$ of its spectrum is defined by

$$x(t) = \frac{1}{2\pi} \int X(\omega) e^{j\omega t} d\omega \tag{2.19}$$

or equivalently,

$$x(t) = \int X(f) e^{j2\pi ft} df. \tag{2.20}$$

We say that the energy signal $x(t)$ and its spectrum $X(\omega)$ form a pair of Fourier transforms *in the ordinary sense*. In the next subsection, we will define the complex FT of power signals (i.e., the Fourier transform *in the limit sense*).

In this book, we will use the following notation: $x(t) \overset{\mathcal{F}}{\leftrightarrow} X(\omega)$ or $x(t) \overset{\mathcal{F}}{\leftrightarrow} X(f)$, that means that $x(t)$ and $X(\omega)$ or $X(f)$ are Fourier transforms. Table 2.1 presents Fourier transforms $X(\omega)$ and $X(f)$ of basic energy and power signals.

2.2.2 *Fourier Transform* in the Limit Sense

Let $x(t)$ be a *power signal* defined as a limit of $y(\alpha, t)$—energy signals with a parameter α going to 0, that is, $x(t) = \lim_{\alpha \to 0} y(\alpha, t)$. For every α, we have a Fourier pair: $y(\alpha, t) \overset{\mathcal{F}}{\leftrightarrow} Y(\alpha, \omega)$. If $\lim_{\alpha \to 0} Y(\alpha, \omega)$ exists, we can denote it with $X(\omega)$, that is, $X(\omega) = \lim_{\alpha \to 0} Y(\alpha, \omega)$. We can say that $x(t)$ and $X(\omega)$ form a Fourier pair *in the limit sense*.

Problem 2.9 Using a limit approach, derive the FTs of the following signals: (a) $x(t) = \delta(t)$, $t \in \mathbb{R}$, (b) $x(t) = 1$, $t \in \mathbb{R}$.

Solution For finding the FT (in the limit sense) of the Dirac delta distribution $x(t) = \delta(t)$ defined by (1.21), we can take $y(\alpha, t) = \frac{1}{\alpha}\Pi\left(\frac{t}{\alpha}\right)$, since $\lim_{\alpha \to 0} \frac{1}{\alpha}\Pi\left(\frac{t}{\alpha}\right) = \delta(t)$. It is easy to verify that $y(\alpha, t)$ satisfies both axioms of (1.21) (see also Table 1.1).

The FT of $\frac{1}{\alpha}\Pi\left(\frac{t}{\alpha}\right)$ is $\mathrm{Sa}\left(\omega \frac{\alpha}{2}\right) = Y(\alpha, \omega)$ (see Table 2.1). We can compute the limit $\lim_{\alpha \to 0} Y(\alpha, \omega) = \lim_{\alpha \to 0} \mathrm{Sa}\left(\omega \frac{\alpha}{2}\right) = 1$ and denote it with $X(\omega)$, that is, $X(\omega) = 1$.

Then according to the limit criterion presented above, we can get the following Fourier pair: $\delta(t) \overset{\mathcal{F}}{\leftrightarrow} 1$.

For the case of the constant signal $x(t) = 1$, $t \in \mathbb{R}$, we propose to choose $y(\alpha, t) = \Pi(\alpha t) = \Pi\left(\frac{t}{1/\alpha}\right)$, since $\lim_{\alpha \to 0} \Pi\left(\frac{t}{1/\alpha}\right) = 1$.

The FT of $\Pi\left(\frac{t}{1/\alpha}\right)$ is $\frac{1}{\alpha}\mathrm{Sa}\left(\frac{\omega}{2\alpha}\right) = Y(\alpha, \omega)$ (see Table 2.1). We need to calculate the limit: $\lim_{\alpha \to 0} Y(\alpha, \omega) = \lim_{\alpha \to 0} \frac{1}{\alpha}\mathrm{Sa}\left(\frac{\omega}{2\alpha}\right) = \lim_{\alpha \to 0} \frac{1}{\alpha}\mathrm{Sa}\left(\frac{\pi f}{\alpha}\right) = \delta(f) = 2\pi\delta(\omega)$ (see Table 1.1) denoted with $X(\omega)$, that is, $X(\omega) = 2\pi\delta(\omega)$.

At last, according to the limit criterion presented above, we obtain the following Fourier pair: $1 \overset{\mathcal{F}}{\leftrightarrow} 2\pi\delta(\omega)$.

The Fourier transformation has many interesting properties listed in Table 2.2. We can observe in what way some signal operations in time-domain make an impact on diverse spectrum properties. A careful reader easily can notice some dual

Table 2.2 Properties of the Fourier transformation

Property	Signal $x(t)$, $t \in \mathbb{R}$	Fourier transform $X(\omega)$, $\omega \in \mathbb{R}$
Dual symmetry	$X(t)$	$2\pi x(-\omega)$
Linearity	$\alpha x_1(t) + \beta x_2(t)$	$\alpha X_1(\omega) + \beta X_2(\omega)$
Reversal	$x(-t)$	$X(-\omega)$
Time-shift	$x(t - t_0)$	$X(\omega)e^{-j\omega t_0}$
Frequency-shift	$x(t)e^{j\omega_0 t}$	$X(\omega - \omega_0)$
Modulation	$x(t)\cos(\omega_0 t)$	$\frac{1}{2}X(\omega+\omega_0) + \frac{1}{2}X(\omega-\omega_0)$
	$x(t)\sin(\omega_0 t)$	$-\frac{1}{2j}X(\omega+\omega_0) + \frac{1}{2j}X(\omega-\omega_0)$
Derivation in time	$\frac{d^n x(t)}{dt^n}$	$(j\omega)^n X(\omega)$
Derivation in frequency	$(-jt)^n x(t)$	$\frac{d^n X(\omega)}{d\omega^n}$
Integration in time	$\int_{-\infty}^{t} x(\tau)d\tau$	$\frac{1}{j\omega}X(\omega) + \pi X(0)\delta(\omega)$
Scaling	$x(a \cdot t)$	$\frac{1}{\lvert a\rvert}X\left(\frac{\omega}{a}\right)$
Time-convolution	$x_1(t) * x_2(t)$	$X_1(\omega) \cdot X_2(\omega)$
Multiplication	$x_1(t) \cdot x_2(t)$	$\frac{1}{2\pi}[X_1(\omega) * X_2(\omega)]$
Plancherel's equality (for energy signals)	$\int_{-\infty}^{\infty} \lvert x(t)\rvert^2 dt = \frac{1}{2\pi}\int_{-\infty}^{\infty} \lvert X(\omega)\rvert^2 d\omega = \int_{-\infty}^{\infty} \lvert X(f)\rvert^2 df$	
Wiener-Khintchine theorem	$R_x(\tau) \overset{\mathcal{F}}{\leftrightarrow} \lvert X(\omega)\rvert^2$	

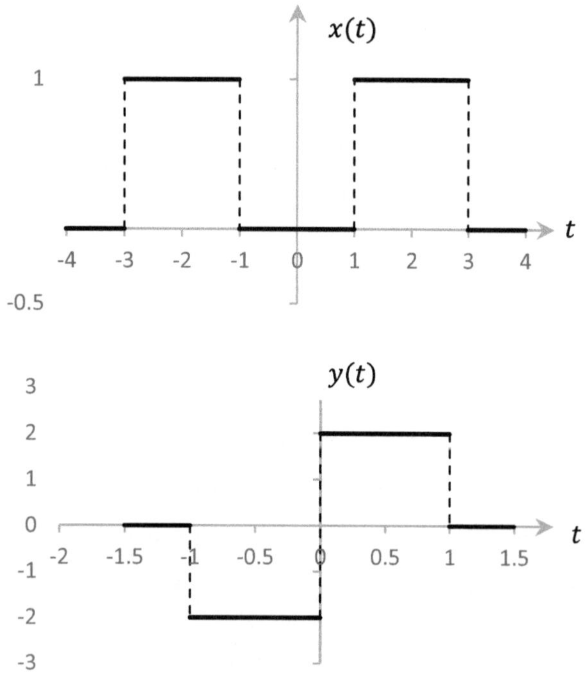

Fig. 2.15 Signals $x(t)$ and $y(t)$ from Problem 2.10

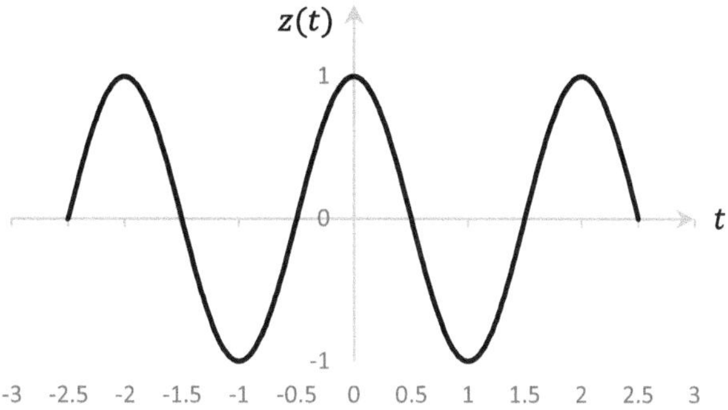

Fig. 2.16 Signal $z(t)$ from Problem 2.11

properties, as a consequence of definitions (2.17) and (2.19). In the next problems, we will learn how to exercise Tables 2.1 and 2.2.

Problem 2.10 Determine the Fourier spectra of signals $x(t)$ and $y(t)$ shown in Fig. 2.15.

Solution We can represent $x(t)$ as the sum of two identical rectangle pulses shifted left and right by ± 2, that is $x(t) = \Pi\left(\frac{t+2}{2}\right) + \Pi\left(\frac{t-2}{2}\right)$. In order to calculate its spectrum, we can apply the linearity and time-shift theorem from Table 2.2. Basing on Table 2.1, we can write $\Pi\left(\frac{t}{2}\right) \overset{\mathcal{F}}{\leftrightarrow} 2\mathrm{Sa}(\omega)$. The Fourier spectrum of $x(t)$ is

$$X(\omega) = 2\mathrm{Sa}(\omega)e^{j2\omega} + 2\mathrm{Sa}(\omega)e^{-j2\omega} = 2\mathrm{Sa}(\omega)\left(e^{j2\omega} + e^{-j2\omega}\right).$$

In parentheses in the above expression, we can recognize the cosine Euler's formula (2.7a). Then

$$X(\omega) = 4\mathrm{Sa}(\omega)\cos(2\omega).$$

Notice that $y(t)$, analogously as above, can be represented as $y(t) = 2\Pi\left(t - \frac{1}{2}\right) - 2\Pi\left(t + \frac{1}{2}\right)$ where $\Pi(t) \overset{\mathcal{F}}{\leftrightarrow} \mathrm{Sa}\left(\frac{\omega}{2}\right)$. The spectrum $Y(\omega)$ is

$$Y(\omega) = 2\mathrm{Sa}\left(\frac{\omega}{2}\right)e^{-j\frac{\omega}{2}} - 2\mathrm{Sa}\left(\frac{\omega}{2}\right)e^{j\frac{\omega}{2}} = 2\mathrm{Sa}\left(\frac{\omega}{2}\right)\left(e^{-j\frac{\omega}{2}} - e^{j\frac{\omega}{2}}\right).$$

We can apply the sine Euler's formula (2.7b) for $Y(\omega)$. Finally, we get

$$Y(\omega) = -4\mathrm{Sa}\left(\frac{\omega}{2}\right)\sin\left(\frac{\omega}{2}\right).$$

Problem 2.11 Given the signal $x(t) = \sum_{n=0}^{49} z(t - 10n)$ where $z(t)$ is a cosine pulse with duration time 5 [s] and amplitude 1, as shown in Fig. 2.16. Compute the Fourier spectrum of $x(t)$.

Solution The signal $x(t)$ is the sum of 50 identical cosine pulses (shifted in time). Because of the linearity of the FT and the time-shift theorem (Table 2.2), its spectrum $X(\omega)$ can be written in the form

$$X(\omega) = \sum_{n=0}^{49} Z(\omega) e^{-j10n\omega},$$

where $Z(\omega)$ is the FT of $z(t)$. Developing the above formula, we can get

$$X(\omega) = Z(\omega) \sum_{n=0}^{49} e^{-j10n\omega} = Z(\omega) \frac{1 - \left(e^{-j10\omega}\right)^{50}}{1 - e^{-j10\omega}} = Z(\omega) \frac{1 - e^{-j500\omega}}{1 - e^{-j10\omega}}$$

Now, let us calculate the spectrum $Z(\omega)$. Since the cosine pulse $z(t)$ can be written as

$$z(t) = \Pi\left(\frac{t}{5}\right) \cos(2\pi f_0 t) = \Pi\left(\frac{t}{5}\right) \cos(\pi t),$$

where $f_0 = \frac{1}{2}$ [Hz] (the fundamental period of the cosine is 2 [s] (see Fig. 2.16)), its spectrum $Z(\omega)$ can be calculated using the modulation property from Table 2.2. We only need to determine the FT of $\Pi\left(\frac{t}{5}\right)$. Using Table 2.1, we can get $\Pi\left(\frac{t}{5}\right) \overset{\mathcal{F}}{\leftrightarrow} 5\mathrm{Sa}\left(\frac{5}{2}\omega\right)$. So,

$$Z(\omega) = \frac{5}{2} \mathrm{Sa}\left(\frac{5}{2}(\omega - \pi)\right) + \frac{5}{2} \mathrm{Sa}\left(\frac{5}{2}(\omega + \pi)\right).$$

Evidently, we can write $Z(\omega)$ in another form, but let us leave it as it is. Finally, we obtain

$$X(\omega) = \frac{5}{2} \left[\mathrm{Sa}\left(\frac{5}{2}(\omega - \pi)\right) + \mathrm{Sa}\left(\frac{5}{2}(\omega + \pi)\right) \right] \frac{1 - e^{-j500\omega}}{1 - e^{-j10\omega}}.$$

Problem 2.12 Using the convolution property of the FT, compute the convolution $y(t) = x_1(t) * x_2(t)$ of two signals: $x_1(t) = 2\Pi\left(\frac{t}{6}\right)$ and $x_2(t) = 3\Pi\left(\frac{t-3}{2}\right)$ from Problem 1.10.

Solution In this problem, we can show an alternative way of calculating a convolution $y(t) = x_1(t) * x_2(t)$. As shown in Table 2.2, the FT $Y(\omega)$ of $y(t)$ is the product of spectra $X_1(\omega)$ and $X_2(\omega)$ of signals $x_1(t)$ and $x_2(t)$. Basing on Table 2.1, we have

$$2\Pi\left(\frac{t}{6}\right) \overset{\mathcal{F}}{\leftrightarrow} 12 \cdot \mathrm{Sa}(3\omega) = X_1(\omega),$$

$$3\Pi\left(\frac{t-3}{2}\right) \overset{\mathcal{F}}{\leftrightarrow} 6 \cdot \mathrm{Sa}(\omega)e^{-j3\omega} = X_2(\omega).$$

The product $Y(\omega) = X_1(\omega) \cdot X_2(\omega)$ is $Y(\omega) = 72 \cdot \mathrm{Sa}(3\omega) \cdot \mathrm{Sa}(\omega)e^{-j3\omega}$. In order to calculate the inverse FT of $Y(\omega)$, we have to perform some trigonometric calculations. First, let us substitute $\mathrm{Sa}(3\omega) \cdot \mathrm{Sa}(\omega)$ using the definition $\mathrm{Sa}(\cdot) = \frac{\sin(\cdot)}{(\cdot)}$ as follows:

$$Y(\omega) = 72 \cdot \frac{\sin(3\omega)}{3\omega} \cdot \frac{\sin(\omega)}{\omega} e^{-j3\omega} = 24 \frac{\sin(3\omega) \cdot \sin(\omega)}{\omega^2} e^{-j3\omega}.$$

Then in the nominator, we can use the identity: $\sin(\alpha) \cdot \sin(\beta) = \frac{1}{2} \times [\cos(\alpha - \beta) - \cos(\alpha + \beta)]$, that is,

$$Y(\omega) = 12 \frac{[\cos(2\omega) - \cos(4\omega)]}{\omega^2} e^{-j3\omega}.$$

Next, in the above expression, for $\cos(2\omega)$ and $\cos(4\omega)$, let us use the formula: $\cos(2\alpha) = 1 - 2\sin^2(\alpha)$ to obtain

$$\begin{aligned} Y(\omega) &= 12 \frac{[1 - 2\sin^2(\omega) - 1 + 2\sin^2(2\omega)]}{\omega^2} e^{-j3\omega} \\ &= [96 \cdot \mathrm{Sa}^2(2\omega) - 24 \cdot \mathrm{Sa}^2(\omega)] e^{-j3\omega}. \end{aligned}$$

Finally, using the time-shift property of the FT and Table 2.1, we can get the convolution $y(t)$:

$$y(t) = 24 \cdot \Lambda\left(\frac{t-3}{4}\right) - 12 \cdot \Lambda\left(\frac{t-3}{2}\right).$$

If we would like to sketch the above signal, we evidently get the identical plot as shown in Fig. 1.40. The reader can easily check this.

2.2.3 Amplitude- and Phase Spectra of Continuous-Time Signals

Definition 2.7 The *amplitude spectrum* of a continuous-time signal $x(t)$ is defined as the absolute value of its spectrum, that is, $|X(\omega)|$. The *phase spectrum* is defined as the argument of the spectrum $X(\omega)$, that is, arg $X(\omega)$. The square of the amplitude spectrum of energy/power signals is called respectively the *energy density/power density spectrum.*

The spectrum $X(\omega)$ of a continuous-time signal $x(t)$ can be expressed in the following form:

$$X(\omega) = |X(\omega)|e^{j\arg X(\omega)}. \tag{2.21}$$

$|X(\omega)|$ is always an even function of ω (of f), while arg $X(\omega)$ is an odd one.

Definition 2.8 The *real spectrum* Re $X(\omega)$ of a continuous-time signal $x(t)$ is defined as

$$\text{Re } X(\omega) = |X(\omega)| \cos(\arg X(\omega)). \tag{2.22}$$

The *imaginary spectrum* Im $X(\omega)$ is

$$\text{Im } X(\omega) = |X(\omega)| \sin(\arg X(\omega)). \tag{2.23}$$

Re $X(\omega)$ is always an even function of ω (of f), while Im $X(\omega)$ is an odd one.

There are the following relations between real-, imaginary-, amplitude-, and phase spectra:

$$|X(\omega)| = \sqrt{[\text{Re } X(\omega)]^2 + [\text{Im } X(\omega)]^2}, \tag{2.24}$$

$$\arg X(\omega) = \text{atan2}(\text{Im } X(\omega), \text{ Re } X(\omega)). \tag{2.25}$$

Problem 2.13 Given the signal $x(t) = \text{Sa}(\pi t) \sin(5\pi t) \cos(10\pi t)$. Determine its spectrum, sketch amplitude- and phase spectra, and calculate the signal energy.

Solution Let us use the trigonometric identity: $\sin(\alpha)\cos(\beta) = \frac{1}{2}[\sin(\alpha + \beta) + \sin(\alpha - \beta)]$ to express $x(t)$ as follows:

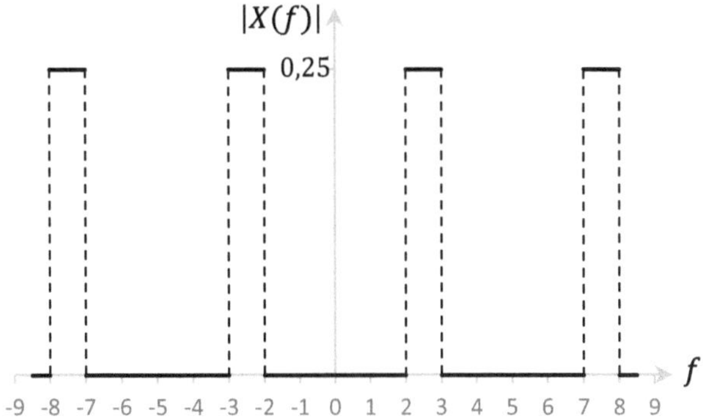

Fig. 2.17 Amplitude spectrum of the signal $x(t)$ from Problem 2.13

$$x(t) = \frac{1}{2}\mathrm{Sa}(\pi t)[\sin(15\pi t) - \sin(5\pi t)],$$

$$x(t) = \frac{1}{2}[\mathrm{Sa}(\pi t)\sin(15\pi t) - \mathrm{Sa}(\pi t)\sin(5\pi t)].$$

To calculate the spectrum of $x(t)$, we should use the linearity and modulation property of the FT (see Table 2.2). From Table 2.1, we can get $\mathrm{Sa}(\pi t) \overset{\mathcal{F}}{\leftrightarrow} \Pi(f)$. The modulation property (see Table 2.2):

$$y(t)\sin(2\pi f_0 t) \overset{\mathcal{F}}{\leftrightarrow} -\frac{1}{2j}Y(f+f_0) + \frac{1}{2j}Y(f-f_0)$$

(notice that we introduced a notation $y(t) \overset{\mathcal{F}}{\leftrightarrow} Y(f)$). In our case, $y(t) = \mathrm{Sa}(\pi t)$ and $Y(f) = \Pi(f)$. We have

$$\mathrm{Sa}(\pi t)\sin(15\pi t) \overset{\mathcal{F}}{\leftrightarrow} -\frac{1}{2j}\Pi(f+7,5) + \frac{1}{2j}\Pi(f-7,5),$$

$$\mathrm{Sa}(\pi t)\sin(5\pi t) \overset{\mathcal{F}}{\leftrightarrow} -\frac{1}{2j}\Pi(f+2,5) + \frac{1}{2j}\Pi(f-2,5).$$

The spectrum $X(f)$ is

$$X(f) = -\frac{1}{4j}\Pi(f+7,5) + \frac{1}{4j}\Pi(f-7,5) - \frac{1}{4j}\Pi(f+2,5) + \frac{1}{4j}\Pi(f-2,5),$$

$$X(f) = \frac{j}{4}\Pi(f+7,5) - \frac{j}{4}\Pi(f-7,5) + \frac{j}{4}\Pi(f+2,5) - \frac{j}{4}\Pi(f-2,5).$$

Notice that $X(f)$ is an imaginary function of frequency (its real part is zero). The imaginary spectrum is

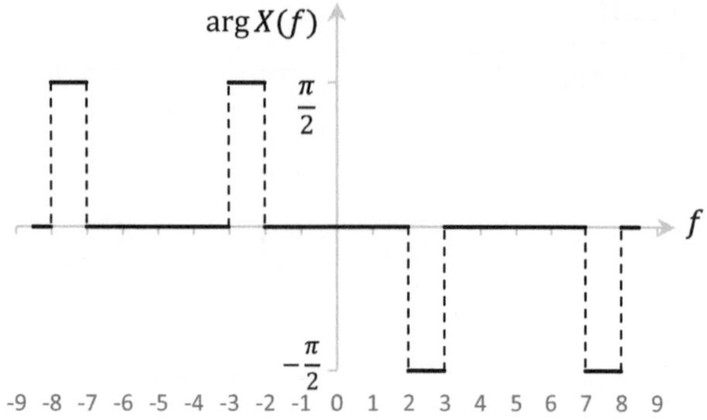

Fig. 2.18 Phase spectrum of the signal $x(t)$ from Problem 2.13

$$\operatorname{Im} X(f) = \frac{1}{4}\Pi(f + 7,5) - \frac{1}{4}\Pi(f - 7,5) + \frac{1}{4}\Pi(f + 2,5) - \frac{1}{4}\Pi(f - 2,5).$$

We can see that it is composed of four non-overlapping rectangle pulses with width 1 [Hz], amplitudes $\pm\frac{1}{4}$, shifted in f-domain by $\pm 2,5$ and $\pm 7,5$ [Hz]. Using (2.24), we can get the amplitude spectrum: $|X(\omega)| = \sqrt{\operatorname{Im}^2 X(\omega)} = |\operatorname{Im} X(\omega)|$. We have (see Fig. 2.17)

$$|X(f)| = \frac{1}{4}\Pi(f + 7,5) + \frac{1}{4}\Pi(f - 7,5) + \frac{1}{4}\Pi(f + 2,5) + \frac{1}{4}\Pi(f - 2,5).$$

Imaginary- and amplitude spectra are related by the formula: Im $X(f) = |X(f)| \sin(\arg X(f))$ (see (2.23)). Comparing the above formulas, we can notice that

$$\sin(\arg X(f)) = \begin{cases} 1 & \text{for } f \in [-8, -7] \cup [-3, -2], \\ -1 & \text{for } f \in [7, 8] \cup [2, 3], \\ 0 & \text{elsewhere.} \end{cases}$$

So, the phase spectrum $\arg X(f)$ (Fig. 2.18) is

$$\arg X(f) = \begin{cases} \frac{\pi}{2} & \text{for } f \in [-8, -7] \cup [-3, -2], \\ -\frac{\pi}{2} & \text{for } f \in [7, 8] \cup [2, 3], \\ 0 & \text{elsewhere.} \end{cases}$$

Finally, we can apply Plancherel's equality (see Table 2.2) to calculate the energy of $x(t)$:

$$E_x = \int_{-\infty}^{\infty} |X(f)|^2 df = 4 \cdot \frac{1}{16} = \frac{1}{4}.$$

Problem 2.14 Using the Wiener-Khintchine theorem, compute the auto-correlation function of the signal $x(t)$ shown in Fig. 1.49. Compare it with the solution of Problem 1.14.

Solution The signal $x(t)$ can be written in the following form:

$$x(t) = \Pi\left(t - \frac{1}{2}\right) + 2\Pi\left(t - \frac{7}{2}\right).$$

The Wiener-Khintchine theorem permits to calculate the auto-correlation function of $x(t)$ as an inverse FT of the energy density spectrum $|X(\omega)|^2$. Let us start with the calculation of $X(\omega)$. To do this, we apply the linearity property and time-shift theorem of the FT from Table 2.2. Using Table 2.1, we get $\Pi(t) \overset{\mathcal{F}}{\leftrightarrow} \mathrm{Sa}\left(\frac{\omega}{2}\right)$. The spectrum $X(\omega)$ is

$$X(\omega) = \mathrm{Sa}\left(\frac{\omega}{2}\right) e^{-j\frac{1}{2}\omega} + 2\mathrm{Sa}\left(\frac{\omega}{2}\right) e^{-j\frac{7}{2}\omega} = \mathrm{Sa}\left(\frac{\omega}{2}\right) e^{-j\frac{1}{2}\omega} \cdot \left[1 + 2e^{-j3\omega}\right]$$
$$= \mathrm{Sa}\left(\frac{\omega}{2}\right) e^{-j\frac{1}{2}\omega} \cdot [1 + 2\cos(3\omega) - 2j\sin(3\omega)].$$

Then we can compute the amplitude spectrum $|X(\omega)|$:

$$|X(\omega)| = \left|\mathrm{Sa}\left(\frac{\omega}{2}\right)\right| \sqrt{(1 + 2\cos(3\omega))^2 + 4\sin^2(3\omega)},$$

where $\left|e^{-j\frac{1}{2}\omega}\right| = 1$. Now, we can compute the energy density spectrum $|X(\omega)|^2$ ($x(t)$ is an energy signal):

$$|X(\omega)|^2 = \mathrm{Sa}^2\left(\frac{\omega}{2}\right) \cdot \left[(1 + 2\cos(3\omega))^2 + 4\sin^2(3\omega)\right]$$
$$= \mathrm{Sa}^2\left(\frac{\omega}{2}\right) \cdot [5 + 4\cos(3\omega)] = 5 \cdot \mathrm{Sa}^2\left(\frac{\omega}{2}\right) + 2 \cdot \mathrm{Sa}^2\left(\frac{\omega}{2}\right)\left(e^{-j3\omega} + e^{j3\omega}\right)$$
$$= 5 \cdot \mathrm{Sa}^2\left(\frac{\omega}{2}\right) + 2 \cdot \mathrm{Sa}^2\left(\frac{\omega}{2}\right) e^{-j3\omega} + 2 \cdot \mathrm{Sa}^2\left(\frac{\omega}{2}\right) e^{j3\omega}.$$

The last step is the calculation of the inverse FT of the above expression, remembering that the auto-correlation function is a function of τ. Using Tables 2.1 and 2.2 (linearity, time-shift), we can get

$$R_x(\tau) = 5 \cdot \Lambda(\tau) + 2 \cdot \Lambda(\tau - 3) + 2 \cdot \Lambda(\tau + 3).$$

Notice that the graph of $R_x(\tau)$ is identical as displayed in Fig. 1.51.

2.2.4 Fourier Transform of Periodic Signals

Let us come back to the formula (2.9) of the exponential Fourier series of a periodic signal $x(t)$. Calculating the FT of (2.9), we can come to the following expression for $X(\omega)$:

$$X(\omega) = \mathcal{F}\left\{\sum_{n\in\mathbb{Z}} c_n e^{jn\omega_0 t}\right\} = \sum_{n\in\mathbb{Z}} c_n \mathcal{F}\left\{e^{jn\omega_0 t}\right\} = 2\pi \sum_{n\in\mathbb{Z}} c_n \delta(\omega - n\omega_0), \tag{2.26}$$

where c_n are coefficients of the exponential Fourier series given by (2.10), $c_n = \frac{1}{T_0} \int\limits_{\langle T_0\rangle} x(t) e^{-jn\omega_0 t} dt$. The formula (2.26) shows that the spectrum of a periodic signal $x(t)$ is *discrete* in domain ω and has a form of peaks (Dirac deltas) with "amplitudes" equal $2\pi c_n$ at $n\omega_0$, $n \in \mathbb{Z}$. The equivalent form of (2.26) in f-domain is

$$X(f) = \sum\nolimits_{n\in\mathbb{Z}} c_n \delta(f - nf_0). \tag{2.27}$$

Moreover, if we compare $c_n = \frac{1}{T_0} \int\limits_{\langle T_0\rangle} x(t) e^{-jn\omega_0 t} dt$ with (2.17) of the FT, we can notice that c_n has also the form of the FT, but:

- In c_n, we integrate within one period T_0, that is, integration limits are $[t_0, t_0 + T_0]$ and t_0 is arbitrary chosen;
- In the interval $[t_0, t_0 + T_0]$, we deal with *a part* of a periodic signal (denoted with $z(t)$ in (1.39)), which repeats every T_0;
- In the exponent, we have $\omega = n\omega_0$;
- The integral is multiplicated by $\frac{1}{T_0}$.

Having all this in mind and denoting the FT of $z(t)$ with $Z(\omega)$, we can write

$$c_n = \frac{1}{T_0} Z(n\omega_0). \tag{2.28}$$

Equivalently, if $z(t) \overset{\mathcal{F}}{\leftrightarrow} Z(f)$, we can calculate coefficients c_n using

$$c_n = \frac{1}{T_0} Z(nf_0). \tag{2.29}$$

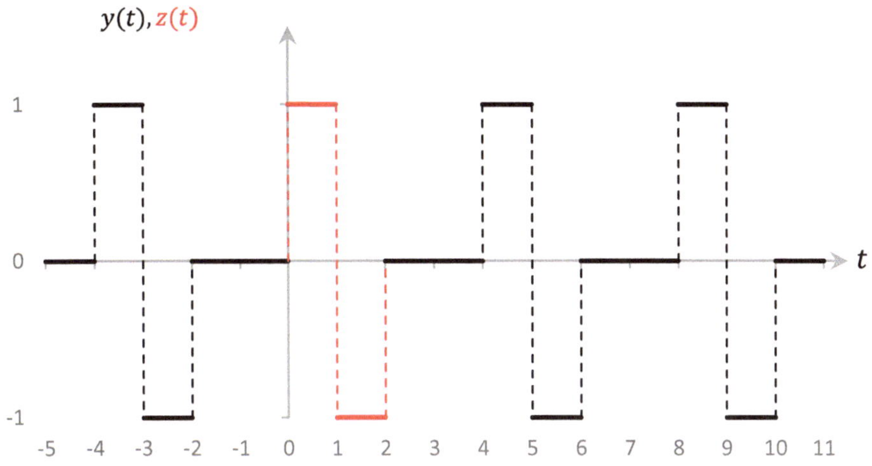

Fig. 2.19 Periodic signal $y(t)$ from Problem 2.15. In red: its elementary segment $z(t)$

Formulas (2.26), (2.27), and (2.28) relate the exponential Fourier series of a periodic signal with its Fourier transform. Let us show how to make profit from these relations solving Problems 2.15, 2.16, 2.17, and 2.18.

Problem 2.15 Determine the Fourier spectrum of the comb distribution $\delta_T(t) = \sum_{n\in\mathbb{Z}} \delta(t - nT)$.

Solution If we look at Fig. 1.26, we can notice that $z(t) = \delta(t)$. Its spectrum $Z(\omega) = 1$. Using (2.28), we can notice that all coefficients $c_n = \frac{1}{T}$, so the spectrum of the comb distribution is

$$\mathcal{F}\{\delta_T(t)\} = \frac{2\pi}{T}\sum_{n\in\mathbb{Z}} \delta(\omega - n\omega_0)$$

and $\omega_0 = \frac{2\pi}{T}$. We can see that the spectrum of the comb distribution with fundamental period T is also the comb distribution (in the domain ω) with period $2\pi/T$. It means that if the delta pulses are closer to each other in the time domain (T gets smaller), the delta pulses in the spectrum move away from each other ($2\pi/T$ gets bigger).

Problem 2.16 Determine the Fourier spectrum of the rectangle pulse train $p(t)$ shown in Fig. 1.4 with amplitude A, fundamental period T_0, and pulse time duration τ.

Solution We have $z(t) = A\Pi(\frac{t}{\tau}) \overset{\mathcal{F}}{\leftrightarrow} Z(\omega) = A\tau \mathrm{Sa}(\frac{\omega\tau}{2})$. Using (2.28), we can get the coefficients c_n:

$$c_n = \frac{1}{T_0} A\tau \mathrm{Sa}\left(\frac{n\omega_0\tau}{2}\right).$$

Then, according to (2.26), the spectrum $P(\omega)$ of $p(t)$ is

$$P(\omega) = 2\pi \frac{A\tau}{T_0} \sum_{n\in\mathbb{Z}} \mathrm{Sa}\left(\frac{n\omega_0\tau}{2}\right) \delta(\omega - n\omega_0).$$

Problem 2.17 Given the signal $y(t) = \sum_{n\in\mathbb{Z}} z(t-4n)$ where $z(t) = \mathbf{1}(t) - 2 \cdot \mathbf{1}(t-1) + \mathbf{1}(t-2)$. Plot $y(t)$ and compute its Fourier spectrum. Determine amplitude- and phase spectra of $y(t)$ in the frequency band $[-5f_0, 5f_0]$, where f_0 is the fundamental frequency of $y(t)$.

Solution The signal $y(t)$ is periodic with fundamental period $T_0 = 4$ [s]. Its plot is displayed in Fig. 2.19. The graph of $z(t)$ is marked in red. We can write $z(t)$ in the following equivalent form

$$z(t) = \Pi\left(t - \frac{1}{2}\right) - \Pi\left(t - \frac{3}{2}\right)$$

and use Tables 2.1 and 2.2 (linearity, time shift) for the calculation of the FT of $z(t)$.

We have

$$Z(\omega) = \mathrm{Sa}\left(\frac{\omega}{2}\right) \cdot \left(e^{-j\frac{\omega}{2}} - e^{-j\frac{3\omega}{2}}\right).$$

Then let us apply (2.28) and calculate exponential Fourier series coefficients c_n ($\omega_0 = \frac{2\pi}{T_0} = \frac{\pi}{2}$ [rad/s]):

$$\begin{aligned} c_n &= \frac{1}{T_0} Z(n\omega_0) = \frac{1}{4} \mathrm{Sa}\left(\frac{n\frac{\pi}{2}}{2}\right) \cdot \left(e^{-j\frac{n\frac{\pi}{2}}{2}} - e^{-j\frac{3n\frac{\pi}{2}}{2}}\right) \\ &= \frac{1}{4} \mathrm{Sa}\left(n\frac{\pi}{4}\right) \cdot \left(e^{-jn\frac{\pi}{4}} - e^{-j3n\frac{\pi}{4}}\right). \end{aligned}$$

The FT of $y(t)$ is given by (2.26), that is,

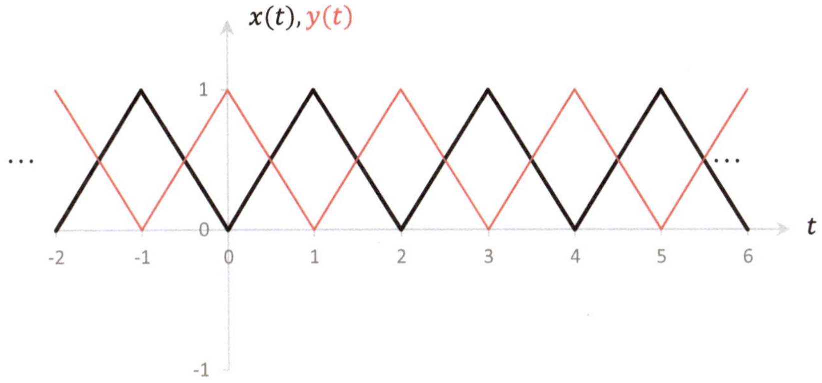

Fig. 2.20 Periodic signals $x(t)$ and $y(t)$ (in red) from Problem 2.16

$$Y(\omega)=2\pi\sum_{n\in\mathbb{Z}}\frac{1}{4}\mathrm{Sa}\left(n\frac{\pi}{4}\right)\cdot\left(e^{-jn\frac{\pi}{4}}-e^{-j3n\frac{\pi}{4}}\right)\cdot\delta\left(\omega-n\frac{\pi}{2}\right).$$

Equivalently,

$$Y(f)=\sum_{n\in\mathbb{Z}}\frac{1}{4}\mathrm{Sa}\left(n\frac{\pi}{4}\right)\cdot\left(e^{-jn\frac{\pi}{4}}-e^{-j3n\frac{\pi}{4}}\right)\cdot\delta\left(f-n\frac{1}{4}\right).$$

Let us recall that the amplitude spectrum of $y(t)$ is defined by $\{|c_n|, n\in\mathbb{Z}\}$ and phase spectrum by $\{\arg c_n, n\in\mathbb{Z}\}$ (see Definition 2.4). We only need to calculate c_n for $n\in\{-5,\ldots,5\}$ as follows:

$$\boldsymbol{c_o}=0,$$

$$\boldsymbol{c_1}=\frac{1}{4}\mathrm{Sa}\left(\frac{\pi}{4}\right)\cdot\left(e^{-j\frac{\pi}{4}}-e^{-j\frac{3\pi}{4}}\right)=\frac{1}{4}\frac{\sin\frac{\pi}{4}}{\frac{\pi}{4}}\cdot\left(\cos\frac{\pi}{4}-j\sin\frac{\pi}{4}-\cos\frac{3\pi}{4}+j\sin\frac{3\pi}{4}\right)$$

$$=\frac{1}{\pi}=\boldsymbol{c_{-1}},$$

$$\boldsymbol{c_2}=\frac{1}{4}\mathrm{Sa}\left(\frac{\pi}{2}\right)\cdot\left(e^{-j\frac{\pi}{2}}-e^{-j\frac{3\pi}{2}}\right)=-\frac{1}{\pi}j,\ \boldsymbol{c_{-2}}=\frac{1}{\pi}j,$$

$$\boldsymbol{c_3}=\frac{1}{4}\mathrm{Sa}\left(\frac{3\pi}{4}\right)\cdot\left(e^{-j\frac{3\pi}{4}}-e^{-j\frac{9\pi}{4}}\right)=-\frac{1}{3\pi}=\boldsymbol{c_{-3}},$$

$$\boldsymbol{c_4}=0=\boldsymbol{c_{-4}},\ \boldsymbol{c_5}=\frac{1}{4}\mathrm{Sa}\left(\frac{5\pi}{4}\right)\cdot\left(e^{-j\frac{5\pi}{4}}-e^{-j\frac{15\pi}{4}}\right)=\frac{1}{5\pi}=\boldsymbol{c_{-5}}.$$

The amplitude spectrum for $n\in\{-5,\ldots,5\}$ is $\left\{\frac{1}{5\pi},0,\frac{1}{3\pi},\frac{1}{\pi},\frac{1}{\pi},0,\frac{1}{\pi},\frac{1}{\pi},\frac{1}{3\pi},0,\frac{1}{5\pi}\right\}$ and the phase spectrum is $\left\{0,0,-\pi,\frac{\pi}{2},0,0,0,-\frac{\pi}{2},\pi,0,0\right\}$.

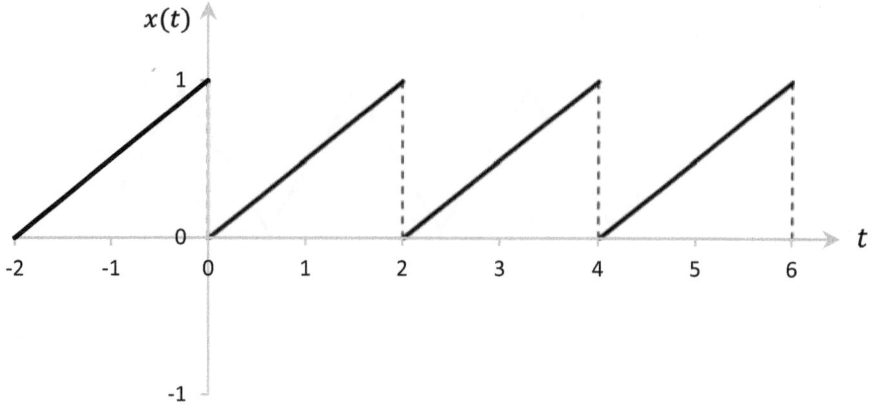

Fig. 2.21 Signal $x(t)$ from Exercise 2.3

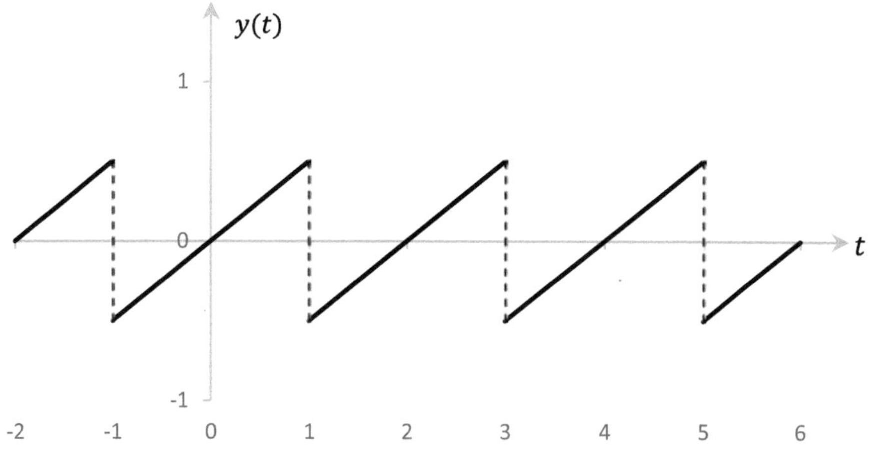

Fig. 2.22 Signal $y(t)$ from Exercises 2.4 and 2.5

Problem 2.18 Given the Fourier spectrum $X(\omega) = \pi e^{j\omega} \sum_{n\in\mathbb{Z}} \mathrm{Sa}^2\left(n\frac{\pi}{2}\right)\delta(\omega - n\pi)$ of a continuous-time signal $x(t)$. Determine $x(t)$ and sketch its graph.

Solution Comparing $X(\omega)$ with (2.26), we can notice that $X(\omega)$ can be expressed as a product $X(\omega) = e^{j\omega}Y(\omega)$ where $Y(\omega) = \pi \sum_{n\in\mathbb{Z}} \mathrm{Sa}^2\left(n\frac{\pi}{2}\right)\delta(\omega - n\pi)$ is the spectrum of a

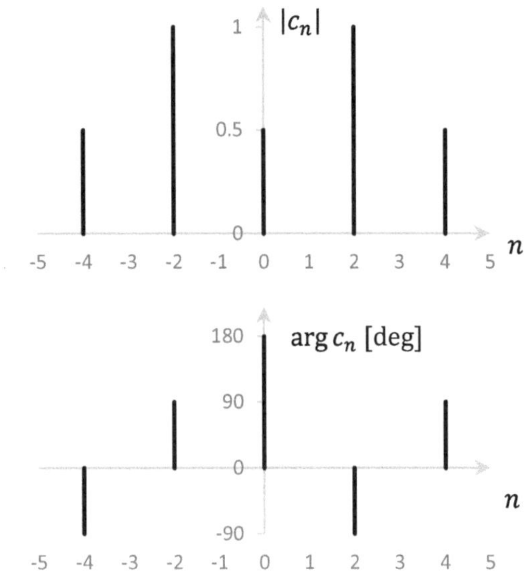

Fig. 2.23 Amplitude- and phase spectra of $x(t)$ from Exercise 2.6

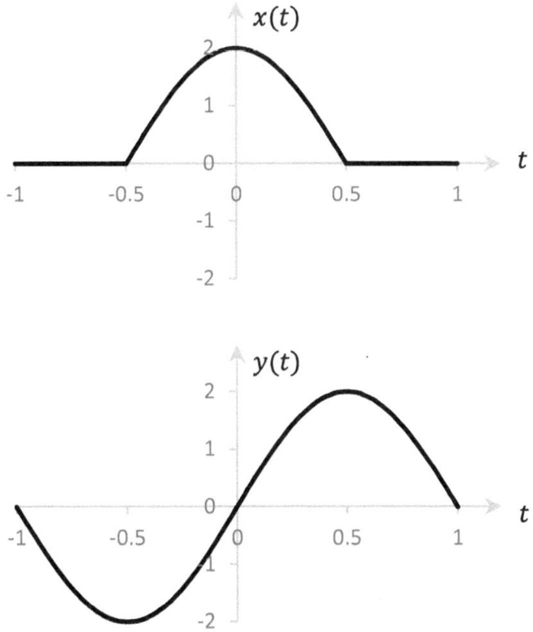

Fig. 2.24 Signals $x(t)$ and $y(t)$ from Exercise 2.8

periodic signal $y(t)$. The idea is to find $y(t)$ and after, using time-shift property (Table 2.2), to express $x(t)$ as $y(t + 1)$.

From (2.26), we have $\omega_0 = \pi$ [rad/s] and $c_n = \frac{1}{2}\mathrm{Sa}^2\left(n\frac{\pi}{2}\right)$. The signal $y(t)$ is periodic with fundamental period $T_0 = 2$ [s]. Now, let us apply (2.28) to get

$Z(n\pi) = \mathrm{Sa}^2\left(n\frac{\pi}{2}\right)$, so $Z(\omega) = \mathrm{Sa}^2\left(\frac{\omega}{2}\right)$. From Table 2.1, we can find $z(t) = \Lambda(t)$, so $y(t) = \sum_{n\in\mathbb{Z}} z(t-2n) = \sum_{n\in\mathbb{Z}} \Lambda(t-2n)$.

Finally, $x(t) = \sum_{n\in\mathbb{Z}} \Lambda(t+1-2n)$ and its plot is displayed in Fig. 2.20.

Exercises

2.1. Given the signal $x(t) = 2 - \frac{\sqrt{2}}{2}\cos(2\pi t) - \frac{\sqrt{2}}{2}\sin(2\pi t) + \frac{1}{2}\cos\left(4\pi t - \frac{\pi}{3}\right)$. Determine and sketch its one-sided amplitude- and phase spectra. Calculate the THD coefficient.

2.2. Given the signal $x(t) = \cos(2\pi t)\sin\left(4\pi t + \frac{\pi}{6}\right)$. Compute its exponential Fourier series and sketch amplitude- and phase spectra. Calculate the average power of $x(t)$.

2.3. Given a saw-tooth waveform with amplitude $A = 1$ and fundamental period $T_0 = 2$ [s] displayed in Fig. 2.21. Compute coefficients of the exponential Fourier series of $x(t)$.

2.4. Using the results of Exercise 2.3, compute coefficients of the exponential Fourier series of the signal $y(t)$ shown in Fig. 2.22.

2.5. Approximate the signal $y(t)$ shown in Fig. 2.22 with the sum $y_N(t)$ of N harmonics, $N = 1, 3, 5, 7, 9$. In each case, plot $y_N(t)$ and calculate the approximation error ε_N.

2.6. Figure 2.23 shows amplitude- and phase spectra of a periodic signal $x(t)$ with fundamental period $T_0 = 0, 1$ [s]. Determine $x(t)$.

2.7. Compute the FT in a limit sense of the following signals: (a) $x(t) = 1, t \in \mathbb{R}$ with $y(\alpha, t) = e^{-\alpha|t|}, \alpha > 0, t \in \mathbb{R}$; (b) $x(t) = \mathrm{sgn}\,(t), t \in \mathbb{R}$ with $y(\alpha, t) = \mathrm{sgn}\,(t)\, e^{-\alpha|t|}, \alpha > 0, t \in \mathbb{R}$.

2.8. Determine Fourier spectra of pulses $x(t)$ (cosine) and $y(t)$ (sine) shown in Fig. 2.24.

2.9. Given the Fourier spectrum $X(\omega) = \frac{1}{j}[\mathrm{Sa}(2\omega - \pi) - \mathrm{Sa}(2\omega + \pi)]$ of a continuous-time signal $x(t)$. Determine $x(t)$.

2.10. Given the signal $x(t) = \sum_{n=0}^{49} z(t-4n)$ where $z(t) = \Lambda(t)$. Determine the Fourier spectrum of $x(t)$.

2.11. Given the signal $x(t) = \mathrm{Sa}\left(\frac{\pi}{2}t\right)\sin(4\pi t)\sin(8\pi t)$. Determine its spectrum, sketch amplitude- and phase spectra and calculate the signal energy.

2.12. Using the convolution theorem for the FT, calculate the auto-convolution of the signal shown in Fig. 1.49. Compare the obtained result with the solution of Exercise 1.8 from Chap. 1.

2.13. Using the Wiener-Khintchine theorem, compute the auto-correlation of the signal $x(t) = A\Pi\left(\frac{t-t_0}{4t_0}\right), A, t_0 \in \mathbb{R}$.

2.14. Compute the Fourier spectrum of a periodic signal $x(t)=\sum_{n\in\mathbb{Z}}z(t-4n)$ where $z(t) = \Lambda(t)$. Determine exponential Fourier series coefficients c_n for $n\in\{-5,\ldots,5\}$.

2.15. Given the Fourier spectrum $X(\omega)=\frac{\pi}{2}e^{-j\omega}\sum_{n\in\mathbb{Z}}\mathrm{Sa}\left(n\frac{\pi}{4}\right)\delta\left(\omega-n\frac{\pi}{2}\right)$ of a continuous-time signal $x(t)$. Determine $x(t)$ and sketch its graph.

Chapter 3
Analysis of Systems in Time Domain

In this chapter, we introduce the notion of a *system* and we analyze system properties in time domain. The reader will get familiar with continuous- and discrete-time systems and their basic features: causality, linearity, and time-invariance. We introduce two functions characterizing a system in time domain, that is, *impulse response* and *step response* of an LTI (linear time-invariant) system. Equations describing LTI continuous- and discrete-time systems are presented. We will show how to compute the system response to the known input signal or vice versa. To make this theory complete, block diagrams of systems in different configurations are considered.

3.1 Basic Notions of Continuous- and Discrete-Time System Theory

Definition 3.1 A *system* is a "black box" representing any device that processes a given input signal (*excitation*) to produce an output signal (*response*).

Notations: In the case of continuous-time systems, the input signal, $x(t)$, $t \in \mathbb{R}$, is a continuous-time signal and the output signal is $y(t)$, $t \in \mathbb{R}$. The input-output relation can be denoted as $x(t) \rightarrow y(t)$. A continuous-time system is often called an *analog system* or an *analog filter*.

In the case of discrete-time systems, the input signal is $x(n)$, $n \in \mathbb{Z}$ and the output signal is $y(n)$, $n \in \mathbb{Z}$, while the input-output relation is $x(n) \rightarrow y(n)$. A discrete-time system is usually called a *digital filter*.

Definition 3.2 A system is called *memoryless* if the output $y(\cdot)$ at a fixed time moment depends only on excitation $x(\cdot)$ at the same moment. If the output $y(\cdot)$ at a fixed time moment is a function of present, past, and/or future values of the input signal, we say that the system is *with memory*.

K. M. Snopek, *Hands-on Signals and Systems Theory*,
https://doi.org/10.1007/978-3-031-56081-1_3

Definition 3.3 A system is called *causal* if its output is a function of only present and past values of the input. If the output depends also on future values of the excitation the system is *noncausal.*

Definition 3.4 A system is *linear* if its output $y(\cdot)$ to a linear combination $\alpha x_1(\cdot) + \beta x_2(\cdot)$, $\alpha, \beta \in \mathbb{R}$ of two input signals $x_1(\cdot)$ and $x_2(\cdot)$ is a linear combination $\alpha y_1(\cdot) + \beta y_2(\cdot)$ of corresponding responses, that is, $x_1(\cdot) \rightarrow y_1(\cdot)$ and $x_2(\cdot) \rightarrow y_2(\cdot)$.

Definition 3.5 A system is *time-invariant* if its response to a time-shifted input is also shifted in time. More exactly, for continuous-time systems we have: $x(t - t_0) \rightarrow y(t - t_0)$, $t_0 \in \mathbb{R}$ and for discrete-time systems: $x(n - n_0) \rightarrow y(n - n_0)$, $n_0 \in \mathbb{Z}$.

Definition 3.6 If the system is linear and time-invariant, we call it an *LTI system* (Linear Time-Invariant).

Important In this book, we always deal with LTI systems, except in Problems 3.1 and 3.2, where system properties are presented in a wider context.

Problem 3.1 Determine features of two systems described by the following input-output relations: (a) $y(t) = A \cdot x(t) \cdot \cos(\omega_0 t)$, $A = \text{const}$, $\omega_0 \in \mathbb{R}$, (b) $y(n) = \frac{1}{2}[x(n) + x(n-1)]$.

Solution The system (a) is *memoryless* and *causal* since the output $y(t)$ depends only on present values of $x(t)$.

Let us check its linearity. Consider two input signals $x_1(t)$ and $x_2(t)$ and the corresponding responses: $y_1(t) = A \cdot x_1(t) \cdot \cos(\omega_0 t)$ and $y_2(t) = A \cdot x_2(t) \cdot \cos(\omega_0 t)$. If we have $\alpha x_1(t) + \beta x_2(t)$ at the input, then at the output appears the following signal:

$$\begin{aligned} A \cdot [\alpha x_1(t) + \beta x_2(t)] \cdot \cos(\omega_0 t) & \\ &= A \cdot \alpha x_1(t) \cdot \cos(\omega_0 t) + A \cdot \beta x_2(t) \cdot \cos(\omega_0 t) \\ &= \alpha[A \cdot x_1(t) \cdot \cos(\omega_0 t)] + \beta[A \cdot x_2(t) \cdot \cos(\omega_0 t)] \\ &= \alpha y_1(t) + \beta y_2(t), \end{aligned}$$

so the system is *linear.*

To check time-invariance of the system (a), we excite it with $x(t - t_0)$. The response is: $A \cdot x(t - t_0) \cdot \cos(\omega_0 t)$ that evidently differs from $y(t - t_0) = A \cdot x(t - t_0) \cdot \cos[\omega_0(t - t_0)]$. The system is *not time-invariant.*

The system (b) is *with memory* and *causal*, since the output $y(n)$ depends only on present and past values of $x(n)$.

To study linearity of the system (b), we calculate its response to the input $\alpha x_1(n) + \beta x_2(n)$, which is $\frac{1}{2}[\alpha x_1(n) + \beta x_2(n) + \alpha x_1(n-1) + \beta x_2(n-1)]$. After simple calculations, we get the response: $\alpha \ \frac{1}{2}[x_1(n) + x_1(n-1)] + \beta \ \ \frac{1}{2} \times$

$[x_2(n)+x_2(n-1)]=\alpha y_1(n)+\beta y_2(n)$ where $y_1(n)=\frac{1}{2}[x_1(n)+x_1(n-1)]$ and $y_2(n)=\frac{1}{2}[x_2(n)+x_2(n-1)]$. The system (b) is *linear*.

At last, we consider the time-shifted input signal $x(n-n_0)$ that produces the response $\frac{1}{2}[x(n-n_0)+x(n-n_0-1)]$ equal $y(n-n_0)$. The system is *time-invariant*.

3.1.1 Impulse and Step Response

Definition 3.7 The *impulse response* $h(t)$ of a continuous-time system is the output signal to the Dirac delta $\delta(t)$. We have: $\delta(t)\rightarrow h(t)$.

The *impulse response* $h(n)$ of a discrete-time system is the output signal to the unit pulse $\delta(n)$. We have: $\delta(n)\rightarrow h(n)$.

Definition 3.8 The *step response* $k(t)$ of a continuous-time system is the output of this system excited with the unit step signal $\mathbf{1}(t)$. We have: $\mathbf{1}(t)\rightarrow k(t)$.

The *step response* $k(n)$ of a discrete-time system is the output of this system excited with the unit step signal $\mathbf{1}(n)$. We have: $\mathbf{1}(n)\rightarrow k(n)$.

Impulse and step responses of LTI systems are related by the following relations:

- For continuous-time systems:

$$h(t)=\frac{d}{dt}k(t) \tag{3.1}$$

$$k(t)=\int_{-\infty}^{t}h(\tau)d\tau \tag{3.2}$$

- For discrete-time systems:

$$h(n)=k(n)-k(n-1) \tag{3.3}$$

Problem 3.2 Given an LTI system defined by two different input-output relations: $x_1(t)=\mathbf{1}(t)\rightarrow y_1(t)=(1-e^{-2t})\cdot\mathbf{1}(t)$ and $x_2(t)=\cos(2t)\rightarrow y_2(t)=\frac{\sqrt{2}}{2}\cos\left(2t-\frac{\pi}{4}\right)$. Find responses $y(t)$ to signals: (a) $x(t)=\delta(t)$, (b) $x(t)=t\cdot\mathbf{1}(t)$, (c) $x(t)=\cos^2(t)$; (d) $x(t)=2\Pi\left(t-\frac{1}{2}\right)$.

Solution First, let us notice that $y_1(t)=k(t)$ (step response).

(a) If the input signal $x(t)=\delta(t)$, the output $y(t)=h(t)$ (impulse response). To determine $y(t)$, we use (3.1) and get

$$y(t) = \frac{d}{dt}y_1(t) = \frac{d}{dt}\{(1 - e^{-2t}) \cdot \mathbf{1}(t)\} = 2e^{-2t} \cdot \mathbf{1}(t) + (1 - e^{-2t}) \cdot \delta(t)$$
$$= 2e^{-2t} \cdot \mathbf{1}(t),$$

since $\frac{d}{dt}\mathbf{1}(t) = \delta(t)$ and $(1 - e^{-2t}) \cdot \delta(t) = (1 - e^{-2 \cdot 0}) \cdot \delta(t) = 0$ (see Table 1.1). So, the impulse response of the system $h(t) = 2e^{-2t} \cdot \mathbf{1}(t)$.

(b) Let us notice that $\frac{d}{dt}\{t \cdot \mathbf{1}(t)\} = \mathbf{1}(t)$, so $t \cdot \mathbf{1}(t) = \int_0^t \mathbf{1}(\tau)d\tau$. Since $\mathbf{1}(t) \to k(t)$, the response $y(t)$ to $x(t) = t \cdot \mathbf{1}(t)$ is equal to the integral

$$y(t) = \int_0^t y_1(\tau)d\tau = \int_0^t [(1 - e^{-2\tau})]d\tau = \left(\tau + \frac{1}{2}e^{-2\tau}\right)\Big|_0^t$$
$$= \left(t - \frac{1}{2} + \frac{1}{2}e^{-2t}\right) \cdot \mathbf{1}(t).$$

(c) First, let us write the input signal in the equivalent form: $x(t) = \frac{1}{2} + \frac{1}{2}\cos(2t)$. We can see that $x(t)$ is a sum of the constant component $\frac{1}{2}$ and the cosine signal $\frac{1}{2}\cos(2t)$. Having in mind the linearity of the system, its response $y(t)$ is a sum of two output signals: $w_1(t)$ and $w_2(t)$, where $\frac{1}{2} \to w_1(t)$ and $\frac{1}{2}\cos(2t) \to w_2(t)$. The response $w_2(t)$ is known: $w_2(t) = \frac{\sqrt{2}}{4}\cos\left(2t - \frac{\pi}{4}\right)$. Let us compute the response $w_1(t)$. We have to recall the definition (1.21) of the Dirac delta: $\frac{1}{2}\int \delta(t)dt = \frac{1}{2}$. Due to the linearity of integration, we can write: $\frac{1}{2}\int h(t)dt = w_1(t)$ where the impulse response calculated in (a) is $h(t) = 2e^{-2t} \cdot \mathbf{1}(t)$. We have

$$w_1(t) = \int_0^\infty e^{-2t}dt = -\frac{e^{-2t}}{2}\Big|_0^\infty = \frac{1}{2}.$$

Finally, $y(t) = \frac{1}{2} + \frac{\sqrt{2}}{4}\cos\left(2t - \frac{\pi}{4}\right)$.

(d) Let us notice that $x(t) = 2\Pi\left(t - \frac{1}{2}\right)$ can equivalently be written as a difference of two unit step signals: $x(t) = 2 \cdot [\mathbf{1}(t) - \mathbf{1}(t - 1)]$. Linearity and time-invariance of the system permit to get the output: $y(t) = 2 \cdot [k(t) - k(t - 1)]$, that is,

$$y(t) = 2(1 - e^{-2t}) \cdot \mathbf{1}(t) - 2\left(1 - e^{-2(t-1)}\right) \cdot \mathbf{1}(t - 1).$$

Problem 3.3 Given the impulse response $h(n) = (0,5)^n \cdot \mathbf{1}(n)$ of a causal discrete-time LTI system. Determine and plot its step response $k(n)$.

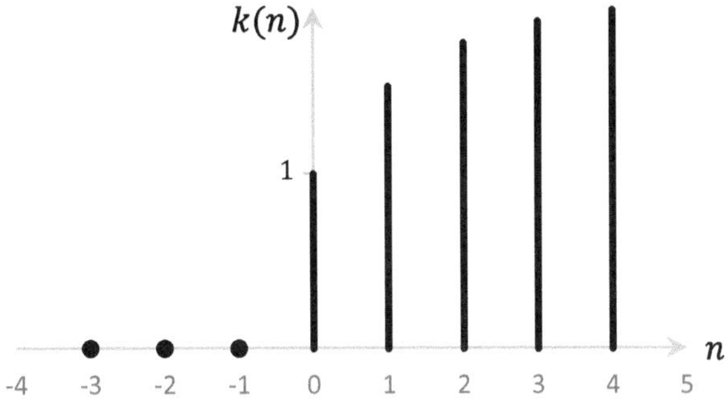

Fig. 3.1 Step response $k(n)$ of the system from Problem 3.3

Solution To determine the step response $k(n)$, we can apply (3.3) to get the following recursive equation: $k(n) = h(n) + k(n-1)$. To solve it, we compute subsequent values of $k(n)$ basing on its precedent value $k(n-1)$ and $h(n)$. For $n = 0, 1, 2, 3, 4$, we have

$k(0) = h(0) = 1$ since $k(-1) = 0$ (causal system),
$k(1) = h(1) + k(0) = 0,5 + 1 = 1,5$,
$k(2) = h(2) + k(1) = 0,25 + 1,5 = 1,75$,
$k(3) = h(3) + k(2) = 0{,}125 + 1,75 = 1{,}875$,
$k(4) = h(4) + k(3) = 0,0625 + 1{,}875 = 1,9375$,

etc. The plot of $k(n)$ is displayed in Fig. 3.1. Let us note that there are other methods of calculation of the step response from the impulse response of a discrete-time LTI system (e.g., see Sect. 8.3).

3.2 Response of the LTI System to a Given Excitation

In the LTI continuous-time system with the impulse response $h(t)$, the output $y(t)$ to a given input signal $x(t)$ is defined as a continuous-time convolution(see (1.35)–(1.36)):

$$y(t) = x(t) * h(t). \tag{3.4}$$

In the LTI discrete-time system with the impulse response $h(n)$, the output $y(n)$ to a given input signal $x(n)$ is defined as a discrete-time convolution (see (1.37)):

$$y(n) = x(n) * h(n). \tag{3.5}$$

Problem 3.4 Given the impulse response $h(t) = e^{-t} \cdot \mathbf{1}(t)$ of a continuous-time LTI system. Find the response to the input signal $x(t) = \Pi\left(t - \frac{1}{2}\right)$.

Solution We will present two different solutions to the considered problem.

1st method. Let us notice that $x(t)$ can be written as $x(t) = \mathbf{1}(t) - \mathbf{1}(t-1)$. The system is linear and time-invariant, so, since $\mathbf{1}(t) \to k(t)$, its response to $x(t)$ is $y(t) = k(t) - k(t-1)$. Using (3.2), we can compute the step response $k(t)$ as follows:

$$k(t) = \int_0^t e^{-\tau} d\tau = -e^{-\tau}\big|_0^t = (1 - e^{-t}) \cdot \mathbf{1}(t).$$

Finally, $y(t) = (1 - e^{-t}) \cdot \mathbf{1}(t) - (1 - e^{-(t-1)}) \cdot \mathbf{1}(t-1)$. Alternatively, $y(t)$ can be expressed as

$$y(t) = \begin{cases} 1 - e^{-t} & \text{for } 0 \le t < 1, \\ e^{-t}(e - 1) & \text{for } t \ge 1, \\ 0 & \text{elsewhere}. \end{cases}$$

Now, we will show that using a convolution method (3.4), we can come to the same result as above.

2nd method. Here, using the graphical method presented in Sect. 1.4 of Chap. 1, we convolve $x(t)$ with $h(t)$ to obtain the response $y(t) = x(t) * h(t)$.

Let us recall this procedure in brief. First, we treat both signals as functions of τ and sketch plots of $h(\tau)$ and of $x(-\tau)$. Then we shift $x(-\tau)$ right with a moving t and integrate the product $h(\tau)x(t-\tau)$ as shown in Fig. 3.2. The result $y(t)$ is exactly the same as above. The system response $y(t)$ is visualized in Fig. 3.3.

Problem 3.5 Determine the response of the system from Problem 3.3 to the input $x(n) = \sin\left(n\frac{\pi}{4}\right) \cdot \mathbf{1}(n)$.

Solution The impulse response of the system is $h(n) = (0,5)^n \cdot \mathbf{1}(n)$. To calculate the response $y(n)$, let us use the convolution relation (3.5):

$$y(n) = \sum_{k=0}^{\infty} x(k)h(n-k) = x(0)h(n) + x(1)h(n-1) + x(2)h(n-2) + \ldots,$$

For subsequent $n = 0, 1, 2, 3, \ldots$, we obtain

$y(0) = x(0)h(0) = 0$,
$y(1) = x(0)h(1) + x(1)h(0) = \sin\left(\frac{\pi}{4}\right) = \frac{1}{\sqrt{2}}$,
$y(2) = x(0)h(2) + x(1)h(1) + x(2)h(0) = \sin\left(\frac{\pi}{4}\right) \cdot \frac{1}{2} + \sin\left(\frac{\pi}{2}\right) = \frac{1}{2\sqrt{2}} + 1$

No.	Graphical computation of $y(t)$
1	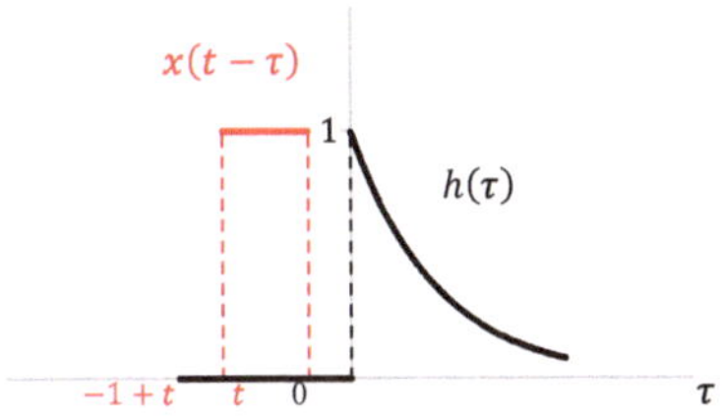
	$t \in (-\infty, 0]$: $\quad y(t) = 0$
2	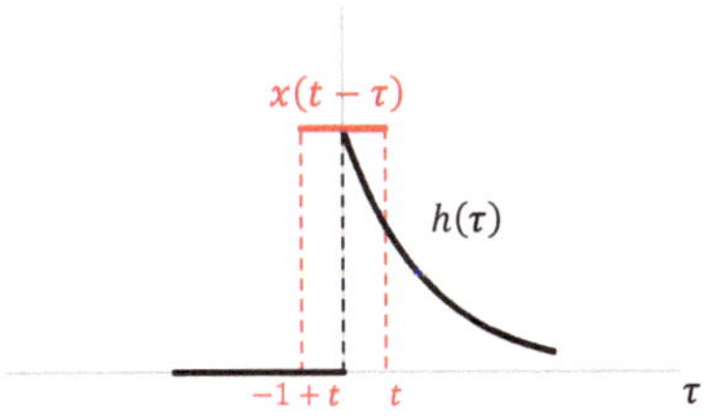
	$t \in (0,1]$: $\quad y(t) = \int_0^t e^{-\tau}\, d\tau = -e^{-\tau}\vert_0^t = 1 - e^{-t}$
3	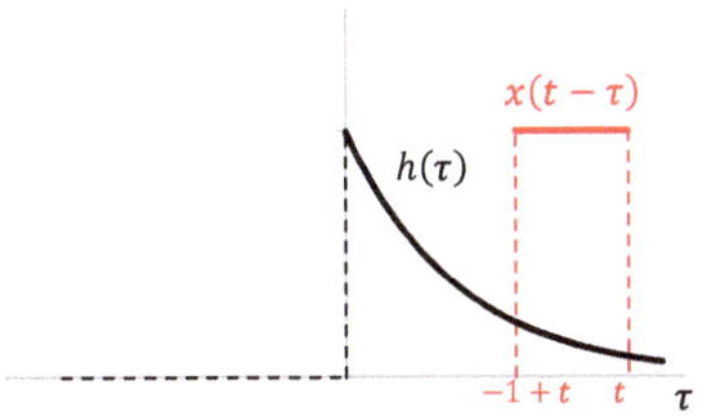
	$t \in (1, \infty)$: $y(t) = \int_{-1+t}^{t} e^{-\tau}\, d\tau = -e^{-\tau}\vert_{-1+t}^{t} = -e^{-t} + e^{-t+1}$ $= e^{-t}(e-1)$

Fig. 3.2 Graphical procedure for computing response $y(t)$ of the system from Problem 3.4

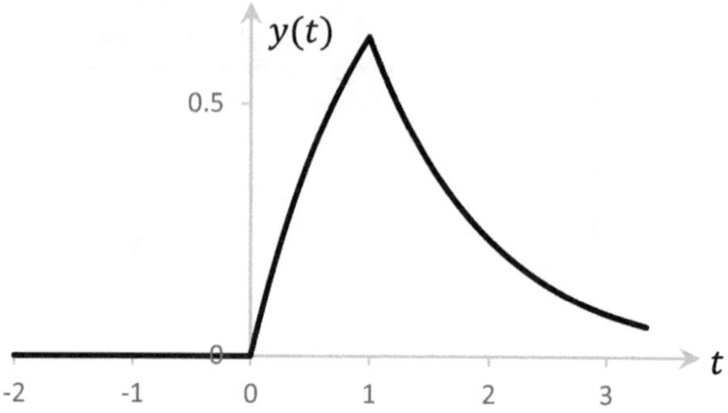

Fig. 3.3 Response $y(t)$ of the system from Problem 3.4

$$y(3) = x(0)h(3) + x(1)h(2) + x(2)h(1) + x(3)h(0) = \sin\left(\frac{\pi}{4}\right) \cdot \frac{1}{4} + \frac{1}{2} + \sin\left(\frac{3\pi}{4}\right)$$
$$= \frac{1}{4\sqrt{2}} + \frac{1}{2} + \frac{1}{\sqrt{2}} = \frac{5}{4\sqrt{2}} + \frac{1}{2}$$

etc.

Let us remember that an alternative way of computation of a discrete-time convolution is the graphical method presented in Sect. 1.4 (Problem 1.11).

3.3 Equations of LTI Systems

A *continuous-time LTI system*, defined by the input-output relation $x(t) \rightarrow y(t)$, is represented by the following *ordinary differential equation*:

$$\begin{aligned} a_m \frac{d^m y(t)}{dt^m} + a_{m-1} \frac{d^{m-1} y(t)}{dt^{m-1}} + \cdots a_1 \frac{dy(t)}{dt} + y(t) \\ = b_l \frac{d^l x(t)}{dt^l} + b_{l-1} \frac{d^{l-1} x(t)}{dt^{l-1}} + \cdots + b_0 x(t), t \geq t_0 \end{aligned} \tag{3.6}$$

with strictly determined *initial conditions* $y(t_0), y'(t_0), \ldots, y^{(m-1)}(t_0)$ and $a_1, \ldots, a_m \in \mathbb{R}, b_0, \ldots, b_l \in \mathbb{R}$. It is often assumed that $t_0 = 0$. The biggest m in (3.6) defines the *order* of a system.

A *discrete-time LTI system*, defined by the "input-output" relation $x(n) \rightarrow y(n)$, is represented by the following *ordinary difference equation*:

$$\begin{aligned} a_m y(n-m) + a_{m-1} y(n-m+1) + \cdots + a_1 y(n-1) + y(n) \\ = b_l x(n-l) + b_{l-1} x(n-l+1) + \cdots + b_1 x(n-1) + b_0 x(n), n \geq n_0 \end{aligned} \tag{3.7}$$

with strictly determined *initial conditions* $y(n_0)$, $y(n_0 - 1)$, ..., $y(n_0 - m + 1)$ and a_1, ..., $a_m \in \mathbb{R}$, b_0, ..., $b_l \in \mathbb{R}$. It is often assumed that $n_0 = 0$. The biggest m in (3.7) defines the *order* of a system.

Definition 3.9 A continuous/discrete-time LTI system defined by (3.6) and (3.7) is *initially at rest* if all initial conditions are zero.

Definition 3.10 A discrete-time LTI system defined by (3.7) is called the *Finite Impulse Response* (FIR) filter if all $a_i = 0$ for $i = 1, \ldots, m$. The Eq. (3.7) is not recursive, and the impulse response $h(n)$ of the FIR filter satisfies

$$h(n) = b_l\delta(n - l) + b_{l-1}\delta(n - l + 1) + \ldots + b_1\delta(n - 1) + b_0\delta(n). \tag{3.8}$$

Definition 3.11 A discrete-time LTI system defined by (3.7) is called the *Infinite Impulse Response* (IIR) filter if at least one of $a_i \neq 0$ for $i = 1, \ldots, m$. In this case, the Eq. (3.7) is recursive.

Our intention is not to present in this book any methods of solving ordinary differential equations in time domain. A reader interested in this subject can find detailed information in [7, 11]. However, in Chaps. 5 and 8, we will show how to solve ordinary differential and difference equations using different transformations (resp. Laplace- and $\mathcal{Z}$-transformations).

Contrary, solving difference equations describing LTI discrete-time systems is rather an easy task. In Problem 3.6, we will compute the digital filter response to a given input signal using a recursive procedure.

Problem 3.6 The IIR filter (LTI) is defined by the equation:
$y(n) + 3 \cdot y(n - 1) + 5 \cdot y(n - 2) = x(n - 2) - 2 \cdot x(n - 1)$. Assuming the system is initially at rest, find its response $y(n)$ to the input $x(n) = \left(\frac{1}{2}\right)^n \cdot \mathbf{1}(n)$.

Solution Let us write the filter equation in the equivalent form

$$y(n) = -2 \cdot x(n - 1) + x(n - 2) - 3 \cdot y(n - 1) - 5 \cdot y(n - 2).$$

and let us apply a recursive procedure for the calculation of $y(n)$.

For $n = 0, 1, 2, 3, 4, \ldots$ we have:

$y(0) = 0$, because $x(n) = 0$ for $n < 0$ and $y(-1) = y(-2) = 0$ (zero initial conditions—system initially at rest),

$$y(1) = -2 \cdot x(0) - 3 \cdot y(0) = -2,$$
$$y(2) = -2 \cdot x(1) + x(0) - 3 \cdot y(1) - 5 \cdot y(0) = -2 \cdot \tfrac{1}{2} + 1 + 6 = 6,$$
$$y(3) = -2 \cdot x(2) + x(1) - 3 \cdot y(2) - 5 \cdot y(1) = -2 \cdot \tfrac{1}{4} + \tfrac{1}{2} - 18 + 10 = -8,$$
$$y(4) = -2 \cdot x(3) + x(2) - 3 \cdot y(3) - 5 \cdot y(2) = -2 \cdot \tfrac{1}{8} + \tfrac{1}{4} + 24 - 30 = -6,$$

etc.

3.4 Block Diagrams of LTI Systems

In the analysis of LTI systems, it is often needed to represent its structure using *block diagrams*. In Table 3.1, we collected basic elements of block diagrams of continuous- and discrete-time systems.

In Problems 3.7, 3.8, 3.9, and 3.10, we will show how to draw block diagrams starting from a system equation and vice versa.

Problem 3.7 Given the equation of the 2nd order continuous-time system: $\frac{d^2y(t)}{dt^2} - 3\frac{dy(t)}{dt} + 2y(t) = \frac{dx(t)}{dt} + x(t)$. Draw the block diagram of this system.

Solution Note that there is more than one solution to this problem.

We propose to write the system equation in the form: $y(t) = \frac{1}{2}\left[\frac{dx(t)}{dt} + x(t) + 3\frac{dy(t)}{dt} - \frac{d^2y(t)}{dt^2}\right]$. In this equality, we can notice two parts:

Table 3.1 Elements of block diagrams of continuous- and discrete-time systems

Element	Continuous-time systems	Discrete-time systems
Pick-off node	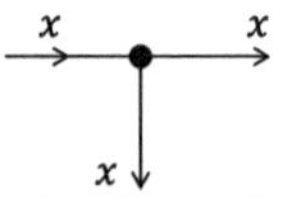	
Adder	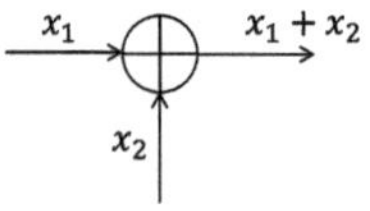	
Multiplier	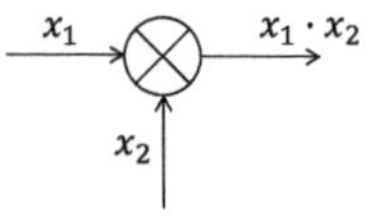	
Constantmultiplier	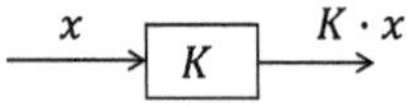	
Shift block	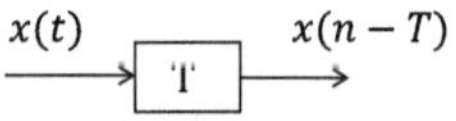	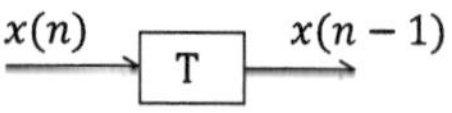
Differentiator block	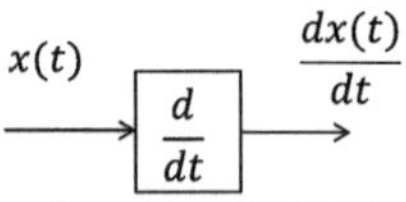	—
Integrating block	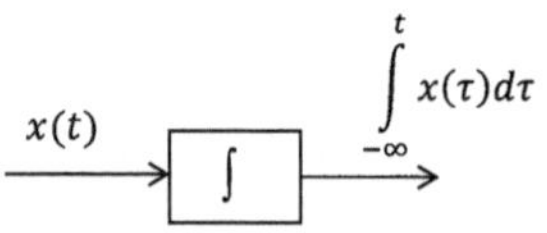	—

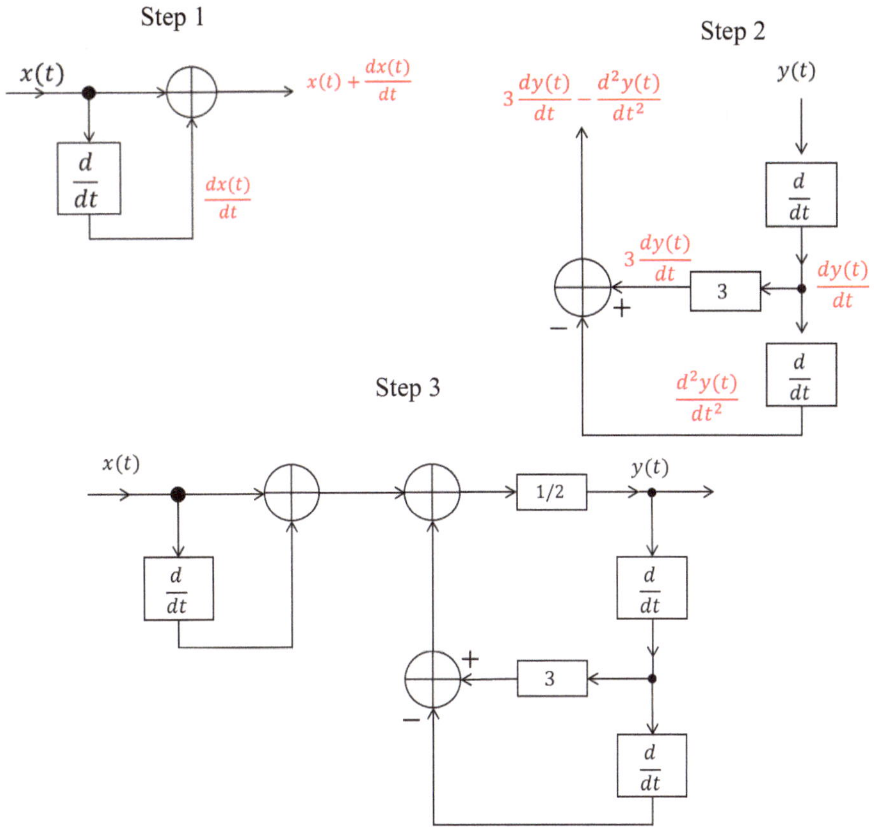

Fig. 3.4 Block diagram of the system from Problem 3.7. In red: signals at the output of corresponding elementary blocks

$\frac{dx(t)}{dt} + x(t)$ depending on $x(t)$ and $3\frac{dy(t)}{dt} - \frac{d^2y(t)}{dt^2}$, both multiplied by $\frac{1}{2}$. We can start with drawing a part of the diagram for $\frac{dx(t)}{dt} + x(t)$ that requires the usage of three elements: a pick-off node, a differentiator block, and an adder (see Fig. 3.4—Step 1). For $3\frac{dy(t)}{dt} - \frac{d^2y(t)}{dt^2}$ at the output, we can use two differentiator blocks, a constant multiplier, and an adder, as shown in Fig. 3.4—Step 2. Let us pay attention to the direction of a signal flow that is from right to left, that is, we apply a *double feedback loop* (more about this topic in the next section concerning interconnections of LTI systems). The final step is to connect both parts by means of an adder and to apply a constant multiplier $\frac{1}{2}$ at its output, as shown in Fig. 3.4—Step 3. Notice that this block diagram is not "economic" w.r.t. a number of used differentiator blocks. In the next section, we will explain a procedure for reducing this number.

Problem 3.8 Given the IIR filter with the equation: $2y(n) - y(n-1) + y(n-3) = x(n) - 5x(n-4)$. Draw its block diagram.

Solution Also in this case, there are more possible solutions to the problem.

Here, we propose to rewrite the above equation as $y(n) = \frac{1}{2}[x(n) - 5x(n-4) + y(n-1) - y(n-3)]$. Similarly as in Problem 3.7, to get the sum $x(n) - 5x(n-4)$, we need to apply a pick-off point, four shift blocks, a constant multiplier, and an adder (Fig. 3.5—Step 1). Notice that here we can apply a different idea to get a subtraction, that is, we mark -5 in the corresponding constant multiplier.

Then to get $y(n-1) - y(n-3)$, we can use a double feedback loop from the output $y(n)$ containing three shift blocks, a pick-off point, and an adder (see Fig. 3.5—Step 2). At the end, we join both parts with an adder followed by a constant multiplier (Fig. 3.5—Step 3). Notice that this block diagram is not "economic" from the point of view of the number of shift blocks used. In the next section, we will explain how this number can be reduced.

Problem 3.9 Figure 3.6 shows the block diagram of a continuous-time LTI system. Find the equation of this system.

Solution A good idea is to introduce an auxiliary signal $w(t)$ marked in red in Fig. 3.6. If we put it at the output of the third differentiator block, we can easily notice that at specific points of the diagram, we have successive derivatives of $w(t)$.

We can write the following equations:

$$x(t) - 2w''(t) - 2w'(t) - w(t) = w'''(t),$$

$$y(t) = \frac{3}{2}w'(t).$$

Let us differentiate the first of the above equations to get

$$x'(t) - 2w'''(t) - 2w''(t) - w'(t) = w^{(4)}(t).$$

From the second relation, we obtain
$w'(t) = \frac{2}{3}y(t)$, so $w''(t) = \frac{2}{3}y'(t)$, $w'''(t) = \frac{2}{3}y''(t)$ and $w^{(4)}(t) = \frac{2}{3}y'''(t)$.
Let us now substitute $w'(t)$, $w''(t)$, $w'''(t)$ and $w^{(4)}(t)$ in the above equation

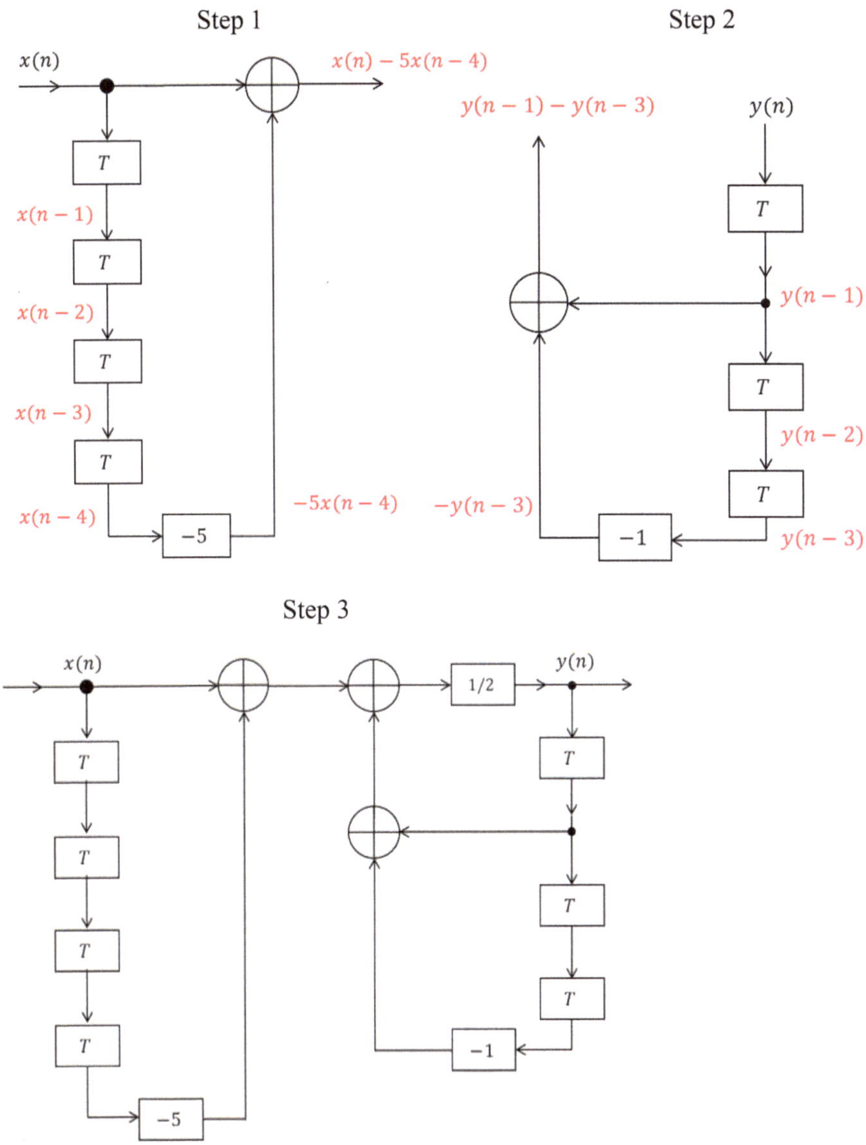

Fig. 3.5 Block diagram of the system from Problem 3.8. In red: signals at output of corresponding elementary blocks

$$x'(t) - \frac{4}{3}y''(t) - \frac{4}{3}y'(t) - \frac{2}{3}y(t) = \frac{2}{3}y'''(t).$$

After ordering the terms, we finally get

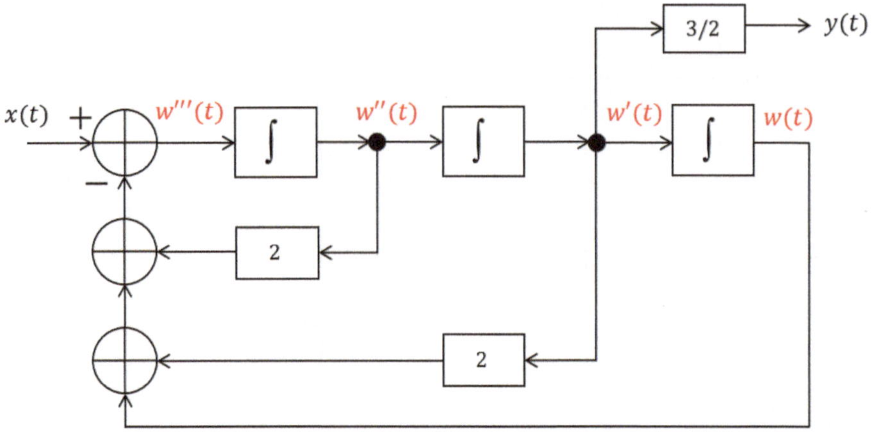

Fig. 3.6 Block diagram of the system from Problem 3.9

$$\frac{2}{3}\frac{d^3y(t)}{dt^3}+\frac{4}{3}\frac{d^2y(t)}{dt^2}+\frac{4}{3}\frac{dy(t)}{dt}+\frac{2}{3}y(t)=\frac{dx(t)}{dt}.$$

Problem 3.10 Figure 3.7 shows the block diagram of a discrete-time LTI system. Find the equation of this system.

Solution Looking at the direction of a signal flow in the block diagram, we can easily determine signals in every branch (marked in red in Fig. 3.7). Now, it is easy to write the equation relating $x(n)$ with $y(n)$. Looking at the top adder, we have

$$y(n)=2x(n)+x(n-1)+y(n-1)-\frac{1}{2}y(n-2).$$

3.5 System Interconnections

There are three basic interconnections of systems: cascade, parallel, and feedback (see Fig. 3.8). In *cascade* interconnection S, the input of the first system S_1 is the input of the whole system S ($x_1=x$), the output of S_1 equals the input of S_2 ($y_1=x_2$), and the output of S_2 is the output of S ($y_2=y$). The impulse response h of a cascade interconnection is equal to a convolution (continuous/discrete-time)

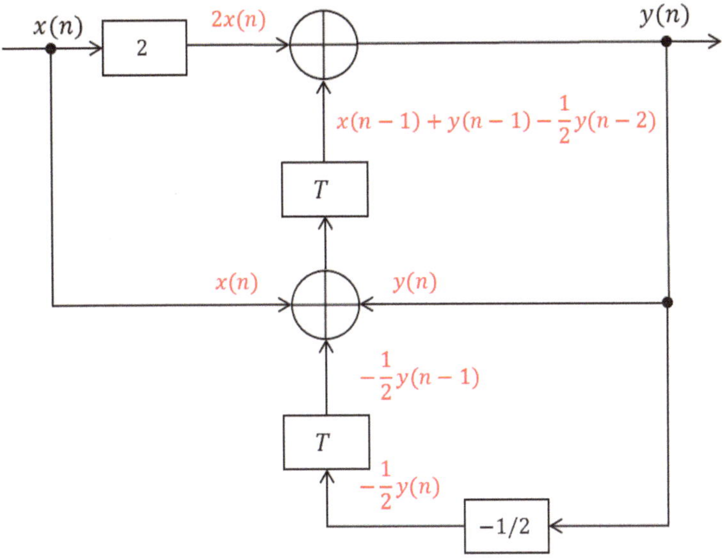

Fig. 3.7 Block diagram of the system from Problem 3.10

$$h = h_1 * h_2, \tag{3.9}$$

where h_1 is the impulse response of S_1 and h_2 is the impulse response of S_2.

In *parallel* interconnection S, subsystems S_1 and S_2 are stimulated with the same input ($x_1 = x_2 = x$) and their outputs y_1 and y_2 are added to form the output of the whole system S ($y_1 + y_2 = y$). The impulse response h of the cascade interconnection is equal to the sum of corresponding impulse responses of S_1 and S_2:

$$h = h_1 + h_2 \tag{3.10}$$

In *feedback* system S, we distinguish two subsystems: S_1—in the main branch, and S_2—in the feedback loop. We can see that the input of S_2 equals the output of S ($x_2 = y$). Its output y_2 is subtracted from x to form the input of S_1 denoted with e, that is, $x - y_2 = e$. The signal e is called the *error* signal. The output of S_1 is y ($y_1 = y$).

Evidently, in each system interconnection, we can have more than two subsystems.

Important Note that in a *cascade interconnection* of two LTI systems S_1 and S_2, the order can be changed, so we have two equivalent interconnections: $S_1 - S_2$ and $S_2 - S_1$. This feature is useful in the realization of LTI systems with a reduced number of elementary blocks, as presented in Problems 3.11 and 3.12.

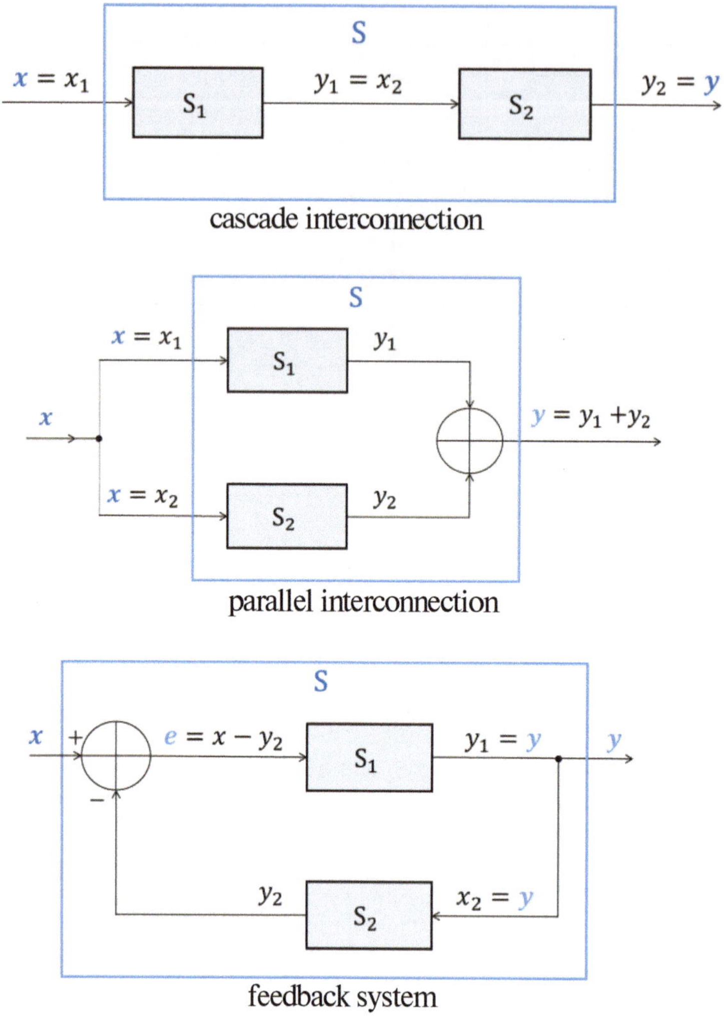

Fig. 3.8 Different interconnections of systems: cascade, parallel, and feedback

> **Problem 3.11** The system from Problem 3.7 is realized as a cascade interconnection of two subsystems. Draw the block diagram with a minimal number of differentiator blocks.

Solution In Fig. 3.9, we marked subsystems S_1 and S_2 of the system from Problem 3.7. Let us change their order into $S_2 - S_1$ as shown in Fig. 3.10. The next step is

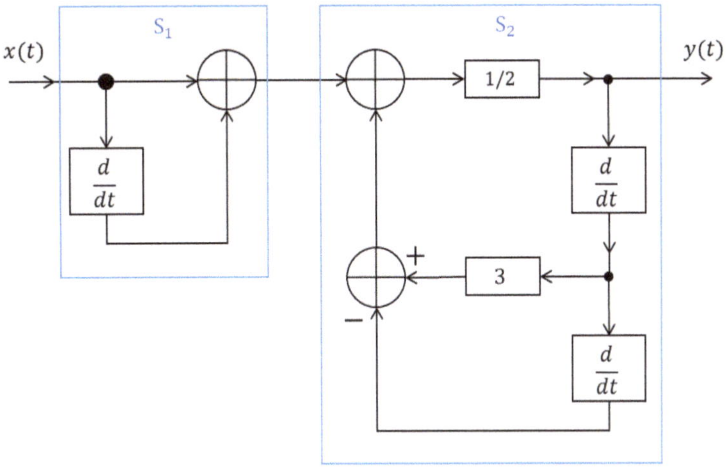

Fig. 3.9 Cascade interconnection $S_1 - S_2$ of the system from Problem 3.7

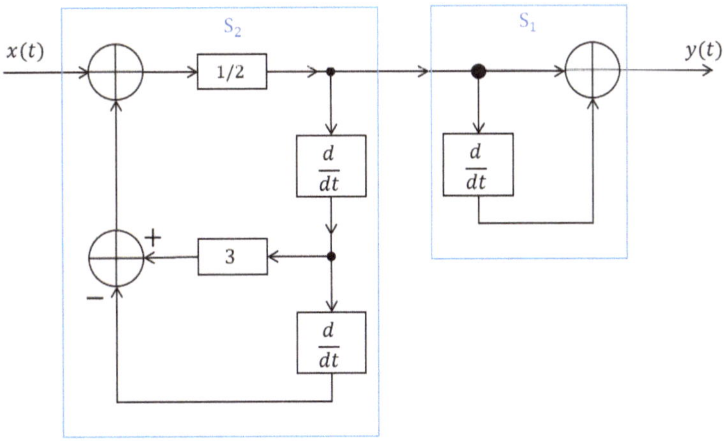

Fig. 3.10 Cascade interconnection $S_2 - S_1$ of the system from Problem 3.7

joining two subsystems as visualized in Fig. 3.11. We can notice that such a realization permits to reduce the number of differentiator blocks from 3 to 2.

Problem 3.12 The system from Problem 3.8 was realized as a cascade interconnection of two subsystems. Draw its block diagram with a minimal number of shift blocks.

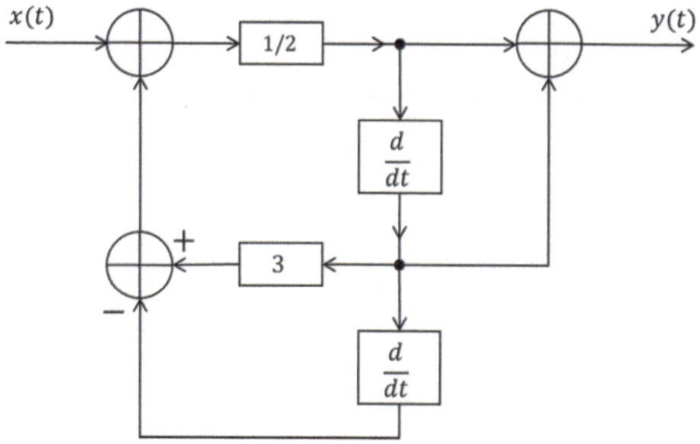

Fig. 3.11 Block diagram of the system from Problem 3.7 with a reduced number of differentiator blocks

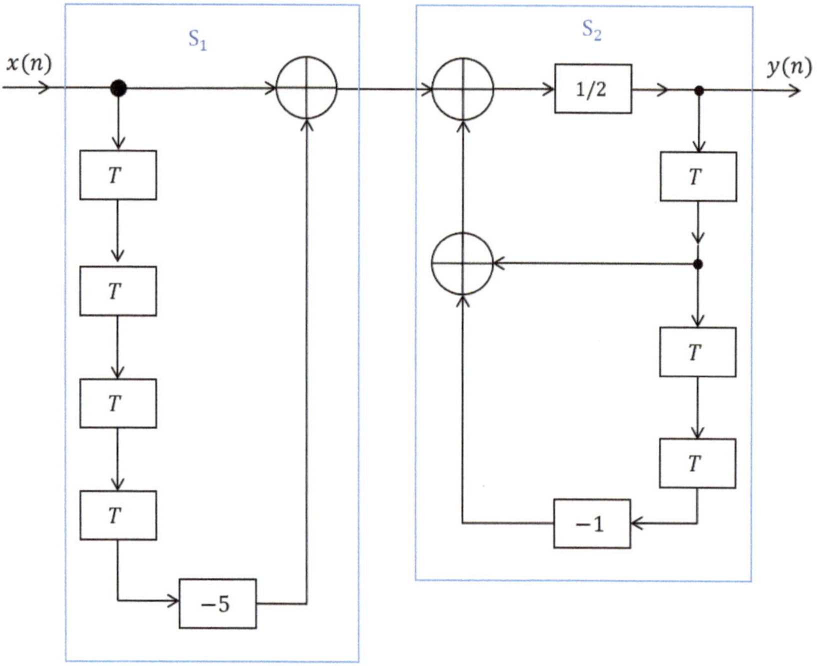

Fig. 3.12 Direct realization type I of the system from Problem 3.8

Solution We can apply the same procedure as in the solution of the precedent Problem 3.11. Figure 3.12 shows the block diagram of the system from Problem 3.8 in the form of a cascade of two subsystems S_1 and S_2. Such an interconnection is called the *direct realization of type I*.

Then we can change the order into $S_2 - S_1$ as shown in Fig. 3.13, and after we can join two subsystems (see Fig. 3.14) to obtain the interconnection called the *direct realization of type II*. We can notice that the number of shift blocks was considerably reduced: from 7 to 4 (see Fig. 3.12).

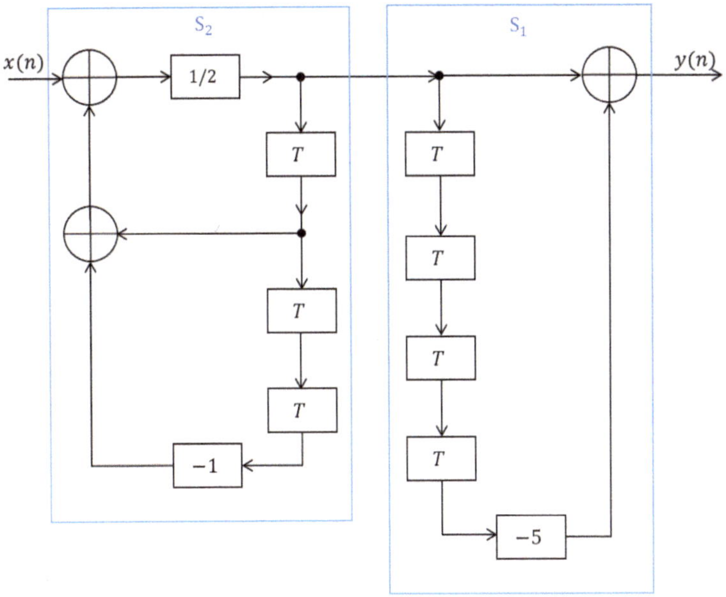

Fig. 3.13 Cascade interconnection $S_2 - S_1$ of the system from Problem 3.8

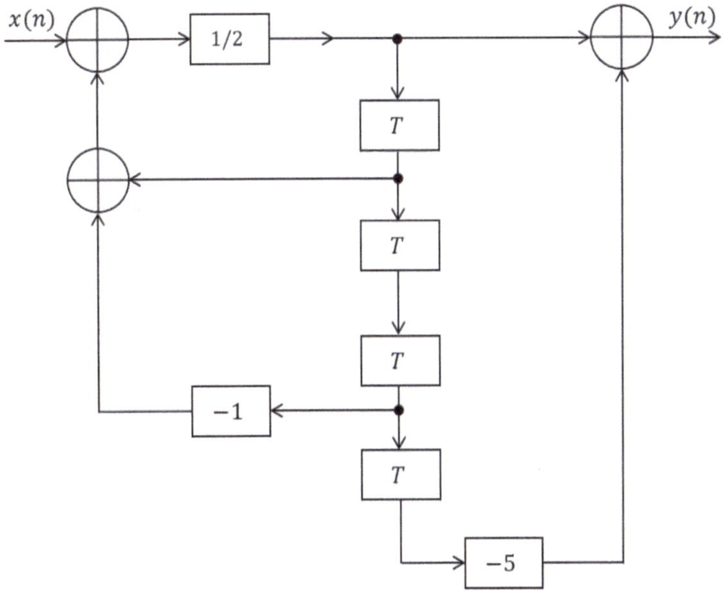

Fig. 3.14 Direct realization of type II of the system from Problem 3.8

Next solved examples concern some interesting problems connected with different systems interconnections.

Problem 3.13 Given the cascade interconnection $S_1 - S_2 - S_3$ of three systems defined by the following input-output relations:

$S_1 : y(n) = \begin{cases} x\left(\frac{n}{2}\right) & \text{for even } n \\ 0 & \text{for odd } n \end{cases};$
$S_2 : y(n) = x(n) + \frac{1}{2}x(n-1) + \frac{1}{4}x(n-2);$
$S_3 : y(n) = x(2n).$

Find the input-output equation of the whole system.

Solution Figure 3.15 shows the structure of the system. The responses of S_1 and S_2 respectively are $y_1(n)$ and $y_2(n)$.

The input-output relations can be rewritten as

$$y_1(n) = \begin{cases} x\left(\frac{n}{2}\right) & \text{for even } n \\ 0 & \text{for odd } n \end{cases},$$

$$y_2(n) = y_1(n) + \frac{1}{2}y_1(n-1) + \frac{1}{4}y_1(n-2),$$

$$y(n) = y_2(2n).$$

We can see that

$$y(n) = y_2(2n) = y_1(2n) + \frac{1}{2}y_1(2n-1) + \frac{1}{4}y_1(2n-2)$$

but $y_1(2n) = x(n)$ (while $2n$ is even), $y_1(2n-1) = 0$ (while $2n-1$ is odd) and $y_1(2n-2) = x(n-1)$ (while $2n-2$ is even).

Finally, we get the input-output equation of the whole system in the form

$$y(n) = x(n) + \frac{1}{4}\,x(n-1),$$

so we deal with a simple FIR filter.

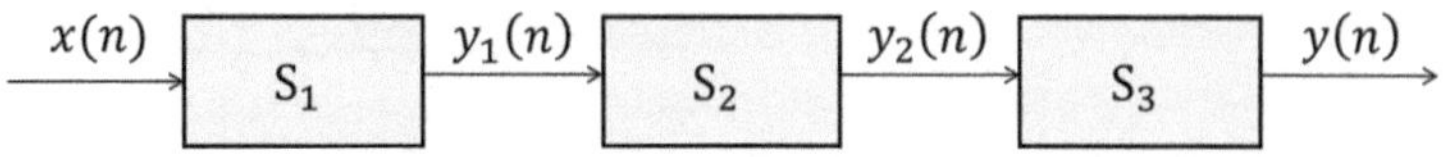

Fig. 3.15 Interconnection of three systems from Problem 3.13

Problem 3.14 A discrete-time LTI system S is realized as: (a) a cascade–, (b) a parallel interconnection, of two systems S_1 and S_2 with impulse responses $h_1(n)$ and $h_2(n) = \mathbf{1}(n) - \mathbf{1}(n-1)$ respectively. The impulse response $h(n)$ of S is $h(n) = \delta(n) + 2\delta(n-1) + 3\delta(n-2) + \delta(n-4) + 3\delta(n-5)$. Determine $h_1(n)$ in each case.

Solution First, let us consider a cascade interconnection of S_1 and S_2. From (3.9), $h(n) = h_1(n) * h_2(n)$ where $h_2(n) = \mathbf{1}(n) - \mathbf{1}(n-1) = \delta(n) + \delta(n-1)$.

We have

$$h(n) = h_1(n) * [\delta(n) + \delta(n-1)] = h_1(n) * \delta(n) + h_1(n) * \delta(n-1).$$

Then

$$h(n) = h_1(n) + h_1(n-1)$$

or equivalently,

$$h_1(n) = h(n) - h_1(n-1).$$

So, we have to solve the following recursive equation:

$$h_1(n) = \delta(n) + 2\delta(n-1) + 3\delta(n-2) + \delta(n-4) + 3\delta(n-5) - h_1(n-1)$$

as follows:

$h_1(0) = \delta(0) = 1,$
$h_1(1) = 2\delta(0) - h_1(0) = 1,$
$h_1(2) = 3\delta(0) - h_1(1) = 2,$
$h_1(3) = -h_1(2) = -2,$
$h_1(4) = \delta(0) - h_1(3) = 3,$
$h_1(5) = 3\delta(0) - h_1(4) = 0$

and finally for $n \geq 6$, $h_1(n) = 0$. The impulse response of S_1 is then

$$h_1(n) = \delta(n) + \delta(n-1) + 2\delta(n-2) - 2\delta(n-3) + 3\delta(n-4).$$

Now, let us consider the parallel interconnection of S_1 and S_2. From (3.10), we know that $h(n) = h_1(n) + h_2(n)$, then

$$h_1(n) = h(n) - h_2(n) = \delta(n-1) + 3\delta(n-2) + \delta(n-4) + 3\delta(n-5).$$

Definition 3.12 A linear time-invariant system is *invertible* if it produces a unique response to each input signal.

Definition 3.13 A linear time-invariant system S^{-1} is called the *inverse* to a given LTI system S if the cascade interconnection S–S^{-1} produces a response equal to the input signal.

For the cascade interconnection S–S^{-1} of two continuous-time LTI systems, we have

$$h(t) * h_1(t) = \delta(t), \tag{3.11}$$

where $h(t)$ is the impulse response of S and $h_1(t)$ is the impulse response of the inverse system S^{-1}.

For the cascade interconnection S–S^{-1} of two discrete-time LTI systems, we can write

$$h(n) * h_1(n) = \delta(n), \tag{3.12}$$

where $h(n)$ is the impulse response of S and $h_1(n)$ is the impulse response of the inverse system S^{-1}.

Problem 3.15 Given a discrete-time LTI system S with the impulse response $h(n) = (1/2)^n \cdot \mathbf{1}(n)$. (a) Find A such that: $h(n) - A \cdot h(n-1) = \delta(n)$. (b) Compute the impulse response $h_1(n)$ of the inverse system S^{-1} to S.

Solution (a) In order to find A, we can substitute $n = 1$ into the relation $h(n) - A \cdot h(n-1) = \delta(n)$ to get $h(1) - A \cdot h(0) = \delta(1) = 0$. Since $h(1) = 1/2$ and $h(0) = 1$, then $A = 1/2$.

(b) We can apply (3.12) in its equivalent form: $\sum_{k=0}^{\infty} h(k)h_1(n-k) = \delta(n)$ (compare with (1.37)). We have

$$\sum_{k-0}^{\infty} h(k)h_1(n-k) = h(0)h_1(n) + h(1)h_1(n-1) + h(2)h_1(n-2) + \ldots,$$

then

$$h(0)h_1(n) + h(1)h_1(n-1) + h(2)h_1(n-2) + \ldots \quad = \delta(n).$$

Let us solve the above difference equation in a recursive way.

For $n = 0$, $h(0)h_1(0) = \delta(0) = 1$, so $h_1(0) = 1$.

For $n = 1$:

$$h(0)h_1(1)+h(1)h_1(0)=\delta(1)=0,$$

$$h_1(1)+\frac{1}{2}h_1(0)=0\Leftrightarrow h_1(1)=-\frac{1}{2}.$$

For $n = 2$:

$$h(0)h_1(2)+h(1)h_1(1)+h(2)h_1(0)=0,$$

$$h_1(2)+\frac{1}{2}h_1(1)+\frac{1}{4}h_1(0)=0,$$

$$h_1(2)+\frac{1}{2}\left(-\frac{1}{2}\right)+\frac{1}{4}=0\Leftrightarrow h_1(2)=0.$$

For $n = 3$:

$$h(0)h_1(3)+h(1)h_1(2)+h(2)h_1(1)+h(3)h_1(0)=0,$$

$h_1(3)+\frac{1}{4}\left(-\frac{1}{2}\right)+\frac{1}{8}=0\Leftrightarrow h_1(3)=0$ and so on.
Finally, since $h_1(0) = 1$ and $h_1(1) = -\frac{1}{2}$, we get $h_1(n)=\delta(n)-\frac{1}{2}\delta(n-1)$.

Exercises

3.1 Study the features of the following continuous- and discrete-time systems described by input-output relations: (a) $y(t) = A + Bx(t)$; $A, B = const$, $A, B \in \mathbb{R}$; (b) $y(t) = x^2(t)$; (c) $y(t)=\int_{t_0}^{t}\tau x(\tau)d\tau, t\geq t_0$; (d) $y(n)=\sum_{i=-\infty}^{\infty}(-0,5)^{n-i}x(i)$; (e) $y(n)=(-0,5)^n\sum_{i=-\infty}^{\infty}x(i)$; (f) $y(n)=x(-n)\cdot x(n-1)$.

3.2 The input-output relation of a *moving average* continuous-time system is

$$y(t)=\frac{1}{T_1+T_2}\int_{t-T_1}^{t+T_2}x(\tau)d\tau; T_1+T_2>0.$$

Determine its features w.r.t. parameters T_1 and T_2.

3.3 The input-output relation of a *moving average* discrete-time system is

$$y(t)=\frac{1}{M+N+1}\sum_{i=-N}^{M}x(n-i); M+N>0$$

Determine its features w.r.t. parameters M and N.

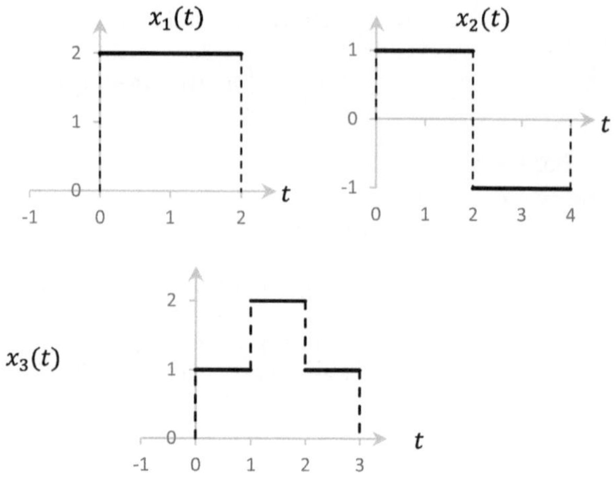

Fig. 3.16 Three different input signals from Exercise 3.4

3.4 The step response of a continuous-time LTI system is $k(t) = \Lambda\left(\frac{t-2}{2}\right)$. Find the output signals $y_1(t)$, $y_2(t)$, $y_3(t)$ to three different input signals $x_1(t)$, $x_2(t)$, $x_3(t)$ shown in Fig. 3.16.

3.5 Given the step response of a continuous-time LTI system: $k(t) = (1 - e^{-t}) \cdot \mathbf{1}(t)$. Find its response to the input $x(t) = \delta(t - 2)$.

3.6 Given the impulse response of a causal discrete-time LTI system: $h(n) = ((0,5)^n - (0,2)^n) \cdot \mathbf{1}(n)$. Determine and sketch its step response.

3.7 Given the input-output relation of a discrete-time system:

$$y(n) = \sum_{k=-\infty}^{\infty} x(k)g(n-2k),$$

where $g(n) = \mathbf{1}(n) - \mathbf{1}(n-3)$. Is this system linear and time-invariant? Find its responses to: (a) $x(n) = \delta(n-2)$; (b) $x(n) = \mathbf{1}(n)$.

3.8 Given the equation of a continuous-time LTI system:

$$2\frac{dy(t)}{dt} + 0,5\,y(t) = x(t) - 0,5\frac{dx(t)}{dt}.$$

Show that an equivalent equation of this system is:

$$y(t) = A\int_{-\infty}^{t} y(\tau)d\tau + Bx(t) + C\int_{-\infty}^{t} x(\tau)d\tau.$$

Determine values of constants A, B, and C. Keeping a minimal number of elementary blocks, draw two different block diagrams of this system.

3.9 Given the equation of a discrete-time LTI system:

$$y(n) - 5y(n-1) + 6y(n-2) = x(n) - x(n-1) - 2x(n-2).$$

Draw the block diagrams as: (a) a direct realization of type I; (b) a direct realization of type II.

3.10 A continuous-time LTI system S is a *cascade connection* S_1—S_2 of two LTI systems. The equation of S_1 is $\frac{dy_1(t)}{dt} = x_1(t)$ and the impulse response of S_2 is $h_2(t) = \mathbf{1}(t)$. Assuming the input signal $x(t) = \mathbf{1}(t) - 2 \cdot \mathbf{1}(t-2) + \mathbf{1}(t-4)$, determine and sketch the response $y(t)$ of S.

3.11 A discrete-time LTI system S is: (a) a *cascade realization*; (b) a *parallel realization* of two systems S_1 and S_2 defined by two difference equations:

$$S_1 : \; y_1(n) = x_1(n) + 2x_1(n-1),$$

$$S_2 : \; y_2(n) = x_2(n-2) + \frac{1}{2}x_2(n-3).$$

Find the difference equation of the system S for cases (a) and (b). Compute and sketch corresponding impulse responses of S.

3.12 Prove that a FIR filter defined by the equation $y(n) = x(n) - x(n-1)$ is the *inverse* of the system S defined by $y(n) = \sum_{k=-\infty}^{n} x(k)$.

Chapter 4
Analysis of Analog Systems in Frequency Domain

In this chapter, we apply the theory of continuous-time Fourier transformation presented in Chap. 2 to study the properties of continuous-time LTI systems (analog filters) in frequency domain. The reader will get to know different types of ideal and real analog filters (low-pass, high-pass, band-pass, and band-stop). The notion of the frequency response function of an analog filter is introduced. We will show how spectrum characteristics of a signal change passing through an analog filter. One will also learn to compute the response of a system to the known input signal. The method of frequency transformation applied in analog filter design is also presented.

4.1 Frequency Response Function

Given a continuous-time signal $x(t)$, $t \in \mathbb{R}$ at the input of a continuous-time LTI system. Its Fourier spectrum is $X(\omega)$. At the output, we can observe the response $y(t)$, $t \in \mathbb{R}$ with the Fourier spectrum $Y(\omega)$. Knowing that $y(t) = x(t) * h(t)$, where $h(t)$ is the impulse response of the system, we can apply to (3.4) the convolution property of the FT and get

$$Y(\omega) = X(\omega)H(\omega), \tag{4.1}$$

where $H(\omega)$ is the FT of the impulse response $h(t)$, that is, $h(t) \overset{\mathcal{F}}{\leftrightarrow} H(\omega)$.

Definition 4.1 The Fourier transform of the impulse response $h(t)$ of a continuous-time LTI system is called its *frequency response function*. In general, it is a complex-valued function expressed as

$$H(\omega) = |H(\omega)|e^{j\arg H(\omega)} = A(\omega)e^{j\varphi(\omega)}, \tag{4.2}$$

K. M. Snopek, *Hands-on Signals and Systems Theory*,
https://doi.org/10.1007/978-3-031-56081-1_4

where $A(\omega) = |H(\omega)|$ is the *magnitude response* and $\varphi(\omega) = \arg H(\omega)$ is the *phase response* of a system. The magnitude characteristic is always an even function of ω, that is, $A(\omega) = A(-\omega)$, while the phase characteristic is an odd function of ω: $\varphi(-\omega) = -\varphi(\omega)$. Evidently, we can consider the frequency response function as a function of *f*.

Definition 4.2 The *real frequency response* $P(\omega)$ is the real part of a complex-valued frequency response function $H(\omega)$. The imaginary part $Q(\omega)$ is called the *imaginary frequency response*. We have

$$H(\omega) = P(\omega) + jQ(\omega). \tag{4.3}$$

From (4.2) and (4.3), we can get the following relations:

$$A(\omega) = \sqrt{[P(\omega)]^2 + [Q(\omega)]^2}, \tag{4.4}$$

$$\varphi(\omega) = \text{atan2}(Q(\omega), P(\omega)). \tag{4.5}$$

The real frequency response is always an even function of ω: $P(\omega) = P(-\omega)$, while the imaginary frequency response is an odd function of ω: $Q(-\omega) = -Q(\omega)$. In Problem 4.1, we will show how to plot different frequency characteristics of a 1st order analog system.

Problem 4.1 Given the impulse response $h(t) = e^{-\alpha t} \cdot \mathbf{1}(t)$ of the1st order analog system. (a) Determine the frequency response function and sketch magnitude and phase characteristics; (b) determine and sketch real and imaginary frequency response characteristics; (c) determine the input-output equation of the system. Consider two cases: $\alpha = \frac{1}{4}$ and $\alpha = 1$.

Solution (a) Using Table 2.1 (Chap. 2), we immediately obtain the frequency response function $H(\omega)$ as the FT of $h(t) = e^{-\alpha t} \cdot \mathbf{1}(t)$, that is, $H(\omega) = \frac{1}{\alpha + j\omega}$. First, we can calculate the magnitude response $A(\omega) = |H(\omega)|$:

$$A(\omega) = \left|\frac{1}{\alpha + j\omega}\right| = \frac{1}{|\alpha + j\omega|} = \frac{1}{\sqrt{\alpha^2 + \omega^2}}.$$

For $\alpha = \frac{1}{4}$: $A(\omega) = \frac{1}{\sqrt{1/16 + \omega^2}}$, for $\alpha = 1$: $A(\omega) = \frac{1}{\sqrt{1 + \omega^2}}$.

Both plots are shown in Fig. 4.1. Notice that the magnitude response of the first filter gets bigger values for lower frequencies than for the second one. In other words, the first filter amplifies “lower frequencies” more than the second one.

Now, let us derive the phase characteristic. We have

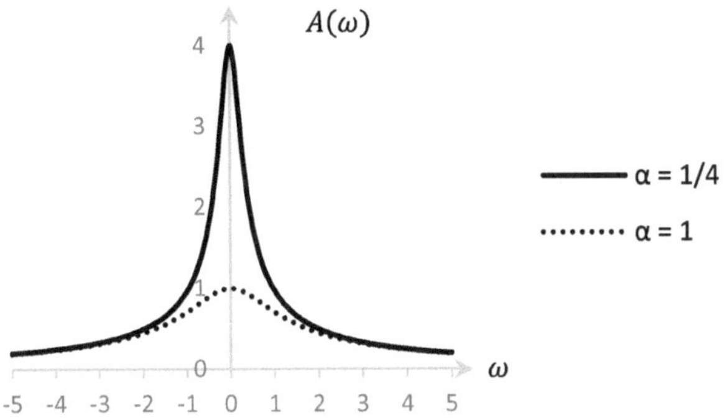

Fig. 4.1 Magnitude response of the analog filter from Problem 4.1

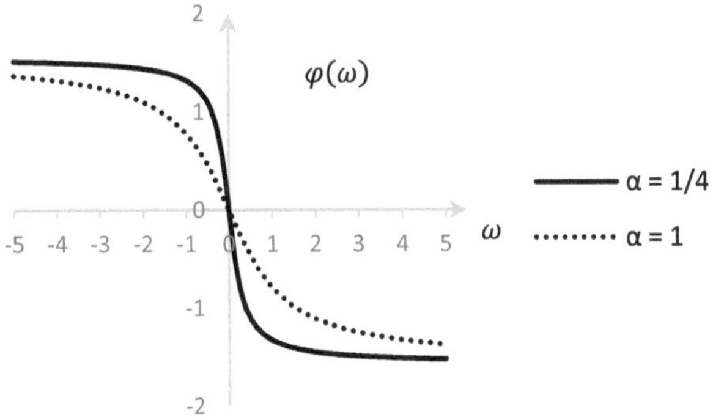

Fig. 4.2 Phase response of the analog filter from Problem 4.1

$$\varphi(\omega) = \arg\left(\frac{1}{\alpha + j\omega}\right) = \arg(1) - \arg(\alpha + j\omega) = 0 - \arg(\alpha + j\omega) = -\operatorname{atan2}(\omega, \alpha)$$
$$= -\operatorname{atan}\left(\frac{\omega}{\alpha}\right).$$

For $\alpha = \frac{1}{4}$: $\varphi(\omega) = -\operatorname{atan}(4\omega)$, for $\alpha = 1$: $\varphi(\omega) = -\operatorname{atan}(\omega)$.

Both plots are presented in Fig. 4.2. We can notice that the system introduces a nonlinear phase shift.

(b) Let us derive real and imaginary parts of $H(\omega)$. We have

$$H(\omega) = \frac{1}{\alpha + j\omega} = \frac{\alpha - j\omega}{(\alpha + j\omega)(\alpha - j\omega)} = \frac{\alpha}{\alpha^2 + \omega^2} + j\frac{-\omega}{\alpha^2 + \omega^2}.$$

Comparing the above expression with (4.3), we get $P(\omega) = \frac{\alpha}{\alpha^2 + \omega^2}$ and $Q(\omega) = \frac{-\omega}{\alpha^2 + \omega^2}$. We can plot both characteristics in the Cartesian plane as shown in Fig. 4.3. However, sometimes it is useful to plot $P(\omega)$ and $Q(\omega)$ simultaneously in the form of the *Nyquist plot* formed of points $(P(\omega), Q(\omega))$, where $\omega \in (-\infty, \infty)$ (Fig. 4.4).

(c) In order to obtain the input-output equation of the system, we should use (4.1), that is,

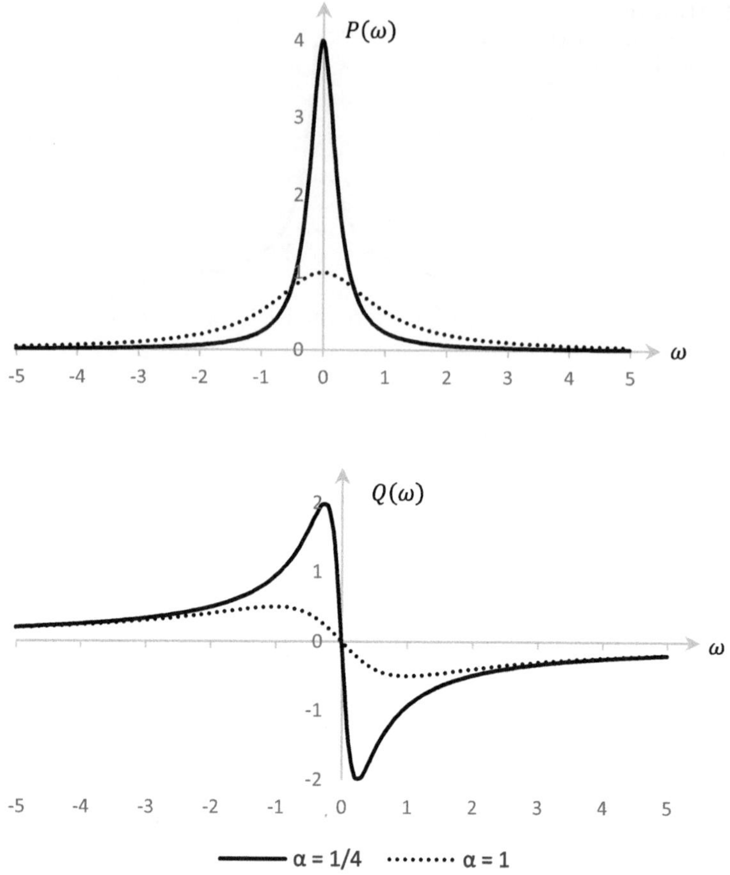

Fig. 4.3 Real and imaginary frequency responses of the analog filter from Problem 4.1

$$Y(\omega) = X(\omega) \cdot \frac{1}{\alpha + j\omega}.$$

Further,

$$Y(\omega)(\alpha + j\omega) = X(\omega),$$
$$\alpha Y(\omega) + j\omega Y(\omega) = X(\omega).$$

We obtained the system equation in frequency domain. Now, let us apply the inverse FT to this equation (derivation in time of FT from Table 2.2) and get

$$\alpha \cdot y(t) + \frac{dy(t)}{dt} = x(t),$$

that is, the 1st order differential equation of the system.

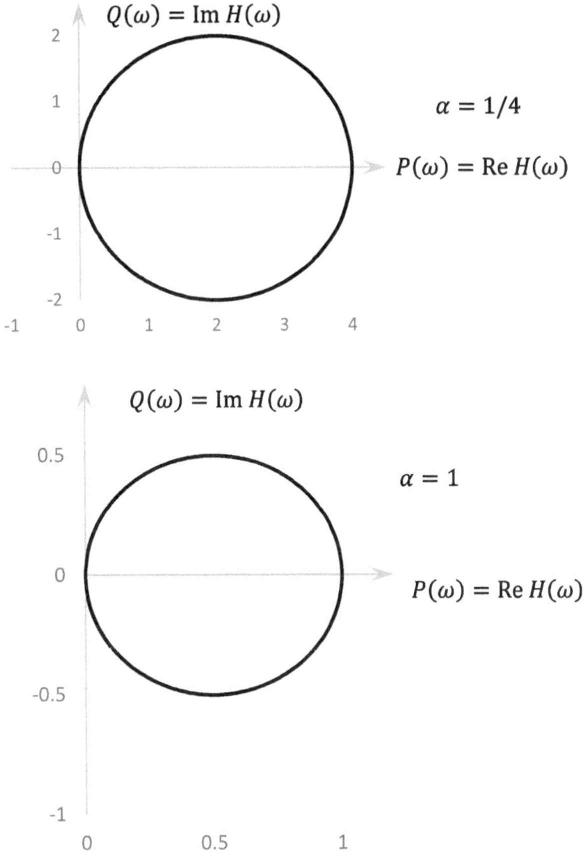

Fig. 4.4 Nyquist plots of the frequency response function of the analog filter from Problem 4.1

4.2 Spectrum Input-Output Relations

Using (4.1) and (4.2), we can derive important relations between amplitude- and phase spectra of input and output signals. The amplitude spectrum $|Y(\omega)|$ of the output signal $y(t)$ is

$$|Y(\omega)| = |X(\omega)| \cdot |H(\omega)| = |X(\omega)| \cdot A(\omega) \tag{4.6}$$

while the phase spectrum arg $Y(\omega)$ is

$$\arg Y(\omega) = \arg X(\omega) + \arg H(\omega) = \arg X(\omega) + \varphi(\omega). \tag{4.7}$$

The relation (4.1) can be very useful if we want to compute a response of a system with known frequency response function $H(\omega)$ excited with the known input signal, as for instance, the complex harmonic signal $e^{j\omega_0 t}$, $\cos(\omega_0 t)$, $\sin(\omega_0 t)$, or any other periodic signal.

4.2.1 Response to a Complex Harmonic Excitation

Let us consider the input signal $x(t) = e^{j\omega_0 t}$ with the spectrum $X(\omega) = 2\pi\delta(\omega - \omega_0)$. The spectrum of the output signal $y(t)$ is

$$Y(\omega) = X(\omega)H(\omega) = 2\pi\delta(\omega - \omega_0)H(\omega) = 2\pi\delta(\omega - \omega_0)H(\omega_0),$$

where we applied the sampling property of the Dirac distribution from Table 1.1. The system response is $y(t) = \mathcal{F}^{-1}\{Y(\omega)\} = e^{j\omega_0 t}H(\omega_0)$. So, we can write the following input-output relation:

$$e^{j\omega_0 t} \to e^{j\omega_0 t}H(\omega_0). \tag{4.8}$$

Important Let us notice that if $\omega_0 = 0$, we can determine the filter response to the constant signal $x(t) = 1$, $t \in \mathbb{R}$, that is, $1 \to H(0)$.

As $H(\omega)$ is the FT of the impulse response $h(t)$, we have $H(0) = \int h(t)dt$. Then $1 \to \int h(t)dt$ (compare with Problem 3.2(c)).

4.2.2 Response to a Harmonic Excitation

Consider the cosine input signal $x(t) = \cos(\omega_0 t) = \frac{1}{2}e^{j\omega_0 t} + \frac{1}{2}e^{-j\omega_0 t}$. Having in mind system linearity and the relation (4.8), we immediately get the system output:

$$y(t) = \frac{1}{2}e^{j\omega_0 t}H(\omega_0) + \frac{1}{2}e^{-j\omega_0 t}H(-\omega_0)$$

Further using (4.2), we have

$$y(t) = \frac{1}{2}e^{j\omega_0 t}A(\omega_0)e^{j\varphi(\omega_0)} + \frac{1}{2}e^{-j\omega_0 t}A(-\omega_0)e^{j\varphi(-\omega_0)}$$

but $A(-\omega_0) = A(\omega_0)$ and $\varphi(-\omega_0) = -\varphi(\omega_0)$.

So,

$$y(t) = A(\omega_0)\left[\frac{1}{2}e^{j(\omega_0 t + \varphi(\omega_0))} + \frac{1}{2}e^{-j(\omega_0 t + \varphi(\omega_0))}\right].$$

Finally, we get the input-output relation

$$\cos(\omega_0 t) \to A(\omega_0)\cos(\omega_0 t + \varphi(\omega_0)) = \text{Re}\left\{e^{j\omega_0 t}H(\omega_0)\right\}. \tag{4.9}$$

The reader can easily develop the response of an LTI analog system to a sine input in the following form:

$$\sin(\omega_0 t) \rightarrow A(\omega_0)\sin(\omega_0 t + \varphi(\omega_0)) = \operatorname{Im}\left\{e^{j\omega_0 t}H(\omega_0)\right\}. \tag{4.10}$$

4.2.3 Response to a Periodic Excitation

Let $x(t)$, $t \in \mathbb{R}$ be a periodic input signal in the form of the exponential Fourier series (2.9): $x(t) = \sum_{n\in\mathbb{Z}} c_n e^{jn\omega_0 t}$. Using (4.8), we can get the output signal $y(t)$ in the form of another exponential Fourier series

$$x(t) = \sum_{n\in\mathbb{Z}} c_n e^{jn\omega_0 t} \rightarrow y(t) = \sum_{n\in\mathbb{Z}} c_n H(n\omega_0) e^{jn\omega_0 t} \tag{4.11}$$

with coefficients $c_n H(n\omega_0)$, $n \in \mathbb{Z}$.

Problem 4.2 Given the filter from Problem 4.1 with the frequency response function $H(\omega) = \frac{1}{\frac{1}{4}+j\omega}$. Consider the input signal $x(t) = \frac{1}{2} + \sin\left(\frac{1}{4}t\right), t \in \mathbb{R}$. Compute the filter response $y(t)$.

Solution Having in mind system linearity, we can determine the response $y_1(t)$ to the constant $x_1(t) = \frac{1}{2}, t \in \mathbb{R}$ and, separately, the response $y_2(t)$ to $x_2(t) = \sin\left(\frac{1}{4}t\right)$, $t \in \mathbb{R}$. The total system response is $y(t) = y_1(t) + y_2(t)$.

Computing $y_1(t)$ is very simple. Since $\frac{1}{2} \rightarrow \frac{1}{2} \cdot H(0) = \frac{1}{2} \cdot \frac{1}{\frac{1}{4}+j0} = 2$, then $y_1(t) = 2$. For computing $y_2(t)$, it is useful to apply (4.10) as follows:

$$\begin{aligned}
\sin\left(\frac{1}{4}t\right) &\rightarrow \operatorname{Im}\left\{e^{j\frac{1}{4}t}H\left(\frac{1}{4}\right)\right\} = \operatorname{Im}\left\{\left(\cos\left(\frac{1}{4}t\right) + j\sin\left(\frac{1}{4}t\right)\right)H\left(\frac{1}{4}\right)\right\} \\
&= \operatorname{Im}\left\{\left(\cos\left(\frac{1}{4}t\right) + j\sin\left(\frac{1}{4}t\right)\right)\frac{1}{\frac{1}{4}+j\frac{1}{4}}\right\} \\
&= \operatorname{Im}\left\{4\left(\cos\left(\frac{1}{4}t\right) + j\sin\left(\frac{1}{4}t\right)\right)\frac{1-j}{(1+j)(1-j)}\right\} \\
&= \operatorname{Im}\left\{2\left(\cos\left(\frac{1}{4}t\right) + j\sin\left(\frac{1}{4}t\right)\right)(1-j)\right\} = 2\sin\left(\frac{1}{4}t\right) - 2\cos\left(\frac{1}{4}t\right).
\end{aligned}$$

So, $y_2(t) = 2\sin\left(\frac{1}{4}t\right) - 2\cos\left(\frac{1}{4}t\right)$. Finally

$$y(t) = 2 + 2\sin\left(\frac{1}{4}t\right) - 2\cos\left(\frac{1}{4}t\right).$$

We can observe that the filter amplifies four times the constant component (filter magnitude response is 4 for $\omega = 0$) and introduces the phase shift equal $-\frac{\pi}{4}$ for $\omega = \frac{1}{4}$.

Let us notice that we can come to the same result applying another form of (4.10), that is,

$$y_2(t) = A\left(\frac{1}{4}\right)\sin\left(\frac{1}{4}t + \varphi\left(\frac{1}{4}\right)\right),$$

where $A\left(\frac{1}{4}\right) = \frac{1}{\sqrt{\frac{1}{16}+\frac{1}{16}}} = \frac{4}{\sqrt{2}} = 2\sqrt{2}$ and $\varphi\left(\frac{1}{4}\right) = -\operatorname{atan}(1) = -\frac{\pi}{4}$ (compare with solution of Problem 4.1). In this case, we have

$$\begin{aligned} y_2(t) = 2\sqrt{2}\sin\left(\frac{1}{4}t - \frac{\pi}{4}\right) &= 2\sqrt{2}\left[\sin\left(\frac{1}{4}t\right)\cos\frac{\pi}{4} - \cos\left(\frac{1}{4}t\right)\sin\frac{\pi}{4}\right] \\ &= 2\sin\left(\frac{1}{4}t\right) - 2\cos\left(\frac{1}{4}t\right). \end{aligned}$$

4.3 Ideal Filters

In filter theory, it is often useful to deal with *ideal filters*. Though not realizable in practice, they permit to understand better an idea of signal filtering. In this part, we present four types of ideal filters: low-pass (LP), high-pass (HP), band-pass (BP), and band-stop (BS). Such nomenclature is connected with filter operating band characterized by frequency parameters collected in Table 4.1.

4.3.1 Low-Pass (LP) Ideal Filter

The frequency response function $H(\omega)$ of the ideal LP filter is given by

Table 4.1 Frequency parameters of ideal filters

Parameter	Description
Bandwidth (width of a *passband*)	Range of frequencies that can "pass" through a filter *unaltered*
Stop bandwidth (width of a *stopband*)	Range of frequencies that are *completely* removed by a filter
Cutoff frequency f_c	Frequency separating a passband from a stopband (LP, HP)
Center frequency f_{center}	Center frequency of filter passband (BP) or filter stopband (BS)

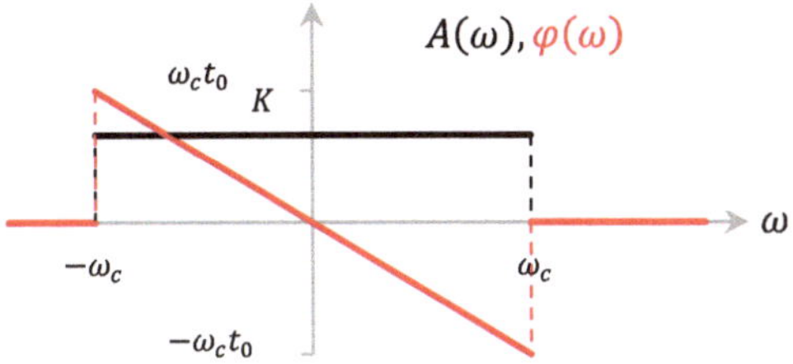

Fig. 4.5 Magnitude and phase responses of an ideal LP filter

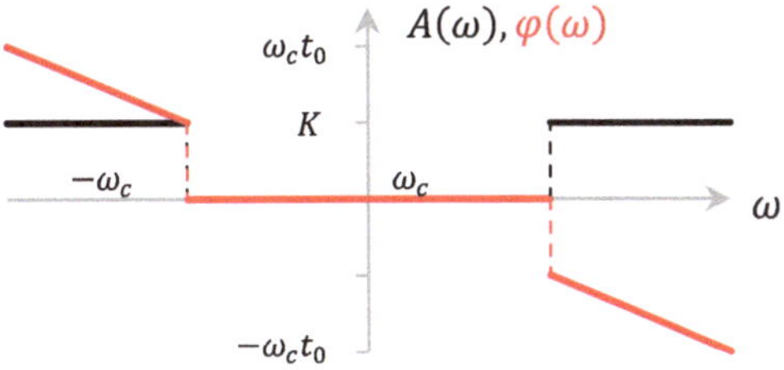

Fig. 4.6 Magnitude and phase responses of an ideal HP filter

$$H(\omega) = \begin{cases} Ke^{-j\omega t_0}, & |\omega| \le \omega_c \\ 0 & \text{elsewhere.} \end{cases} \tag{4.12}$$

We can notice that the magnitude response $A(\omega) = |H(\omega)|$ is constant and equal K in the passband $|\omega| \le \omega_c$ and 0 in the stopband $|\omega| > \omega_c$. The phase response $\varphi(\omega) = \arg H(\omega)$ is linear in the passband (and equals $-\omega t_0$). In the stopband, $\varphi(\omega) = 0$. The cutoff frequency $f_c = \frac{\omega_c}{2\pi}$ (in Hz). Both characteristics are presented in Fig. 4.5.

4.3.2 *High-Pass (HP) Ideal Filter*

The frequency response function $H(\omega)$ of the ideal HP filter is given by

$$H(\omega) = \begin{cases} Ke^{-j\omega t_0}, & |\omega| > \omega_c \\ 0 & \text{elsewhere.} \end{cases} \tag{4.13}$$

The magnitude response $A(\omega)$ is constant and equal K in the passband $|\omega| > \omega_c$ and 0 in the stopband $|\omega| \le \omega_c$. The phase response $\varphi(\omega)$ is linear in the passband (and equals $-\omega t_0$), while in the stopband, $\varphi(\omega) = 0$. The cutoff frequency $f_c = \frac{\omega_c}{2\pi}$ (in Hz). Both characteristics are presented in Fig. 4.6.

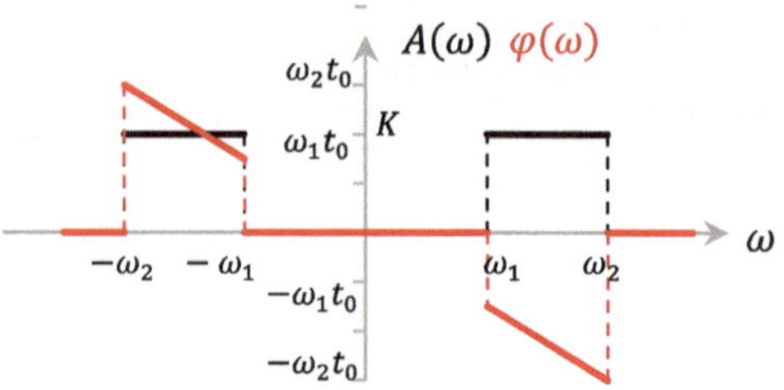

Fig. 4.7 Magnitude and phase responses of an ideal BP filter

4.3.3 Band-Pass (BP) Ideal Filter

The frequency response function $H(\omega)$ of the ideal BP filter is given by

$$H(\omega) = \begin{cases} Ke^{-j\omega t_0}, & \omega_1 \leq |\omega| \leq \omega_2 \\ 0 & \text{elsewhere.} \end{cases} \tag{4.14}$$

The magnitude response $A(\omega)$ equals K in the passband $\omega_1 \leq |\omega| \leq \omega_2$ and 0 -in the stopband $\omega \in (-\infty, -\omega_2) \cup (-\omega_1, \omega_1) \cup (\omega_2, +\infty)$. The phase response $\varphi(\omega)$ is linear in the passband (and equals $-\omega t_0$) while in the stopband, $\varphi(\omega) = 0$. The passband center frequency $f_{\text{center}} = \frac{\omega_1 + \omega_2}{4\pi}$ (in Hz). Both characteristics are presented in Fig. 4.7.

4.3.4 Band-Stop (BS) Ideal Filter

The frequency response function $H(\omega)$ of the ideal BS filter is given by

$$H(\omega) = \begin{cases} Ke^{-j\omega t_0}, & \omega \in (-\infty, -\omega_2) \cup (-\omega_1, \omega_1) \cup (\omega_2, +\infty) \\ 0 & \text{elsewhere.} \end{cases} \tag{4.15}$$

The magnitude response $A(\omega)$ equals K in the passband $\omega \in (-\infty, -\omega_2) \cup (-\omega_1, \omega_1) \cup (\omega_2, +\infty)$ and 0—in the stopband $|\omega| \leq \omega_1$. The phase response $\varphi(\omega)$ is linear in the passband (and equals $-\omega t_0$) while in the stopband, $\varphi(\omega) = 0$. The stopband center frequency $f_{\text{center}} = \frac{\omega_1 + \omega_2}{4\pi}$ (in Hz). We can notice that an ideal BS filter is a superposition of ideal LP- and HP filters. Magnitude and phase characteristics are presented in Fig. 4.8.

In Problems 4.3 and 4.4, we will show how to determine the response of an ideal filter to a given input signal.

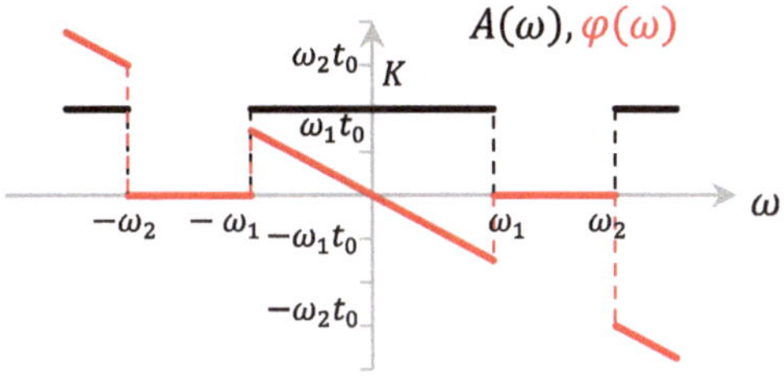

Fig. 4.8 Magnitude and phase responses of an ideal BS filter

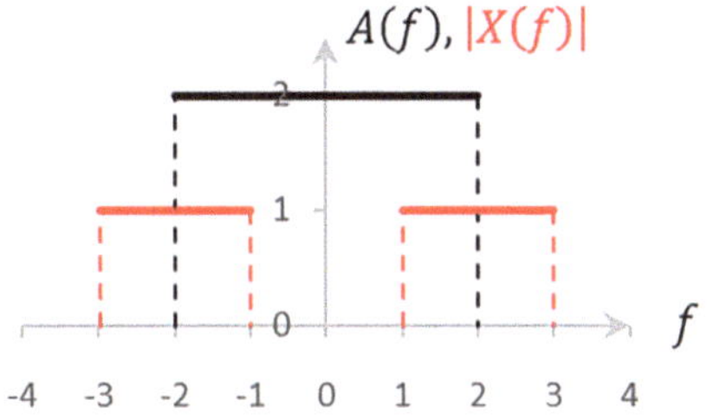

Fig. 4.9 Magnitude response $A(f)$ of the ideal LP filter and amplitude spectrum $|X(f)|$ of the input signal (b) from Problem 4.3

Problem 4.3 Given the frequency response of an LP ideal filter:

$$H(f) = \begin{cases} 2e^{-j8\pi f}, |f| \leq 2 \\ 0 \qquad \text{elsewhere.} \end{cases}$$

Compute the system output if the input is: (a) $x(t) = \cos(3\pi t)$, $t \in \mathbb{R}$; (b) $x(t) = 4\text{Sa}(2\pi t)\cos(4\pi t)$, $t \in \mathbb{R}$; (c) $x(t)$—a periodic bipolar rectangle waveform of even symmetry with amplitude $A = 1$, angular frequency $\omega_0 = 2\pi$ [rad/s] and duty factor $d = \frac{1}{2}$.

Solution The magnitude response $A(f) = 2$ in the passband $|f| \leq 2$ and 0 in the stopband $|f| > 2$ (see Fig. 4.9). The phase response $\varphi(f) = -8\pi f$ in the passband, 0 elsewhere.

(a) The input $x(t) = \cos(3\pi t)$, $t \in \mathbb{R}$. In order to determine $y(t)$, we can apply (4.9) and get

$$y(t) = A(3\pi)\cos[3\pi t + \varphi(3\pi)] = 2\cos(3\pi t - 12\pi) = 2\cos[3\pi(t - 4)].$$

We can see that the filter response is also a cosine signal with the same angular frequency 3π [rad/s], amplitude 2, and initial phase -12π [rad] (cosine signal shifted right by 4).

(b) In this case, we can compute $y(t)$ as the inverse FT of $Y(f)$. First, we need to determine the spectrum $X(f)$ of the input $x(t) = 4\text{Sa}(2\pi t)\cos(4\pi t)$. Let us apply the modulation property from Table 2.2. From Table 2.1, we have.

$$4\text{Sa}(2\pi t) \overset{\mathcal{F}}{\leftrightarrow} 2\Pi\left(\frac{f}{2}\right), \text{ then } X(f) = \Pi\left(\frac{f+2}{2}\right) + \Pi\left(\frac{f-2}{2}\right).$$

The amplitude spectrum $|X(f)| = X(f)$, while the phase spectrum $\arg X(f) = 0$. In Fig. 4.9, we present the filter magnitude response $A(f)$ and amplitude spectrum $|X(f)|$ of the input signal.

Now, let us apply (4.6) to compute the amplitude spectrum $|Y(f)|$ of the filter response as follows:

$$|Y(f)| = |X(f)| \cdot A(f) = 2\left[\Pi\left(f + \frac{3}{2}\right) + \Pi\left(f - \frac{3}{2}\right)\right].$$

The phase spectrum $\arg Y(f)$ of the output signal equals $\varphi(\omega) = -8\pi f$ (compare with (4.7)). Finally, the system response spectrum is

$$Y(f) = |Y(f)|e^{j\arg Y(f)} = 2\left[\Pi\left(f + \frac{3}{2}\right) + \Pi\left(f - \frac{3}{2}\right)\right]e^{-j8\pi f}.$$

To compute $y(t)$, we have to apply the inverse FT for $Y(f)$ and get

$$y(t) = 4\text{Sa}(\pi(t-4))\cos(3\pi(t-4)).$$

(b) In order to determine $Y(f)$, we have to apply again (4.1). First, we should compute the spectrum of a periodic even-symmetric bipolar rectangle waveform with amplitude $A = 1$, angular frequency $\omega_0 = 2\pi$ [rad/s] and duty factor $d = \frac{1}{2}$. Notice that the fundamental period $T_0 = 1$ [s] and the fundamental frequency $f_0 = 1$ [Hz]. For convenience, we have plotted the input signal in Fig. 4.10.

In Fig. 4.10 in red, we marked the signal $z(t)$ (similarly as in Problem 2.15). We have

$$z(t) = \Pi\left(\frac{t}{0,5}\right) - \Pi\left(\frac{t-0,5}{0,5}\right)$$

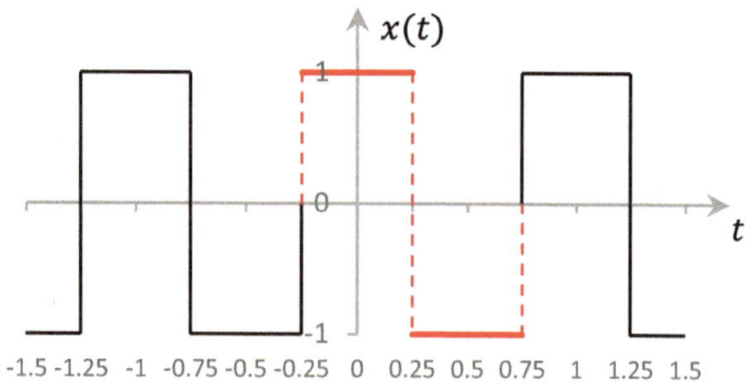

Fig. 4.10 Periodic input rectangle waveform from Problem 4.3

and its FT is $Z(f) = \frac{1}{2}\mathrm{Sa}\left(\frac{\pi f}{2}\right)\left(1 - e^{-j\pi f}\right)$ (see Table 2.2—linearity and time-shift property of FT).

The input signal spectrum is given by $X(f) = \sum_{n\in\mathbb{Z}} c_n \delta(f - nf_0)$ with coefficients $c_n = \frac{1}{T_0} Z(nf_0)$ (see (2.27) and (2.29)). We have

$$c_n = Z(nf_0) = \frac{1}{2}\mathrm{Sa}\left(\frac{\pi n f_0}{2}\right)\left(1 - e^{-j\pi n f_0}\right) = \frac{1}{2}\mathrm{Sa}\left(\frac{\pi n}{2}\right)\left(1 - e^{-j\pi n}\right),$$

then

$$X(f) = \sum_{n\in\mathbb{Z}} \frac{1}{2}\mathrm{Sa}\left(\frac{\pi n}{2}\right)\left(1 - e^{-j\pi n}\right)\delta(f - n\cdot 1).$$

We can notice that the spectrum $X(f)$ of the input signal is discrete in frequency and the only components at frequencies n [Hz] are weighted delta pulses.

Now, in order to get the filter response, we have to refer once again to the filter passband defined by the cutoff frequency $f_c = 2$ [Hz]. It is evident that the ideal LP filter eliminates all frequencies bigger than f_c. Then at filter output, we have

$$Y(f) = \left(\sum_{n=-2}^{2} \frac{1}{2}\mathrm{Sa}\left(\frac{\pi n}{2}\right)\left(1 - e^{-j\pi n}\right)\delta(f - n)\right)e^{-j8\pi f}$$

(do not forget about the nonzero phase response of the filter!). Then, since for $n = 0$ and for $n = \pm 2$, $\mathrm{Sa}\left(\frac{\pi n}{2}\right) = 0$, we get

$$\begin{aligned} Y(f) &= \frac{1}{2}\left[\mathrm{Sa}\left(-\frac{\pi}{2}\right)\left(1 - e^{j\pi}\right)\delta(f+1) + \mathrm{Sa}\left(\frac{\pi}{2}\right)\left(1 - e^{-j\pi}\right)\delta(f-1)\right]e^{-j8\pi f} \\ &= \mathrm{Sa}\left(\frac{\pi}{2}\right)[\delta(f+1) + \delta(f-1)]e^{-j8\pi f} = \frac{\sin\frac{\pi}{2}}{\frac{\pi}{2}}[\delta(f+1) + \delta(f-1)]e^{-j8\pi f} \\ &= \frac{2}{\pi}[\delta(f+1) + \delta(f-1)]e^{-j8\pi f}. \end{aligned}$$

Finally, we can compute the response $y(t)$ as the inverse FT of $Y(f)$, that is,

$$y(t) = \frac{4}{\pi}\cos\left(2\pi(t-4)\right).$$

We can see that only the first harmonic of the periodic input rectangle waveform passed through our LP filter.

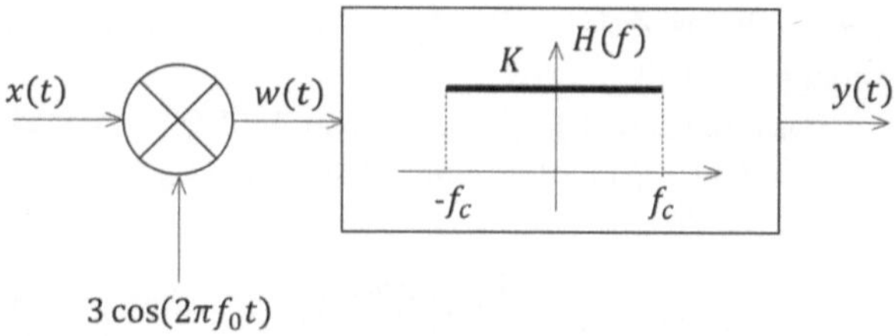

Fig. 4.11 Signal processing system from Problem 4.4

Problem 4.4 Given the signal processing system shown in Fig. 4.11. The input $x(t)$ is a cosine signal with amplitude $A_x = 2$ and frequency $f_x = 60$ Hz. It is multiplied by 3 cos $(2\pi f_0 t)$ to get $w(t)$ and then $w(t)$ is passing through an ideal LP filter with gain K and cutoff frequency f_c. At the output, we observe a cosine signal $y(t)$ with amplitude $A_y = 12$ and frequency $f_x = 600$ Hz. Determine values of parameters f_0, f_c, K.

Solution We have to study what happens at each point of the considered processing system. It is easier to perform calculations in frequency domain.

First, we need to determine the FT of the input signal. From Table 2.1, we get

$$x(t) = 2\cos(2\pi \cdot 60 \cdot t) \overset{\mathcal{F}}{\leftrightarrow} X(f) = \delta(f + 60) + \delta(f - 60).$$

Then we have to determine the spectrum of $w(t) = x(t) \cdot 3$ cos $(2\pi f_0 t)$ applying modulation property of FT (Table 2.2). We obtain

$$w(t) = x(t) \cdot 3\cos(2\pi f_0 t) \overset{\mathcal{F}}{\leftrightarrow} W(f) = \frac{3}{2}[X(f + f_0) + X(f - f_0)].$$

Then

$$W(f) = \frac{3}{2}[\delta(f + 60 + f_0) + \delta(f - 60 + f_0) + \delta(f + 60 - f_0) + \delta(f - 60 - f_0)],$$

which is visualized in Fig. 4.12. The spectrum $W(f)$ consists of four Dirac delta pulses at frequencies $\pm(f_0 - 60)$ and $\pm(f_0 + 60)$.

The input-output relation of the considered ideal LP filter is: $w(t) \to y(t)$ and the output signal spectrum $Y(f) = W(f)H(f)$. The system response $y(t) = 12$ cos $(2\pi \cdot 600 \cdot t)$ is known and its spectrum $Y(f) = 6 \cdot [\delta(f + 600) + \delta(f - 600)]$ consists of two delta peaks at frequencies ± 600 Hz.

The ideal LP filter should pass only two peaks of $W(f)$ at $\pm(f_0 - 60)$ and delete those at $\pm(f_0 + 60)$. So, the only solution to this problem is $f_0 - 60 = 600$, that is, $f_0 = 660$ Hz.

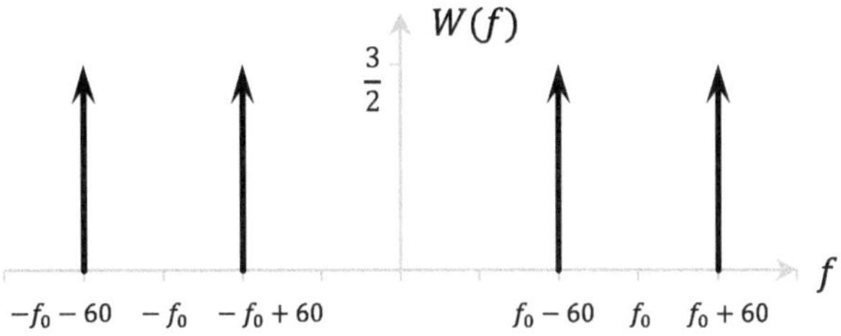

Fig. 4.12 Spectrum of $w(t)$ from Problem 4.4

Table 4.2 Frequency parameters of real filters

Parameter	Description
3dB bandwidth (width of a *passband*)	Range of frequencies that can "pass" through a filter with small changes
Stop bandwidth (width of a *stopband*)	Range of frequencies that are *considerably attenuated* by a filter
Cutoff frequency f_c	Frequency separating a 3 dB passband from a stopband (LP, HP)
Center frequency f_{center}	Center frequency of the 3 dB passband (BP filter) or stopband (BS filter)

Let us now determine the cutoff frequency f_c. It should assume values between $(f_0 - 60)$ and $(f_0 + 60)$, that is, $600 < f_c < 720$ [Hz].

Finally, since $Y(f) = W(f)H(f)$, amplitudes of spectrum peaks should satisfy $\frac{3}{2} \cdot K = 6$, that is, $K = 4$.

4.4 Real Filters

In the precedent section, we considered ideal filters that are not realizable in practice due to their non-causal impulse responses. In real applications, we deal with filters having magnitude characteristics approximating an "ideal" case. Here, we would like to focus on two problems strictly connected to frequency properties of analog filters. First, we will define the 3 dB bandwidth of a real filter and then we will present the popular method of analog filter design known as the *frequency transformation*. In Chap. 5, we will come back to real filters presenting in detail Butterworth and Chebyshev (type I and II) analog filters.

Let us present in Table 4.2 the frequency parameters of real filters. A reader can compare it with Table 4.1.

4.4.1 3 dB Bandwidth

Definition 4.3 The *3 dB bandwidth of an LP filter* is the range $[-\omega_c, \omega_c]$, where ω_c is the angular frequency for which the magnitude response $A(\omega)$ of a filter decreases by 3 dB from its maximum value $A_{max} = A(0)$, that is, $20 \log A(\omega_c) = 20 \log A_{max} - 3$, which is equivalent to

$$A(\omega_c) = \frac{A_{max}}{\sqrt{2}}. \tag{4.16}$$

Evidently, we can define the 3 dB bandwidth of an LP filter in the domain f (in Hz) as $[-f_c, f_c]$, where $f_c = \frac{\omega_c}{2\pi}$.

Definition 4.4 The *3 dB bandwidth of a BP filter* is the range $[\omega_1, \omega_2]$, where ω_1 and ω_2 are angular frequencies for which the magnitude response $A(\omega)$ of a filter decreases by 3 dB from its maximum value $A_{max} = A(\omega_{center})$ and $\omega_{center} = \frac{\omega_1+\omega_2}{2}$, that is, $20 \log A(\omega_1) = 20 \log A(\omega_2) = 20 \log A_{max} - 3$, which is equivalent to

$$A(\omega_1) = A(\omega_2) = \frac{A_{max}}{\sqrt{2}}. \tag{4.17}$$

Let us note that we can define the 3 dB bandwidth of a BP filter in the domain f (in Hz) as $[f_1, f_2]$, where $f_1 = \frac{\omega_1}{2\pi}$ and $f_2 = \frac{\omega_2}{2\pi}$.

In Problem 4.5, we will show how to calculate the 3 dB bandwidth of a real LP filter.

Problem 4.5 Determine the 3 dB bandwidth of the filter from Problem 4.1 with the frequency response function $H(\omega) = \frac{1}{\frac{1}{4}+j\omega}$.

Solution In Problem 4.1, we have already determined the filter magnitude response $A(\omega) = \frac{1}{\sqrt{\frac{1}{16}+\omega^2}}$ and plotted it in Fig. 4.1. We can see that $A(\omega)$ gets its maximum value for $\omega = 0$, that is, $A_{max} = A(0) = 4$. We can apply (4.16) to get

$$A(\omega_c) = \frac{1}{\sqrt{\frac{1}{16} + \omega_c^2}} = \frac{4}{\sqrt{2}}.$$

Further,

$$4\sqrt{\frac{1}{16} + \omega_c^2} = \sqrt{2} \Leftrightarrow 16\left(\frac{1}{16} + \omega_c^2\right) = 2 \Leftrightarrow 16\omega_c^2 = 1 \Leftrightarrow \omega_c = \frac{1}{4}.$$

The 3 dB bandwidth is $[-\omega_c, \omega_c] = \left[-\frac{1}{4}, \frac{1}{4}\right]$ [rad/sec] as marked in red in Fig. 4.13.

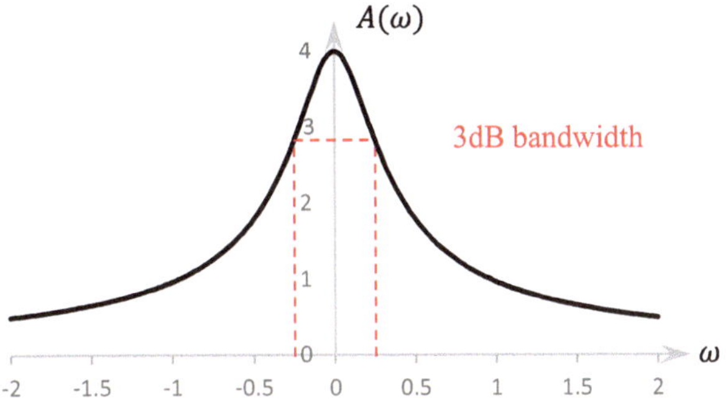

Fig. 4.13 Magnitude response of the filter from Problem 4.1. In red: the 3 dB bandwidth calculated in Problem 4.5

Table 4.3 Frequency transformation formulas for analog filter design

Type of a designed filter	Frequency response function of the designed filter
LP (ω_{c_2}—3 dB angular frequency)	$H_{\mathrm{LP}}(\omega) = H\left(\omega\frac{\omega_{c_1}}{\omega_{c_2}}\right)$
HP (ω_{c_2}—3 dB angular frequency)	$H_{\mathrm{HP}}(\omega) = H(-\frac{\omega_{c_1}\omega_{c_2}}{\omega})$
BP ($[\omega_1, \omega_2]$—3 dB passband)	$H_{\mathrm{BP}}(\omega) = H\left(\omega_{c_1}\frac{\omega^2 - \omega_1\omega_2}{\omega(\omega_2 - \omega_1)}\right)$
BS ($[\omega_1, \omega_2]$—3 dB stopband)	$H_{\mathrm{BS}}(\omega) = H\left(\omega_{c_1}\frac{\omega(\omega_2 - \omega_1)}{\omega_1\omega_2 - \omega^2}\right)$

4.4.2 *Frequency Transformation*

The *frequency transformation* is a known analog filter design method. Let us explain its successive steps. Starting from a prototype LP filter with the known frequency response function $H(\omega)$ and strictly determined 3 dB cutoff angular frequency denoted with ω_{c_1}, we can design analog filters of different types using formulas collected in Table 4.3. We only need to replace a variable ω in $H(\omega)$ with a corresponding mathematical formula. An example of analog filter design will be presented in Problem 4.6.

Problem 4.6 Given a prototype LP filter with the frequency response function $H(\omega) = \frac{10}{10+j\omega}$. Using the frequency transformation method, design the following filters: (a) an LP filter with the 3 dB angular frequency $\omega_{c_2} = 20$ rad/s; (b) an HP filter with the 3 dB angular frequency $\omega_{c_2} = 20$ rad/s; (c) a BP filter with 3 dB passband $[5, 15]$ rad/s; (d) a BS filter with 3 dB stopband $[5, 15]$ rad/s. In each case, plot the frequency response function of the designed filter.

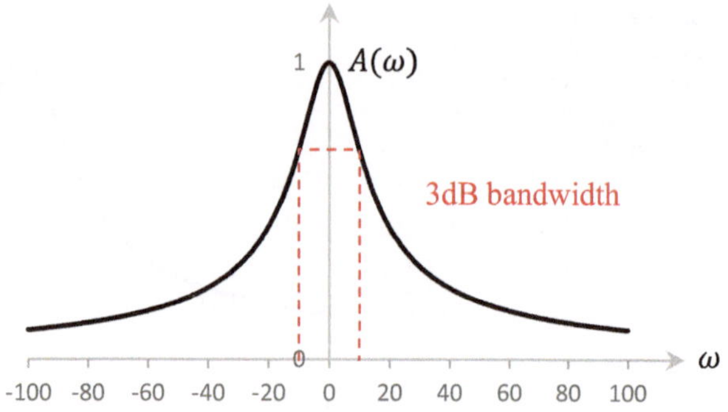

Fig. 4.14 Magnitude response of the LP prototype filter from Problem 4.6. In red: the 3 dB bandwidth

Solution The prototype LP filter frequency response function is $H(\omega) = \frac{10}{10+j\omega}$. First, we have to compute its 3 dB cutoff angular frequency ω_{c_1}. The magnitude response is $A(\omega) = \frac{10}{\sqrt{100+\omega^2}}$ and its maximum $A_{\max} = A(0) = 1$. Using (4.16), we get $A(\omega_{c_1}) = \frac{10}{\sqrt{100+\omega_{c_1}^2}} = \frac{1}{\sqrt{2}} \Leftrightarrow \omega_{c_1} = 10$ [rad/s].

In Fig. 4.14, we show the magnitude response of the prototype LP filter with its 3 dB bandwidth marked in red.

(a) Now we want to design an LP filter with $\omega_{c_2} = 20$ rad/s, so we have to replace ω in $H(\omega) = \frac{10}{10+j\omega}$ with $\omega\frac{\omega_{c_1}}{\omega_{c_2}} = \omega\frac{10}{20} = \frac{1}{2}\omega$ getting

$$H_{\mathrm{LP}}(\omega) = \frac{10}{10 + j\frac{1}{2}\omega} = \frac{20}{20 + j\omega}.$$

The magnitude response of the "new" low-pass filter is

$$A(\omega) = |H_{\mathrm{LP}}(\omega)| = \frac{20}{\sqrt{400 + \omega^2}}.$$

Its plot is presented in Fig. 4.15. It is twice as large as that of the prototype filter.

(b) Now, our task is to design an HP filter with the 3 dB angular frequency $\omega_{c_2} = 20$ rad/s. We have to apply the formula $H_{\mathrm{HP}}(\omega) = H\left(-\frac{\omega_{c_1}\omega_{c_2}}{\omega}\right) = H\left(-\frac{200}{\omega}\right)$ from Table 4.3 to get

$$H_{\mathrm{HP}}(\omega) = \frac{10}{10 + j\left(-\frac{200}{\omega}\right)} = \frac{\omega}{\omega - 20j}.$$

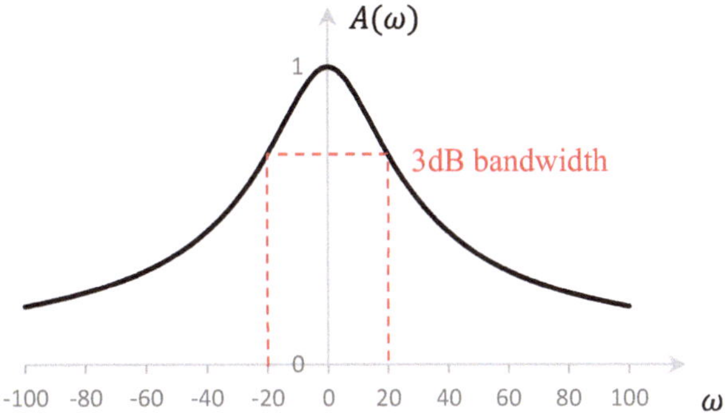

Fig. 4.15 Magnitude response of the designed LP filter from Problem 4.6. In red: the 3 dB bandwidth

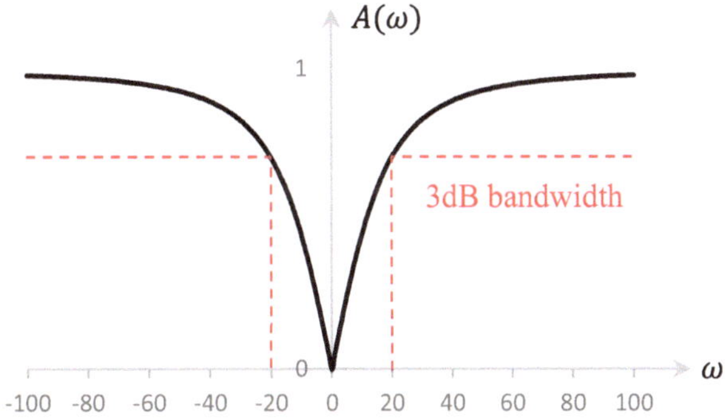

Fig. 4.16 Magnitude response of the designed HP filter from Problem 4.6. In red: the 3 dB bandwidth

The magnitude response of the designed HP filter (see Fig. 4.16) is:

$$A(\omega) = |H_{\mathrm{HP}}(\omega)| = \frac{|\omega|}{\sqrt{\omega^2 + 400}}.$$

In order to design a BP filter with the 3 dB passband $[\omega_1, \omega_2] = [5, 15]$ rad/s, we apply the formula $H_{\mathrm{BP}}(\omega) = H\left(\omega_{c_1} \frac{\omega^2 - \omega_1\omega_2}{\omega(\omega_2 - \omega_1)}\right) = H\left(10 \frac{\omega^2 - 75}{\omega \cdot 10}\right) = H\left(\frac{\omega^2 - 75}{\omega}\right)$ from Table 4.3 and get

$$H_{\mathrm{BP}}(\omega) = \frac{10}{10 + j\left(\frac{\omega^2 - 75}{\omega}\right)} = \frac{10\omega}{10\omega + j(\omega^2 - 75)}.$$

The magnitude response (see Fig. 4.17) of the "new" BP filter is given by

$$A(\omega) = |H_{\mathrm{BP}}(\omega)| = \frac{10|\omega|}{\sqrt{100\omega^2 + (\omega^2 - 75)^2}} = \frac{10|\omega|}{\sqrt{\omega^4 - 50\omega^2 + 5625}}.$$

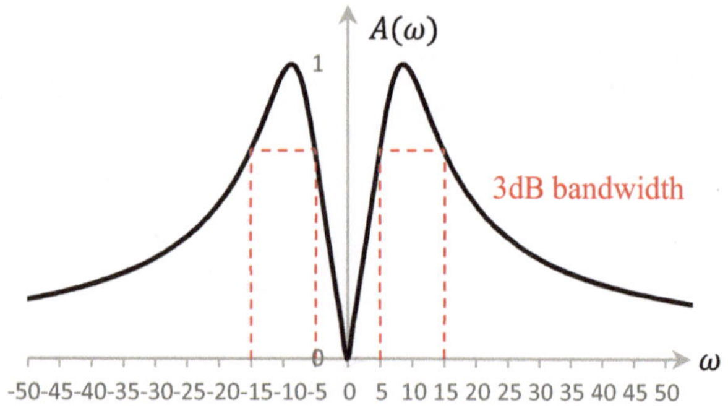

Fig. 4.17 Magnitude response of the designed BP filter from Problem 4.6. In red: the 3 dB bandwidth

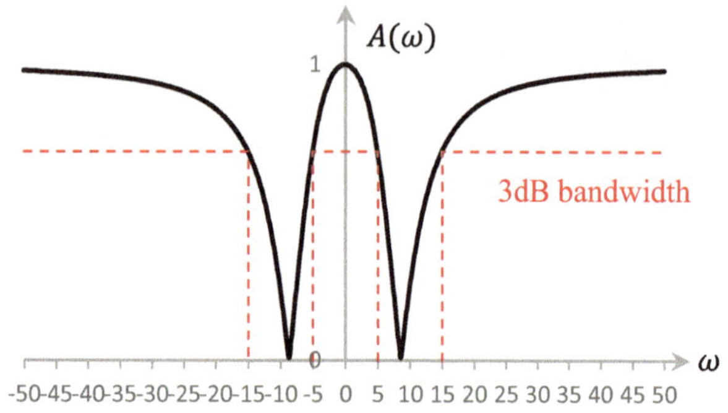

Fig. 4.18 Magnitude response of the designed BS filter from Problem 4.6. In red: the 3 dB bandwidth

(c) At last, we design a BS filter with the 3 dB stopband $[\omega_1, \omega_2] = [5, 15]$ rad/s. Let us apply the formula $H_{\mathrm{BS}}(\omega) = H\left(\omega_{c_1} \frac{\omega(\omega_2 - \omega_1)}{\omega_1\omega_2 - \omega^2}\right) = H\left(100\frac{\omega}{75 - \omega^2}\right)$ from Table 4.3 and have

$$H_{\mathrm{BP}}(\omega) = \frac{10}{10 + j\left(100\frac{\omega}{75 - \omega^2}\right)} = \frac{1}{1 + j\left(\frac{10\omega}{75 - \omega^2}\right)} = \frac{75 - \omega^2}{75 - \omega^2 + j \cdot 10\omega}.$$

The magnitude response (see Fig. 4.18) of the designed BS filter is as follows:

$$A(\omega) = |H_{\mathrm{BS}}(\omega)| = \frac{\left|75 - \omega^2\right|}{\sqrt{(75 - \omega^2)^2 + 100\omega^2}} = \frac{\left|75 - \omega^2\right|}{\sqrt{\omega^4 - 50\omega^2 + 5625}}.$$

In Figs. 4.14, 4.15, 4.16, 4.17, and 4.18, we can observe that the magnitude responses of real filters considerably differ from their ideal analogs. All considered filters have magnitude frequency plots with slopes, as a consequence of a low order of the filter (here the order equals 1). In Chap. 5, we will see that when a filter order gets bigger, the slopes get steeper.

Exercises

4.1 Given the frequency response function of an ideal LP filter with the cutoff frequency $5f_s$: $H(f) = \begin{cases} 1 & \text{for } |f| \le 5f_s \\ 0 & \text{elsewhere.} \end{cases}$, $f_s = \frac{1}{T_s}$. Determine the filter responses $y_1(t)$ and $y_2(t)$ to the following input signals: (a) $x_1(t) = \delta(t)$; (b) $x_2(t) = \sum_{n\in\mathbb{Z}} \delta(t - nT_s)$.

4.2 Given an ideal LP filter with $H(\omega) = \begin{cases} e^{j\omega} & \text{for } |\omega| \le 5 \\ 0 & \text{elsewhere.} \end{cases}$. Sketch its magnitude and phase characteristics. Determine its response $y(t)$ to the input signal $x(t) = \left(1 + \frac{\sin 2t}{t}\right)\cos(4t), t \in \mathbb{R}$.

4.3 Given an ideal BP filter with $H(\omega) = \begin{cases} 4e^{-j3\omega} & \text{for } 2 \le |\omega| \le 4 \\ 0 & \text{elsewhere.} \end{cases}$. Sketch its magnitude and phase characteristics. Determine filter responses $y_1(t)$ and $y_2(t)$ to the following input signals: (a) $x_1(t) = \mathrm{Sa}(3t)$; (b) $x_2(t) = \mathrm{Sa}(2t)\cos(3t)$, $t \in \mathbb{R}$.

4.4 Given the signal processing system shown in Fig. 4.19. The input signal$x(t)$ is a cosine with amplitude $A_x = 1$ and frequency $f_x = 50$ Hz. It is multiplied by $2\cos(2\pi f_0 t)$, getting $w(t)$, and then $w(t)$ is passing through an ideal HP filter with gain K and cutoff frequency f_c. At the output, we observe a cosine signal $y(t)$ with amplitude $A_y = 5$ and frequency $f_x = 500$ Hz. Determine values of parameters f_0, f_c, K.

4.5 Given the signal processing system shown in Fig. 4.20, where

$$H_1(\omega) = \begin{cases} 3 & \text{for } |\omega| \le 2 \\ 0 & \text{elsewhere} \end{cases} \text{ and } H_2(\omega) = \begin{cases} e^{-j\omega} & \text{for } |\omega| \ge \frac{3}{2} \\ 0 & \text{elsewhere.} \end{cases}$$

Find the response $y(t)$ to the input $x(t) = \frac{1}{\pi}\mathrm{Sa}(t)\cos(2t), t \in \mathbb{R}$.

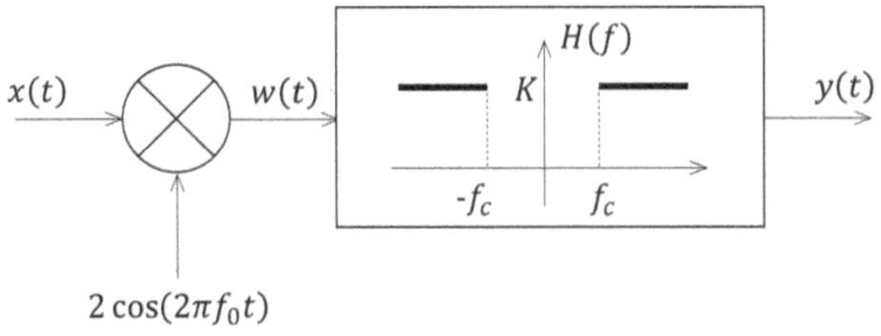

Fig. 4.19 Signal processing system from Exercise 4.4

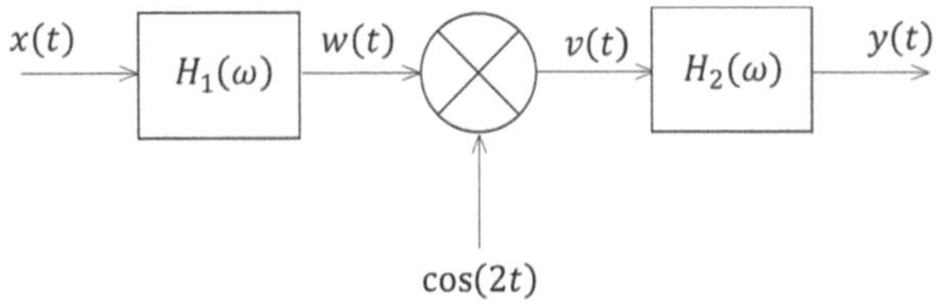

Fig. 4.20 Signal processing system from Exercise 4.5

4.6 Given an ideal LP filter with $H(\omega) = \begin{cases} 1 & \text{for } |\omega| \le \omega_c \\ 0 & \text{elsewhere} \end{cases}$

Compute the filter response $y(t)$ in case when the input signal is a periodic bipolar rectangle waveform with fundamental period T_0 defined within one period as

$$x(t) = \begin{cases} 1 & \text{for } 0 < t \le T_0/2 \\ -1 & T_0/2 < t \le T_0. \end{cases}$$

and $\omega_c \cdot T_0 = 10$ [rad].

4.7 The input signal of an ideal LP filter is an even-symmetric periodic unipolar rectangle waveform with amplitude 1, fundamental period $T_0 = 1$ [s] and duration time of a rectangle pulse $\tau = 0,\ 5$ [s]. The frequency response function of the filter is

$$H(f) = \begin{cases} 1 & \text{for } |f| \le f_c \\ 0 & \text{elsewhere.} \end{cases}$$

Determine the filter response for two different cutoff frequencies of the filter: (a) $f_c = 2,\ 5$; (b) $f_c = 3,\ 5$ [Hz].

4.8 Given the signal $x(t) = [1 + \cos(2\pi t)] \cos(20\pi t)$, $t \in \mathbb{R}$ at the input of a filter with frequency response function:

$$H(f) = \begin{cases} -\frac{1}{2}|f - 10| + 1 & \text{for } 8 \le f \le 12 \\ -\frac{1}{2}|f + 10| + 1 & \text{for } -12 \le f \le -8 \\ 0 & \text{elsewhere.} \end{cases}$$

Determine the filter response $y(t)$ and sketch its spectrum $Y(f)$.

4.9 Given three different analog filters with corresponding impulse responses: (a) $h_1(t) = \mathbf{1}(t)$; (b) $h_2(t) = -2\delta(t) + 5e^{-2t}\mathbf{1}(t)$; (c) $h_3(t) = 2t \cdot e^{-t}\,\mathbf{1}(t)$. Determine the frequency response function of each filter and find its response to the input signal $x(t) = \cos t$, $t \in \mathbb{R}$.

4.10 Given a real filter with the frequency response function $H(\omega) = \frac{5}{5 + j\omega}$. Compute its impulse response $h(t)$ and its step response $k(t)$. Sketch magnitude and phase responses and determine the 3 dB bandwidth. Find the filter response to the input signal $x(t) = 2 + 2\cos(10t + \pi/4)$, $t \in \mathbb{R}$.

4.11 Given a real filter with the frequency response function $H(\omega) = \frac{j\omega}{5 + j\omega}$. Compute its impulse response $h(t)$ and its step response $k(t)$. Sketch magnitude and phase responses and determine the 3 dB bandwidth. Find the filter response to the input signal $x(t) = \Pi(t - 1)$.

4.12 Given the input-output differential equation of an LTI system

$$4\frac{dy(t)}{dt} + y(t) = x(t).$$

Compute the frequency response function $H(\omega)$ and the impulse response $h(t)$. Sketch the magnitude response and compute the 3 dB bandwidth. What is the filter response to the input signal $x(t) = e^{-2t} \cdot \mathbf{1}(t)$, $t \in \mathbb{R}$?

4.13 Given the input-output differential equation of a LTI system

$$\frac{d^2y(t)}{dt^2} + 4\frac{dy(t)}{dt} + 4y(t) = x(t).$$

Determine the frequency response function $H(\omega)$ and the impulse response $h(t)$. Sketch the magnitude response and calculate the 3 dB bandwidth. What is the filter response to the input signal $x(t) = e^{-2t} \cdot \mathbf{1}(t)$, $t \in \mathbb{R}$?

4.14 Given the frequency response function $H(\omega) = \frac{1}{1+j\frac{\omega}{2}\tau}$, where $\tau > 0$, $\tau \in \mathbb{R}$. The input signal is $x(t) = \cos(2t) + \cos^2(2t)$, $t \in \mathbb{R}$. The filter response is $y(t)$. (a) Calculate the ratio p_{in} of the average power of the 2nd harmonic to the average power of the 1st harmonic of $x(t)$; (b) calculate the value of τ such that the ratio p_{out} of the average power of the 2nd harmonic to the average power of the 1st harmonic of $y(t)$ is $p_{\text{out}} = 0,4p_{\text{in}}$; (c) determine $y(t)$ for the value of τ calculated in (b).

4.15 Given a prototype LP filter with the frequency response function $H(\omega) = \frac{2}{1+4j\omega}$ and 3 dB angular frequency ω_1. (a) Compute and sketch its magnitude response and determine the 3 dB bandwidth; (b) design an HP filter with the 3 dB angular frequency $\omega_2 = 4\omega_1$ and sketch its magnitude response; (d) design an LP filter with the 3 dB angular frequency $\omega_2 = 4\omega_1$ and sketch its magnitude response.

4.16 Given a prototype LP filter defined by the input-output relation

$$y(t) = \int_0^t [2x(\tau) - 3y(\tau)]d\tau.$$

(a) Compute and plot its magnitude response and determine the 3 dB angular frequency ω_1; (b) design a BP filter with the 3 dB passband [4, 12] rad/s and plot its magnitude response; (c) design a BS filter with the 3 dB stopband [4, 12] rad/s and plot its magnitude response.

Chapter 5
Laplace Transformation in Continuous-Time LTI Systems

In this chapter, we deal with the one-sided Laplace transformation, widely applied in analog systems. We recall in brief the definition and basic properties of the Laplace transformation. We prove its usefulness in solving differential input-output equations of LTI systems. The concepts of zero-input and zero-state responses are also presented. Next, using the convolution theorem, we introduce a notion of the *transfer function* of an analog system and present formulas for transfer functions of different system interconnections introduced in Chap. 3. We also show the relation between the transfer function and the frequency response function of a system defined in Chap. 4. A special section is dedicated to elements of the theory of *stability* of analog systems. In the last section, we present in detail three classes of real filters: Butterworth and Chebyshev (type I and II), with their transfer functions and magnitude responses.

5.1 The One-Sided Laplace Transformation

Definition 5.1 The *unilateral (one-sided) Laplace transform* $X(s)$, $s = \sigma + j\omega$, $s \in \mathbb{C}$ of a continuous-time signal $x(t)$, $t \in \mathbb{R}$ is defined as

$$X(s) = \int_0^\infty x(t)e^{-st}dt. \tag{5.1}$$

The transformation (5.1) converts a continuous-time signal $x(t)$ into a function $X(s)$ of a complex variable s. We denote the pair of Laplace transforms as $x(t) \overset{\mathcal{L}}{\leftrightarrow} X(s)$. In Table 5.1, we present some useful pairs of one-sided Laplace transforms.

The set of values for which the integral (5.1) converges absolutely is called the *region of convergence* (ROC).

K. M. Snopek, *Hands-on Signals and Systems Theory*,
https://doi.org/10.1007/978-3-031-56081-1_5

Table 5.1 Laplace transforms of chosen causal signals

$x(t),\ t \in \mathbb{R}$	$X(s),\ s \in \mathbb{C}$
$\delta(t)$	1
$\mathbf{1}(t)$	$\frac{1}{s}$
$t \cdot \mathbf{1}(t)$	$\frac{1}{s^2}$
$t^n \cdot \mathbf{1}(t),\ n \in N$	$\frac{n!}{s^{n+1}}$
$e^{-\alpha t} \cdot \mathbf{1}(t)$	$\frac{1}{s+\alpha}$
$t^n e^{-\alpha t} \cdot \mathbf{1}(t)$	$\frac{n!}{(s+\alpha)^{n+1}}$
$\cos(\omega_0 t) \cdot \mathbf{1}(t)$	$\frac{s}{s^2+\omega_0^2}$
$\sin(\omega_0 t) \cdot \mathbf{1}(t)$	$\frac{\omega_0}{s^2+\omega_0^2}$
$t \cdot \cos(\omega_0 t) \cdot \mathbf{1}(t)$	$\frac{s^2-\omega_0^2}{\left(s^2+\omega_0^2\right)^2}$
$t \cdot \sin(\omega_0 t) \cdot \mathbf{1}(t)$	$\frac{2\omega_0 s}{\left(s^2+\omega_0^2\right)^2}$
$e^{-\alpha t} \cdot \cos(\omega_0 t) \cdot \mathbf{1}(t)$	$\frac{s+\alpha}{(s+\alpha)^2+\omega_0^2}$
$e^{-\alpha t} \cdot \sin(\omega_0 t) \cdot \mathbf{1}(t)$	$\frac{\omega_0}{(s+\alpha)^2+\omega_0^2}$

We often deal with the problem of calculating the *inverse Laplace transform* of a given function $X(s)$. Usually in system analysis, $X(s)$ is a rational function in the form of a quotient $X(s) = \frac{N(s)}{D(s)}$, where $N(s)$ and $D(s)$ are polynomials of the variable s. For such functions, there are some algebraic methods of calculation of the inverse Laplace transform presented below in solved problems in Sect. 5.2.

The properties of the Laplace transformation are collected below in Table 5.2.

5.2 Zero-Input and Zero-State Responses

Let us recall the general form (3.6):

$$a_m \frac{d^m y(t)}{dt^m} + a_{m-1} \frac{d^{m-1} y(t)}{dt^{m-1}} + \ldots a_1 \frac{dy(t)}{dt} + y(t) = b_l \frac{d^l x(t)}{dt^l} + b_{l-1} \frac{d^{l-1} x(t)}{dt^{l-1}} + \ldots + b_0 x(t), t \geq t_0 \tag{5.1}$$

of the input output differential equation with *initial conditions* $y(t_0), y'(t_0), \ldots, y^{(m-1)}(t_0)$. The solution of (5.1) is the system response $y(t)$ expressed as the sum $y(t) = y_{0i}(t) + y_{0s}(t)$ of two components: $y_{0i}(t)$ called the *zero-input response* (*free response*) and $y_{0s}(t)$—the *zero-state response* (*forced response*).

The zero-input response $y_{0i}(t)$ is the solution of (5.1) when *the input signal $x(t)$ is zero*, that is,

Table 5.2 Properties of the one-sided Laplace transformation

Property	Signal $x(t)$, $t \in \mathbb{R}$	Laplace transform $X(s)$, $s \in \mathbb{C}$
Linearity	$\alpha x_1(t) + \beta x_2(t)$	$\alpha X_1(s) + \beta X_2(s)$
Time-shift	$x(t - t_0) \cdot \mathbf{1}(t - t_0)$, $t_0 > 0$	$X(s)e^{-st_0}$
s-shift	$x(t)e^{-at}$	$X(s + a)$
Modulation	$x(t) \cos(\omega_0 t)$	$\frac{1}{2}X(s+j\omega_0)+\frac{1}{2}X(s-j\omega_0)$
	$x(t) \sin(\omega_0 t)$	$-\frac{1}{2j}X(s+j\omega_0)+\frac{1}{2j}X(s-j\omega_0)$
Derivation in time	$\frac{d^n x(t)}{dt^n}$	$s^n X(s) - s^{n-1}x(0^+) - s^{n-2}x'(0^+) - \ldots - x^{(n-1)}(0^+)$
Derivation in frequency	$t^n x(t)$	$(-1)^n \frac{d^n X(s)}{ds^n}$
Integration in time	$\int_{-\infty}^{t} x(\tau)d\tau$	$\frac{X(s)}{s}$
Scaling	$x(a \cdot t)$, $a > 0$	$\frac{1}{a}X\left(\frac{s}{a}\right)$
Time-convolution	$x_1(t) * x_2(t)$	$X_1(s) \cdot X_2(s)$
Multiplication	$x_1(t) \cdot x_2(t)$	$\frac{1}{2\pi j}[X_1(s) * X_2(s)]$
Initial value	$\lim_{t\to 0^+} x(t) = \lim_{s\to\infty} sX(s)$	
Final value	$\lim_{t\to\infty} x(t) = \lim_{s\to 0} sX(s)$	

$$a_m \frac{d^m y(t)}{dt^m} + a_{m-1}\frac{d^{m-1}y(t)}{dt^{m-1}} + \ldots a_1 \frac{dy(t)}{dt} + y(t) = 0. \tag{5.2}$$

The zero-state response $y_{0s}(t)$ is the solution of (5.1) with *zero initial conditions,* that is, $y(t_0) = 0$, $y'(t_0) = 0, \ldots, y^{(m-1)}(t_0) = 0$.

Let us now explain how to solve differential equations of LTI systems using the Laplace transformation and how to determine zero-input– and zero-state responses. First, using the differentiation property of the Laplace transformation (see Table 5.2), we transform the eq. (5.1) into the system equation in the s-domain. Then, we compute the Laplace transform $Y(s)$ of the output signal. At last, we determine $y(t)$ as the inverse Laplace transform of $Y(s)$.

Let us demonstrate this method solving Problem 5.1, in which we will also determine zero-input and zero-state components of the output signal.

Problem 5.1 Given the input-output equation of an LTI analog system: $\frac{d^2y(t)}{dt^2} + 5\frac{dy(t)}{dt} + 6y(t) = x(t)$ excited with $x(t) = e^{-t} \cdot \mathbf{1}(t)$. Assuming initial conditions $y(0^+) = 1$, $y'(0^+) = -2$, compute its zero-input and zero-state responses $y_{0i}(t)$ and $y_{0s}(t)$.

Solution Using Table 5.2, let us write the system equation in the s-domain:

$$s^2Y(s) - sy(0^+) - y'(0^+) + 5[sY(s) - y(0^+)] + 6Y(s) = X(s).$$

After reordering the terms, we get

$$(s^2 + 5s + 6)Y(s) = X(s) + (s + 5)y(0^+) + y'(0^+),$$
$$Y(s) = \frac{X(s)}{s^2 + 5s + 6} + \frac{(s + 5)y(0^+) + y'(0^+)}{s^2 + 5s + 6}.$$

We can see that $Y(s)$ has two components: $\frac{X(s)}{s^2+5s+6}$ (depending on the input) and $\frac{(s+5)y(0^+)+y'(0^+)}{s^2+5s+6}$ (depending on initial conditions). The first component, denoted with $Y_{0s}(s)$, is the Laplace transform of the zero-state response $y_{0s}(t)$: $y_{0s}(t) \overset{\mathcal{L}}{\leftrightarrow} Y_{0s}(s)$. The second component $Y_{0i}(s)$ is the Laplace transform of the zero-input response $y_{0i}(t)$: $y_{0i}(t) \overset{\mathcal{L}}{\leftrightarrow} Y_{0i}(s)$. We have $Y(s) = Y_{0s}(s) + Y_{0i}(s)$.

Substituting in the above equation the initial conditions $y(0^+) = 1$, $y'(0^+) = -2$, we obtain

$$Y(s) = \frac{X(s)}{s^2 + 5s + 6} + \frac{s + 3}{s^2 + 5s + 6}.$$

As $s^2 + 5s + 6 = (s + 2)(s + 3)$, we have

$$Y(s) = \frac{X(s)}{(s + 2)(s + 3)} + \frac{1}{s + 2} = Y_{0s}(s) + Y_{0i}(s).$$

First, let us calculate the zero-state response $y_{0s}(t)$. From Table 5.1, we can get the Laplace transform pair: $x(t) = e^{-t} \cdot \mathbf{1}(t) \overset{\mathcal{L}}{\leftrightarrow} X(s) = \frac{1}{s+1}$. So, $Y_{0s}(s) = \frac{1}{(s+1)(s+2)(s+3)}$ and we can compute its inverse Laplace transform using the *partial fraction decomposition* of $Y_{0s}(s)$ as follows:

$$Y_{0s}(s) = \frac{A}{s + 1} + \frac{B}{s + 2} + \frac{C}{s + 3}$$
$$= \frac{A(s + 2)(s + 3) + B(s + 1)(s + 3) + C(s + 1)(s + 2)}{(s + 1)(s + 2)(s + 3)},$$

where A, B, and C are real constants. Comparing with $\frac{1}{(s+1)(s+2)(s+3)}$, we get

$$A(s + 2)(s + 3) + B(s + 1)(s + 3) + C(s + 1)(s + 2) = 1.$$

We can, of course, solve the set of three equations but there is an easier way of calculation of constants A, B, C known as the *method of residues* [4, 17]. Notice that, if we substitute in the above equation $s = -1$, we get $2A = 1 \Leftrightarrow A = \frac{1}{2}$. For $s = -2$, we obtain $B = 1$ and for $s = -3$, we have $2C = 1 \Leftrightarrow C = \frac{1}{2}$. The Laplace transform of the zero-state response $y_{0s}(t)$ is a sum of three components as follows:

$$Y_{0s}(s) = \frac{1}{2} \cdot \frac{1}{s+1} + \frac{1}{s+2} + \frac{1}{2} \cdot \frac{1}{s+3}$$

so $y_{0s}(t)$ can easily be determined basing on Table 5.2 in the form

$$y_{0s}(t) = \left(\frac{1}{2}e^{-t} + e^{-2t} + \frac{1}{2}e^{-3t}\right) \cdot \mathbf{1}(t).$$

Now, we will compute the zero-input response $y_{0i}(t)$. Since $Y_{0i}(s) = \frac{1}{s+2}$, we immediately get

$$y_{0i}(t) = e^{-2t} \cdot \mathbf{1}(t).$$

The system response $y(t) = y_{0i}(t) + y_{0s}(t)$, then

$$y(t) = \left(\frac{1}{2}e^{-t} + 2e^{-2t} + \frac{1}{2}e^{-3t}\right) \cdot \mathbf{1}(t).$$

5.3 Transfer Function

In Chap. 3 we stated that the response $y(t)$ of a LTI analog system to a given input signal $x(t)$ is a convolution of $x(t)$ with the impulse response $h(t)$ (see (3.4)). Applying the convolution theorem for the Laplace transformation (see Table 5.2), we obtain the following equality in the s-domain:

$$Y(s) = X(s)H(s), \tag{5.3}$$

where $y(t) \overset{\mathcal{L}}{\leftrightarrow} Y(s)$, $x(t) \overset{\mathcal{L}}{\leftrightarrow} X(s)$ and $h(t) \overset{\mathcal{L}}{\leftrightarrow} H(s)$.

Definition 5.2 The *transfer function* $H(s)$ of a LTI continuous-time system is the Laplace transform of its impulse response $h(t)$: $h(t) \overset{\mathcal{L}}{\leftrightarrow} H(s)$.

In particular, for the input-output relation $x(t) = \mathbf{1}(t) \to y(t) = k(t)$, we can write the following input-output relation in the s-domain: $X(s) = \frac{1}{s} \to Y(s) = K(s)$. Notice that the transfer function $H(s)$ is related to the Laplace transform $K(s)$ of the step response $k(t)$ by

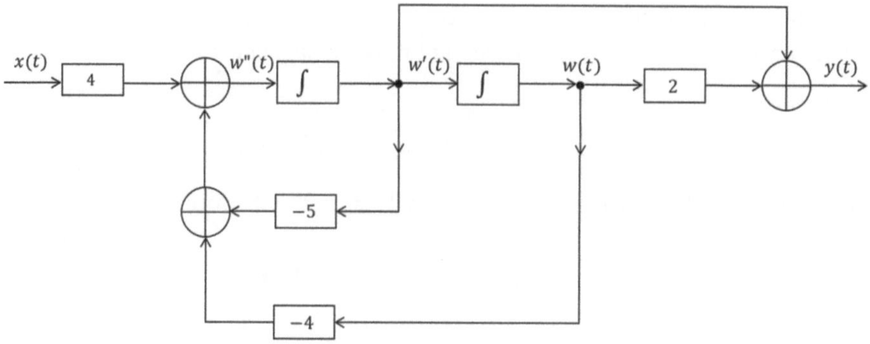

Fig. 5.1 Block diagram of the system from Problem 5.2

$$H(s) = \frac{K(s)}{1/s} = sK(s). \tag{5.4}$$

From (5.3), we can see that the transfer function $H(s)$ equals the ratio of the Laplace transform $Y(s)$ of the output signal to the Laplace transform $X(s)$ of the input signal under an assumption that *all initial conditions are zero*. In other words, the transfer function is the ratio of the Laplace transform $Y_{0s}(s)$ of the *zero-state response* to the Laplace transform $X(s)$. We have

$$H(s) = \left.\frac{Y(s)}{X(s)}\right|_{\text{initial conditions}=0} = \frac{Y_{0s}(s)}{X(s)}. \tag{5.5}$$

In Problem 5.2, we will compute the transfer function of a given system and its impulse response using (5.2). We will also show how to get the system equation starting from $H(s)$.

Problem 5.2 Given an analog system with the block diagram shown in Fig. 5.1 excited with $x(t) = (1 - e^{-2t}) \cdot \mathbf{1}(t)$. (a) Compute the transfer function of the system and its impulse response; (b) determine the input-output equation.

Solution (a) We proceed analogously as in Problem 3.9 and formulate the set of equations defining the system as follows:

$$\begin{cases} 4x(t) - 5w'(t) - 4w(t) = w''(t) \\ w'(t) + 2w(t) = y(t) \end{cases},$$

where $w(t)$ is an internal signal. For calculation of the transfer function, we have to assume zero initial conditions and apply the Laplace transformation to the above equations ($x(t) \overset{\mathcal{L}}{\leftrightarrow} X(s), y(t) \overset{\mathcal{L}}{\leftrightarrow} Y(s), w(t) \overset{\mathcal{L}}{\leftrightarrow} W(s)$). We get

$$\begin{cases} 4X(s) - 5sW(s) - 4W(s) = s^2W(s) \\ sW(s) + 2W(s) = Y(s). \end{cases}$$

After reordering the terms, we have

$$\begin{cases} X(s) = \frac{1}{4}W(s)\left(s^2 + 5s + 4\right) \\ Y(s) = W(s)(s+2). \end{cases}$$

Dividing the second equation by the first one, we can get the transfer function

$$H(s) = 4\frac{s+2}{s^2 + 5s + 4} = 4\frac{s+2}{(s+1)(s+4)}.$$

Now, we can use the following partial fraction decomposition of $H(s)$:

$$H(s) = 4\left(\frac{A}{s+1} + \frac{B}{s+4}\right) = 4\frac{As + 4A + Bs + B}{(s+1)(s+4)} = 4\frac{(A+B)s + 4A + B}{(s+1)(s+4)}$$

to get $\begin{cases} A+B=1 \\ 4A+B=2 \end{cases} \Leftrightarrow A = \frac{1}{3}, B = \frac{2}{3}$.

We have

$$H(s) = \frac{4}{3}\frac{1}{s+1} + \frac{8}{3}\frac{1}{s+4}$$

so the impulse response

$$h(t) = \frac{4}{3}\left(e^{-t} + 2e^{-4t}\right) \cdot \mathbf{1}(t).$$

(b) If we want to determine the input-output equation, we should start from the transfer function

$$H(s) = \frac{Y(s)}{X(s)} = 4\frac{s+2}{s^2 + 5s + 4}$$

that is equivalent to

$$s^2Y(s) + 5sY(s) + 4Y(s) = 4sX(s) + 2X(s).$$

The inverse Laplace transformation of the above equation permits to get the system equation in the time domain

$$\frac{d^2y(t)}{dt^2} + 5\frac{dy(t)}{dt} + 4y(t) = 4\frac{dx(t)}{dt} + 2x(t).$$

5.3.1 Relation Between the Transfer Function and the Frequency Response of a Continuous-Time LTI System

Let us notice that the Fourier transform $X(\omega)$ of a causal signal $x(t)$, $t \in \mathbb{R}$ is

$$X(\omega) = \int_0^\infty x(t)e^{-j\omega t}dt \tag{5.6}$$

(see (2.17)). Notice that (5.6) is a special case of the one-sided Laplace transform (5.1), where $s = j\omega$. Having this in mind, we can write the following relation between the transfer function $H(s)$ and the frequency response $H(\omega)$:

$$H(\omega) = H(s)|_{s=j\omega}. \tag{5.7}$$

5.3.2 Transfer Functions of System Interconnections

In Sect. 3.5, we presented formulas (3.9) and (3.10) for impulse responses of cascade and parallel system interconnections. Using properties of the Laplace transformation presented in Table 5.2, we can easily derive formulas for transfer functions of the following system interconnections:

$$\text{cascade}: H(s) = H_1(s) \cdot H_2(s), \tag{5.8}$$

$$\text{parallel}: H(s) = H_1(s) + H_2(s). \tag{5.9}$$

For the negative feedback system presented in Fig. 3.8, we can write the following relations in the s-domain:

$$Y(s) = [X(s) - Y_2(s)] \cdot H_1(s),$$
$$Y_2(s) = Y(s) \cdot H_2(s).$$

Substituting $Y_2(s)$ from the second equation to the first one, we obtain

$$Y(s) = [X(s) - Y(s) \cdot H_2(s)] \cdot H_1(s)$$

After reordering the terms, we have

$$Y(s) + \mathrm{Y(s)} \cdot H_1(s) \cdot H_2(s) = X(s) \cdot H_1(s).$$

The transfer function of the *feedback* system is given by

$$H(s) = \frac{H_1(s)}{1 + H_1(s) \cdot H_2(s)} \tag{5.10}$$

The usefulness of formulas (5.8), (5.9) and (5.10) is demonstrated in Problems 5.3 and 5.4.

Problem 5.3 Two systems S_1 and S_2 are connected in parallel, as shown in Fig. 3.8. Given the input-output equation of S_1: $\frac{dy_1(t)}{dt} = x_1(t)$ and the step response of S_2: $k_2(t) = (1 - e^{-t}) \cdot \mathbf{1}(t)$, (a) determine the impulse response $h(t)$ of the whole system; (b) calculate the response $y(t)$ to the input signal $x(t) = (t - 1) \cdot \mathbf{1}(t - 1)$.

Solution (a) Let us calculate the impulse response $h(t)$ as the inverse Laplace transform of the transfer function $H(s)$ given by (5.9). We need to calculate $H_1(s)$ and $H_2(s)$. From the equation of the system S_1, we can see that

$$sY_1(s) = X_1(s) \Rightarrow H_1(s) = \frac{Y_1(s)}{X_1(s)} = \frac{1}{s}.$$

The transfer function $H_2(s)$ can be computed from (5.4): $H_2(s) = sK_2(s)$, where $K_2(s) = \frac{1}{s} - \frac{1}{s+1}$ (see Table 5.1). We have $H_2(s) = 1 - \frac{s}{s+1} = \frac{1}{s+1}$. Since

$$H(s) = H_1(s) + H_2(s) = \frac{1}{s} + \frac{1}{s+1},$$

the impulse response $h(t) = (1 + e^{-t}) \cdot \mathbf{1}(t)$ (see Definition 5.2).

(b) The output signal $y(t)$ can be computed as the inverse Laplace transform of $Y(s)$, where $Y(s) = X(s)H(s)$. First, we need to calculate the Laplace transform of the

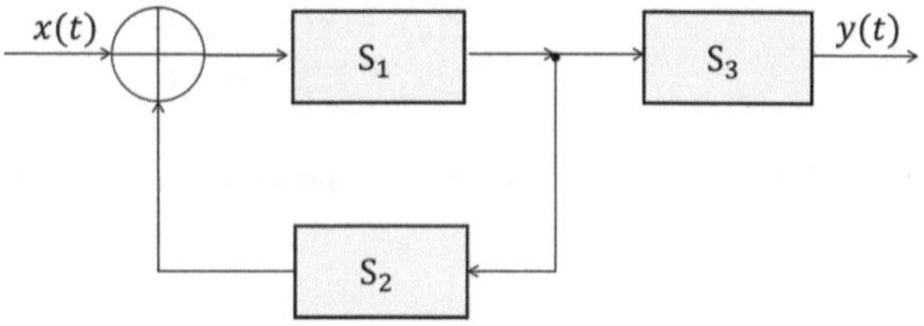

Fig. 5.2 Block diagram of the system from Problem 5.4

input signal $x(t) = (1 - t) \cdot \mathbf{1}(t-1)$. Applying the time-shift property from Table 5.2, we get $X(s) = \frac{1}{s}e^{-s}$. Then

$$Y(s) = \frac{1}{s}e^{-s}\left(\frac{1}{s} + \frac{1}{s+1}\right) = \frac{1}{s^2}e^{-s} + \frac{1}{s(s+1)}e^{-s} = \frac{1}{s^2}e^{-s} + \frac{s+1-s}{s(s+1)}e^{-s}$$
$$Y(s) = \frac{1}{s^2}e^{-s} + \frac{1}{s}e^{-s} - \frac{1}{s+1}e^{-s}.$$

Due to the linearity of the Laplace transformation, we finally obtain the system response

$$y(t) = (t-1) \cdot \mathbf{1}(t-1) + \mathbf{1}(t-1) - e^{-(t-1)} \cdot \mathbf{1}(t-1) = (t - e^{-(t-1)}) \cdot \mathbf{1}(t-1).$$

Problem 5.4 Given an analog LTI system with a block diagram presented in Fig. 5.2. The transfer functions of corresponding systems are:

$$H_1(s) = \frac{s-1}{s^2+4s+3}, \qquad H_2(s) = \frac{s^3+3s^2+4s+2}{s^2+2s-3}, \quad H_3(s) = \frac{1}{s+2}.$$

(a) Compute the transfer function $H(s)$ of the whole system, (b) determine and sketch its impulse response $h(t)$, (c) determine the input-output equation, (d) compute and sketch the magnitude response of the system.

Solution (a) Notice that the considered system is a cascade interconnection of two systems: the first one is a feedback system composed of S_1 and S_2 and the second one is S_3. Then the transfer function of the whole system is a product of two transfer functions: $\frac{H_1(s)}{1-H_1(s)\cdot H_2(s)}$ and $H_3(s)$ as follows:

$$H(s) = \frac{H_1(s)}{1 - H_1(s) \cdot H_2(s)} \cdot H_3(s),$$

where the minus sign in the denominator is a consequence of the positive feedback (compare with (5.10) and Fig. 3.8).

Let us substitute all transfer functions into the above equality:

$$H(s) = \frac{\frac{s-1}{s^2+4s+3}}{1-\frac{s-1}{s^2+4s+3}\frac{s^3+3s^2+4s+2}{s^2+2s-3}} \cdot \frac{1}{s+2}.$$

Simplifying the above multilevel fraction gives

$$H(s) = \frac{\frac{s-1}{(s+1)(s+3)}}{1-\frac{s-1}{(s+1)(s+3)}\frac{(s+1)(s^2+2s+2)}{(s-1)(s+3)}} \cdot \frac{1}{s+2},$$

.

$$H(s) = \frac{\frac{s-1}{(s+1)(s+2)(s+3)}}{1-\frac{(s^2+2s+2)}{(s+3)^2}} = \frac{\frac{s-1}{(s+1)(s+2)}}{\frac{(s+3)^2-(s^2+2s+2)}{s+3}}$$

and finally, we get the transfer function

$$H(s) = \frac{(s-1)(s+3)}{(s+1)(s+2)(4s+7)}.$$

(b) In order to compute the impulse response $h(t)$ as the inverse Laplace transform of the $H(s)$, we have first to decompose $H(s)$ into the sum

$$H(s) = \frac{A}{s+1} + \frac{B}{s+2} + \frac{C}{4s+7}.$$

Applying the *residues method* for the calculation of constants A, B, and C gets

$$\begin{aligned}
A &= \lim_{s\to -1}(s+1)H(s) = \lim_{s\to -1}\frac{(s-1)(s+3)}{(s+2)(4s+7)} = -\frac{4}{3},\\
B &= \lim_{s\to -2}(s+2)H(s) = \lim_{s\to -2}\frac{(s-1)(s+3)}{(s+1)(4s+7)} = -3,\\
C &= \lim_{s\to -\frac{7}{4}}(4s+7)H(s) = \lim_{s\to -\frac{7}{4}}\frac{(s-1)(s+3)}{(s+1)(s+2)} = \frac{55}{3}.
\end{aligned}$$

Then

$$H(s) = -\frac{4}{3}\frac{1}{s+1} - \frac{3}{s+2} + \frac{55}{12}\frac{1}{(s+\frac{7}{4})}.$$

Using Table 5.1, we get the impulse response (see Fig. 5.3):

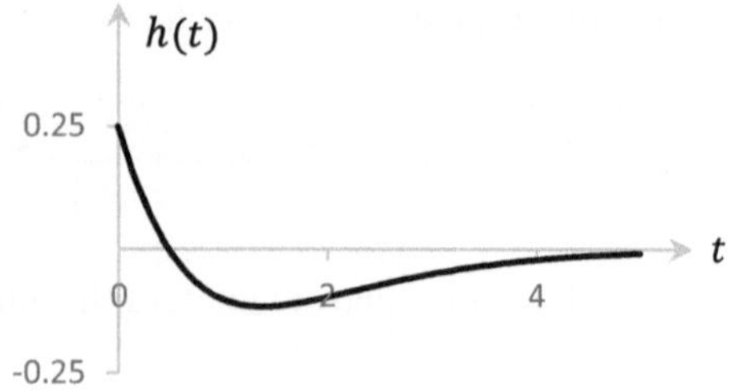

Fig. 5.3 Impulse response $h(t)$ of the system from Problem 5.4

$$h(t)=\left(-\frac{4}{3}e^{-t}-3e^{-2t}+\frac{55}{12}e^{-\frac{7}{4}t}\right)\cdot\mathbf{1}(t).$$

(c) To determine the input-output differential equation of the system, we should start from the system equation in the s-domain. Thus,

$$H(s)=\frac{Y(s)}{X(s)}=\frac{(s-1)(s+3)}{(s+1)(s+2)(4s+7)}=\frac{s^2+2s-3}{(s^2+3s+2)(4s+7)}$$
$$=\frac{s^2+2s-3}{4s^3+19s^2+29s+14}$$

which is equivalent to

$$\left(4s^3+19s^2+29s+14\right)Y(s)=\left(s^2+2s-3\right)X(s),$$

$$4s^3Y(s)+19s^2Y(s)+29sY(s)+14Y(s)=s^2X(s)+2sX(s)-3X(s).$$

The input-output equation in the time domain is

$$4\frac{d^3y(t)}{dt^3}+19\frac{d^2y(t)}{dt^2}+29\frac{dy(t)}{dt}+14y(t)=\frac{d^2x(t)}{dt^2}+2\frac{dx(t)}{dt}-3x(t).$$

(d) Using (4.2), the magnitude response is

$$A(\omega)=\left|\frac{(j\omega-1)(j\omega+3)}{(j\omega+1)(j\omega+2)(4j\omega+7)}\right|=\frac{\sqrt{(1+\omega^2)(9+\omega^2)}}{\sqrt{(1+\omega^2)(4+\omega^2)(49+16\omega^2)}},$$

$$A(\omega)=\sqrt{\frac{9+\omega^2}{(4+\omega^2)(49+16\omega^2)}}.$$

Its plot is displayed in Fig. 5.4. We can notice that the considered system is a low-pass filter.

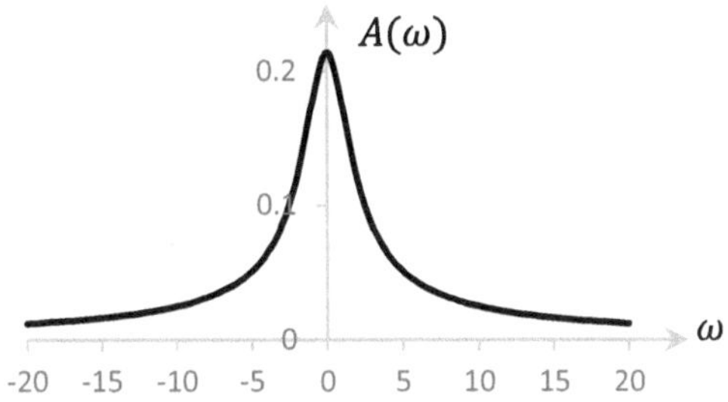

Fig. 5.4 Magnitude response $A(\omega)$ of the system from Problem 5.4

5.3.3 *Zeros and Poles of the Transfer Function*

The transfer function $H(s)$ of an analog LTI system is usually the rational function of the form $\frac{N(s)}{D(s)}$, where $N(s)$ and $D(s)$ are polynomials of a variable s.

Definition 5.3 If the degree of $N(s)$ is less than or equal to the degree of $D(s)$, we say that the transfer function is *proper.* If the degree of $N(s)$ is less than the degree of $D(s)$, the transfer function is *strictly proper.*

Definition 5.4 The *zeros* of the transfer function $H(s)$ are roots of $N(s)$, while the roots of $D(s)$ are called the *poles* of $H(s)$.

The *pole-zero plot* is the graphical representation of poles and zeros of the transfer function $H(s)$ in domain of the complex variable s. It is very useful in the study of the stability of a system.

5.4 Stability of Analog LTI Systems

Definition 5.5 We say that a continuous-time LTI system, defined by the input-output relation $x(t) \to y(t)$, is *BIBO–stable* (BIBO—Bounded Input Bounded Output) if the following condition is fulfilled:

$$\forall t \in \mathbb{R}, \exists C_1, C_2 \in \mathbb{R} : |x(t)| \leq C_1 < \infty \Rightarrow |y(t)| \leq C_2 < \infty. \tag{5.11}$$

The condition (5.11) is equivalent to the *absolute integrability* of the impulse response $h(t)$ expressed as

$$\int_{-\infty}^{\infty} |h(t)| dt < \infty. \tag{5.12}$$

For causal systems, the condition (5.12) gets the form $\int_{0}^{\infty} |h(t)|dt < \infty$. The system BIBO–stable can be simply called "*stable*".

Theorem 5.1 An analog LTI system is BIBO–stable if and only if all poles of its transfer function $H(s)$ are *in the strict left half of the s-plane* and $H(s)$ is proper.

To study the stability of analog systems with rational transfer functions $H(s) = \frac{N(s)}{D(s)}$, we have to calculate the roots of $D(s)$. It is an easy task for polynomials of degree 1 or 2. For higher–order polynomials, we can apply some algebraic criteria, for example, the *Hurwitz criterion*, the *Hurwitz-Liénard criterion,* or the *Routh-Hurwitz criterion*.

5.4.1 Hurwitz Criterion

Definition 5.6 The polynomial $D(s) = a_m s^m + a_{m-1} s^{m-1} + \ldots + a_1 s + a_0$, $a_m > 0$, $a_i \in \mathbb{R}$, $i = 0, \ldots, m$, of degree m is called *the Hurwitz polynomial* if *all* its roots are in the strict left half of the s-plane: Re $s < 0$.

Theorem 5.2 The polynomial $D(s) = a_m s^m + a_{m-1} s^{m-1} + \ldots + a_1 s + a_0$, $a_m > 0$, $a_i \in \mathbb{R}$, $i = 0, \ldots, m$, is the Hurwitz polynomial if and only if all the leading principal minors Δ_i, $i = 1, \ldots, m$ of the Hurwitz matrix $\mathcal{H}_m$ are *positive*.

The *Hurwitz matrix* $\mathcal{H}_m$ is a square matrix $m \times m$ of the following structure:

$$\mathcal{H}_m = \begin{bmatrix} a_{m-1} & a_m & 0 & \cdots & \\ a_{m-3} & a_{m-2} & a_{m-1} & & \\ a_{m-5} & a_{m-4} & a_{m-3} & \ddots & \vdots \\ & \vdots & & & \\ 0 & 0 & 0 & \cdots & a_0 \end{bmatrix} \tag{5.13}$$

The first three leading principal minors of the matrix (5.13) are

$$\Delta_1 = a_{m-1}, \quad \Delta_2 = \begin{vmatrix} a_{m-1} & a_m \\ a_{m-3} & a_{m-2} \end{vmatrix}, \quad \Delta_3 = \begin{vmatrix} a_{m-1} & a_m & 0 \\ a_{m-3} & a_{m-2} & a_{m-1} \\ a_{m-5} & a_{m-4} & a_{m-3} \end{vmatrix} \quad \ldots \tag{5.14}$$

and the last one is $\Delta_m = a_0 \Delta_{m-1}$. If we need to calculate matrix determinants with order >3, we can apply the Laplace decomposition described, for example, in [11].

Let us apply the Hurwitz criterion for checking the stability of the system from Problem 5.4.

Problem 5.5 Consider the system from Problem 5.4 with the transfer function $H(s) = \frac{s^2+2s-3}{4s^3+19s^2+29s+14}$. Check its BIBO–stability using the Hurwitz criterion.

Solution We have to check if $D(s)$ (in the denominator of $H(s)$) is the Hurwitz polynomial. We have $D(s) = 4s^3 + 19s^2 + 29s + 14$, that is, $a_3 = 4$, $a_2 = 19$, $a_1 = 29$, $a_0 = 14$.

Let us construct the Hurwitz matrix $\mathcal{H}_3$ of dimension 3 × 3:

$$\mathcal{H}_3 = \begin{bmatrix} a_2 & a_3 & 0 \\ a_0 & a_1 & a_2 \\ 0 & 0 & a_0 \end{bmatrix} = \begin{bmatrix} 19 & 4 & 0 \\ 14 & 29 & 19 \\ 0 & 0 & 14 \end{bmatrix}$$

and calculate the first three leading principal minors as follows:

$$\Delta_1 = a_2 = 19, \quad \Delta_2 = \begin{vmatrix} a_2 & a_3 \\ a_0 & a_1 \end{vmatrix} = \begin{vmatrix} 19 & 4 \\ 14 & 29 \end{vmatrix} = 495, \quad \Delta_3 = a_0 \Delta_2 = 6930.$$

We can see that all minors Δ_i, $i = 1, 2, 3$ are positive, so the considered system is BIBO–stable.

On the other hand, if we looked once again at the plot of the impulse response from Fig. 5.3, we could easily notice that a condition concerning its absolute integrability is also fulfilled.

5.4.2 *Hurwitz-Liénard Criterion*

Theorem 5.3 The polynomial $D(s) = a_m s^m + a_{m-1} s^{m-1} + \ldots + a_1 s + a_0$, $a_i > 0$, $i = 0, \ldots, m$, is the Hurwitz polynomial if and only if the leading principal minors of the Hurwitz matrix $\mathcal{H}_m$ satisfy: $\Delta_1 > 0, \Delta_3 > 0, \ldots$ or $\Delta_2 > 0, \Delta_4 > 0, \ldots$

5.4.3 *Routh-Hurwitz Criterion*

Theorem 5.4 The polynomial $D(s) = a_m s^m + a_{m-1} s^{m-1} + \ldots + a_1 s + a_0$, $a_i > 0$, $i = 0, \ldots, m$, is the Hurwitz polynomial if and only if all elements of the left column of the Routh array are *positive*.

The *Routh array* structure with all necessary mathematical formulas is shown in Fig. 5.5. Let us illustrate this method in Problem 5.6.

Problem 5.6 Given an LTI analog system with the transfer function

$$H(s) = \frac{s+1}{s^4 + 2s^3 + ks^2 + 4s + 5}.$$

Using the Routh-Hurwitz criterion, indicate values of the parameter k for which the system is stable.

a_m	a_{m-2}	a_{m-4}	a_{m-6}	$\cdots$
a_{m-1}	a_{m-3}	a_{m-5}	a_{m-7}	$\cdots$
b_1	b_2	b_3	b_4	$\cdots$
c_1	c_2	c_3	$\cdots$	
d_1	d_2	$\cdots$		
e_1				

$$b_1 = \frac{\begin{vmatrix} a_m & a_{m-2} \\ a_{m-1} & a_{m-3} \end{vmatrix}}{-a_{m-1}}, b_2 = \frac{\begin{vmatrix} a_m & a_{m-4} \\ a_{m-1} & a_{m-5} \end{vmatrix}}{-a_{m-1}}, b_3 = \frac{\begin{vmatrix} a_m & a_{m-6} \\ a_{m-1} & a_{m-7} \end{vmatrix}}{-a_{m-1}}, \ldots$$

$$c_1 = \frac{\begin{vmatrix} a_{m-1} & a_{m-3} \\ b_1 & b_2 \end{vmatrix}}{-b_1}, c_2 = \frac{\begin{vmatrix} a_{m-1} & a_{m-5} \\ b_1 & b_3 \end{vmatrix}}{-b_1}, c_3 = \frac{\begin{vmatrix} a_{m-1} & a_{m-7} \\ b_1 & b_4 \end{vmatrix}}{-b_1}, \ldots$$

$$d_1 = \frac{\begin{vmatrix} b_1 & b_2 \\ c_1 & c_2 \end{vmatrix}}{-c_1}, d_2 = \frac{\begin{vmatrix} b_1 & b_3 \\ c_1 & c_3 \end{vmatrix}}{-c_1}, \ldots$$

$$e_1 = \frac{\begin{vmatrix} c_1 & c_2 \\ d_1 & d_2 \end{vmatrix}}{-d_1}, \ldots$$

Fig. 5.5 Routh array for checking the stability of an LTI system

Solution We have $D(s) = s^4 + 2s^3 + ks^2 + 4s + 5$, i.e., $a_4 = 1, a_3 = 2, a_2 = k, a_1 = 4, a_0 = 5$. The first two rows of the Routh array are

$$\begin{matrix} a_4 = 1 & a_2 = k & a_0 = 5 \\ a_3 = 2 & a_1 = 4 & 0 \end{matrix}$$

Let us calculate b_1 and b_2:

$$b_1 = \frac{\begin{vmatrix} a_4 & a_2 \\ a_3 & a_1 \end{vmatrix}}{-a_3} = \frac{\begin{vmatrix} 1 & k \\ 2 & 4 \end{vmatrix}}{-2} = \frac{4-2k}{-2} = k-2,$$

$$b_2 = \frac{\begin{vmatrix} a_4 & a_0 \\ a_3 & 0 \end{vmatrix}}{-a_3} = \frac{\begin{vmatrix} 1 & 5 \\ 2 & 0 \end{vmatrix}}{-2} = 5.$$

Then we can calculate c_1 as follows:

$$c_1 = \frac{\begin{vmatrix} a_3 & a_1 \\ b_1 & b_2 \end{vmatrix}}{-b_1} = \frac{\begin{vmatrix} 2 & 4 \\ k-2 & 5 \end{vmatrix}}{2-k} = \frac{18-4k}{2-k} = \frac{2(9-2k)}{2 \quad k}.$$

According to Theorem 5.3, the system is stable, if and only if all elements of the left column of the Routh array are positive, that is, $a_4 > 0$, $a_3 > 0$, $b_1 > 0$ and $c_1 > 0$. Evidently, $a_4 > 0$ and $a_3 > 0$. The system will be stable if

$$b_1 = k - 2 > 0 \Leftrightarrow k > 2$$

and

$$c_1 = \frac{2(9-2k)}{2-k} > 0 \Leftrightarrow (9-2k)(2-k) > 0 \Leftrightarrow k < 2 \text{ or } k > 4,5.$$

Concluding, the system is stable for $k > 4, 5$.

5.5 Chosen Classes of Real Filters

In Chap. 4, we considered different types of ideal filters: low-pass, high-pass, band-pass, and band-stop and we got acquainted with the frequency transformation method of analog filter design. In this Section, we focus on three important classes of real filters: Butterworth and Chebyshev (type I and II) filters and study their magnitude responses depending on a filter order.

In Fig. 5.6, we present regions (marked white on a grey background) in which a magnitude response of a real filter (LP, HP, BP, BS) should be located. We can observe that in the passband, it should assume values between $(1 - \delta_p)$ and $(1 + \delta_p)$ and in the stopband between 0 and δ_s (sometimes it can oscillate or ripple). Then, values of ω_p and ω_s (also with subscripts) indicate regions of transition. The angular frequency ω_p is often chosen as the 3 dB angular frequency and ω_s is the edge angular frequency of the stopband.

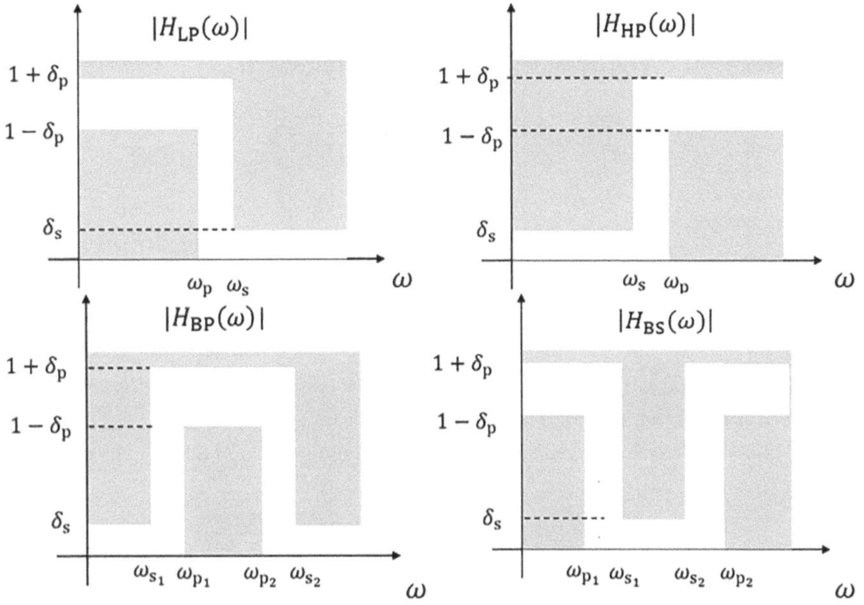

Fig. 5.6 Requirements for magnitude responses of different types of real filters (LP, HP, BP, BS)

In filter design, we always need to determine the values of parameters: δ_p, δ_s, ω_p and ω_s. Other interesting parameters are: the minimal gain in the passband and the maximal attenuation in the stopband.

The *minimal gain* in the passband (expressed in decibels) is defined as

$$A_p = -20\log\left(1-\delta_p\right). \tag{5.15}$$

The *maximal attenuation* (expressed in decibels) is given by

$$A_s = -20\log(\delta_s). \tag{5.16}$$

Now, let us describe the chosen classes of real filters with different orders. We present their transfer functions, poles-diagrams, and magnitude responses and discuss their behavior.

5.5.1 Butterworth Filters

The filter was called after Stephen Butterworth (1885–1958)—a famous British physicist, constructor, and brilliant mathematician who, in 1930, in his paper "On the Theory of Filter Amplifiers," proposed a maximally flat low-pass filter. Let ω_c be its 3 dB cutoff angular frequency. The transfer function *H*(*s*) of the Butterworth filter with order *n* is

For odd n : $H(s)$

$$= \frac{\omega_c}{s+\omega_c}\prod\nolimits_{i=1}^{(n-1)/2}\frac{\omega_c^2}{s^2+2\cos(\theta_i)\omega_c s+\omega_c^2}, \quad \theta_i = \frac{i\cdot 180^0}{n}, \tag{5.17}$$

Foreven n : $$H(s) = \prod\nolimits_{i=1}^{n/2}\frac{\omega_c^2}{s^2+2\cos\,(\theta_i)\omega_c s+\omega_c^2}, \quad \theta_i = \frac{(i-0,5)\cdot 180^0}{n}. \tag{5.18}$$

Table 5.3 contains formulas of polynomials *D*(*s*) in the denominator of *H*(*s*) for Butterworth filters with order 1,. . .,5.

The poles λ_k, $k = 1, 2, \ldots, n$ of the transfer function (5.17) and (5.18) are

$$\lambda_k = j\omega_c \exp\left(j\frac{\pi+2(k-1)\pi}{2n}\right), k = 1, 2, \ldots, n \tag{5.19}$$

and the magnitude response of the *n*th order Butterworth filter

Table 5.3 Polynomials $D(s)$ for the Butterworth filters with order 1,. . .,5

order n	$D(x),\ x=\frac{s}{\omega_c}$
1	$1 + x$
2	$x^2 + \sqrt{2}x + 1$
3	$(x + 1)(x^2 + x + 1)$
4	$(x^2 + 0,765x + 1)(x^2 + 1,848x + 1)$
5	$(x + 1)(x^2 + 0,618x + 1)(x^2 + 1,618x + 1)$

$$A(\omega)=\frac{1}{\sqrt{1+\left(\frac{\omega}{\omega_c}\right)^{2n}}}\ . \tag{5.20}$$

Problem 5.6 Determine transfer functions and poles of the Butterworth filters with orders 1, 2, and 3 and the 3 dB cutoff angular frequency $\omega_c = 1$ rad/s. Plot their magnitude responses.

Solution First, let us consider the Butterworth filter with order $n = 1$. Its transfer function is

$$H(s)=\frac{\omega_c}{s+\omega_c}=\frac{1}{s+1}.$$

We can see that there is only one pole at $s = -1$, which is in accordance with (5.19), that is, $\lambda_1 = j\omega_c \exp\left(j\frac{\pi}{2}\right) = -1$.

Next, let us consider the Butterworth filter with order 2. Using (5.18) or Table 5.3, we get its transfer function

$$H(s)=\frac{1}{s^2+\sqrt{2}s+1}$$

with $\theta_1 = 45^\circ$. We can easily calculate two poles of the transfer function:

$$\lambda_1=-\frac{\sqrt{2}}{2}+j\frac{\sqrt{2}}{2},\lambda_2=-\frac{\sqrt{2}}{2}-j\frac{\sqrt{2}}{2}.$$

Their plot is presented in Fig. 5.7 (left).

Finally, from (5.17) and Table 5.3, we get the transfer function of the 3rd order Butterworth filter

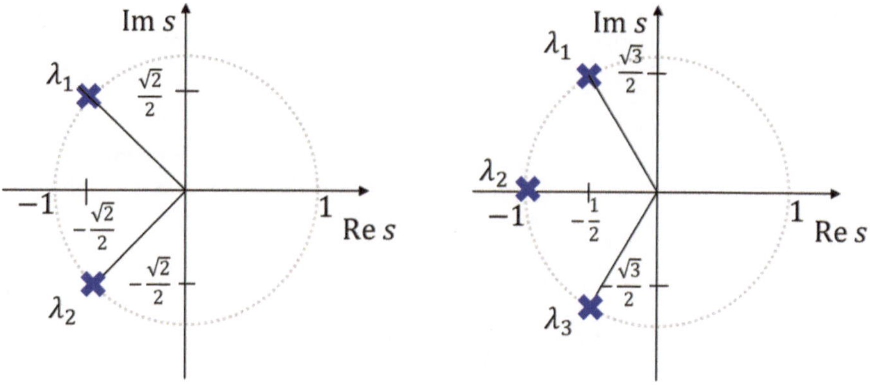

Fig. 5.7 Pole diagrams of Butterworth filters with order 2 (left) and 3 (right)

$$H(s) = \frac{1}{s+1} \cdot \frac{1}{s^2 + s + 1}$$

with $\theta_1 = 60^\circ$. There are three poles defined as

$$\lambda_k = j\omega_c \exp\left(j\frac{\pi + 2(k-1)\pi}{6}\right), k = 1, 2, 3,$$

that is, $\lambda_1 = -\frac{1}{2} + j\frac{\sqrt{3}}{2}, \lambda_2 = -1, \lambda_3 = -\frac{1}{2} - j\frac{\sqrt{3}}{2}$, which is visualized in Fig. 5.7 (right).

Now, let us determine the magnitude responses of Butterworth filters with order 1,2, and 3. Evidently, we can calculate $A(\omega)$ using (5.7) or directly from (5.20).

The magnitude response of the first order Butterworth filter: $A(\omega) = \frac{1}{\sqrt{1+\omega^2}}$ (compare with Problem 4.2). For the Butterworth filter with order 2, $A(\omega) = \frac{1}{\sqrt{1+\omega^4}}$ and for the order $n = 3$: $A(\omega) = \frac{1}{\sqrt{1+\omega^6}}$. The plots of magnitude responses of all considered filters are displayed in Fig. 5.8. We can observe that magnitude characteristics are always flat in the passband and that their slopes in the transition region become steeper for bigger values of n.

In filter design, we often have to determine the order n of the Butterworth filter with given values of ω_c and ω_s and the maximal attenuation δ_s in the stopband. We can write

$$A(\omega_s) = \delta_s = \frac{1}{\sqrt{1 + (\frac{\omega_s}{\omega_c})^{2n}}} thickmathspace \Leftrightarrow thickmathspace 1 + (\frac{\omega_s}{\omega_c})^{2n}$$

$$= \frac{1}{\delta_s^2} \Leftrightarrow n = \frac{1}{2} \frac{\log\left(\frac{1}{\delta_s^2} - 1\right)}{\log\left(\frac{\omega_s}{\omega_c}\right)}$$

For small values of δ_s, that is, $\delta_s \ll 1$, we get the following rule for the order n:

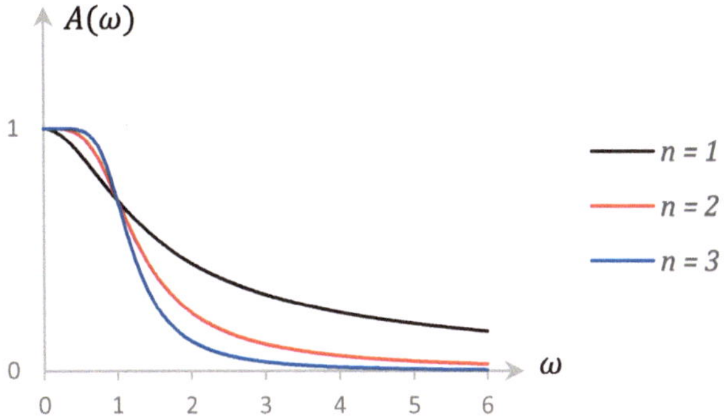

Fig. 5.8 Magnitude responses of Butterworth filters with order $n = 1, 2, 3$

$$n \approx \frac{\log\left(\frac{1}{\delta_s}\right)}{\log\left(\frac{\omega_s}{\omega_c}\right)} = \frac{-\log(\delta_s)}{\log\left(\frac{\omega_s}{\omega_c}\right)}. \tag{5.21}$$

Problem 5.7 Determine the order of the Butterworth filter with the 3 dB cutoff frequency $f_c = 0{,}5$ kHz, the limit frequency of the stopband $f_s = 2$ kHz and the maximal attenuation in the stopband $A_s = 50$ dB.

Solution From (5.16), we get

$$50 = -20\log(\delta_s) \Leftrightarrow -\log(\delta_s) = 2,5.$$

Then using (5.21), we can calculate the ratio

$$\frac{-\log(\delta_s)}{\log\left(\frac{\omega_s}{\omega_c}\right)} = \frac{2,5}{\log\left(\frac{2000}{500}\right)} \approx 4,152.$$

We choose the order $n = 5$ of the Butterworth filter that satisfies all conditions.

5.5.2 *Chebyshev Filters (Type I)*

The transfer function of the nth order Chebyshev filter (type I) is

Table 5.4 Chebyshev polynomials with order 1,. . .,5

Order n	$C_n(x)$
1	x
2	$2x^2 - 1$
3	$4x^3 - 3x$
4	$8x^4 - 8x^2 + 1$
5	$16x^5 - 20x^3 + 5x$

$$H(s) = \frac{H_0}{\prod_{k=1}^{n}\left(1 - \frac{s}{\lambda_k}\right)}, \tag{5.22}$$

where

$$H_0 = \begin{cases} 1 & \text{for } n \text{ odd} \\ \frac{1}{\sqrt{1+\varepsilon^2}} & \text{for } n \text{ even} \end{cases}$$

and λ_k, $k = 1, 2, \ldots, n$ are poles lying on an ellipse of semi-minor axis r and semi-major axis R, defined with the poles $\widehat{\lambda}_k$ of the transfer function of the Butterworth filter of the same order n given by

$$\lambda_k = r \cdot \operatorname{Re}\left(\widehat{\lambda}_k\right) + jR \cdot \operatorname{Im}\left(\widehat{\lambda}_k\right) \tag{5.23}$$

$$r = \sinh\left(\sinh^{-1}\left(\frac{1}{\varepsilon}\right)/n\right), R = \cosh\left(\sinh^{-1}\left(\frac{1}{\varepsilon}\right)/n\right) \tag{5.24}$$

and ε is a ripple factor.

The magnitude response of the nth order Chebyshev filter (type I) is defined by Chebyshev polynomials $C_n(\cdot)$ (see Table 5.4):

$$A(\omega) = \frac{1}{\sqrt{1 + \varepsilon^2 C_n^2\left(\frac{\omega}{\omega_p}\right)}} \tag{5.25}$$

Definition 5.7 The *Chebyshev polynomials* can be determined using the following recursive formula:

$$C_0(x) = 1, \quad C_1(x) = x, \quad C_n(x) = 2xC_{n-1}(x) - C_{n-2}(x) \tag{5.26}$$

or, equivalently, as

$$C_n(x) = \begin{cases} \cos(n\cos^{-1}x) & \text{for} \quad 0 \le x \le 1, \\ \cosh\left(n\cosh^{-1}x\right) & \text{for} \quad x > 1. \end{cases} \tag{5.27}$$

Problem 5.8 Determine the transfer functions of the 1st- and 2nd order Chebyshev filters (type I) with the passband cutoff angular frequency $\omega_p = 1$ rad/s. Calculate the poles and plot their diagram in the complex plane. Sketch the corresponding magnitude responses. Assume the ripple factor $\varepsilon = 0, 5$.

Solution Consider the Chebyshev filter (type I) with order $n = 2$. In Problem 5.6, we have already calculated the poles of the 2nd order Butterworth filter: $\widehat{\lambda}_1 = -\frac{\sqrt{2}}{2} + j\frac{\sqrt{2}}{2}, \widehat{\lambda}_2 = -\frac{\sqrt{2}}{2} - j\frac{\sqrt{2}}{2}$.

Using (5.24), we can calculate the parameters of an ellipse as follows:

$$\begin{aligned} r = & \quad \sinh\left(\sinh^{-1}(2)/2\right) \cong 0,7862, \\ R = & \quad \cosh\left(\sinh^{-1}(2)/2\right) \cong 1,272. \end{aligned}$$

Then from (5.23), the poles λ_1 and λ_2 of the transfer function are

$$\begin{aligned} \lambda_1 = & \quad 0,7862 \cdot \left(-\frac{\sqrt{2}}{2}\right) + j \cdot 1,272 \cdot \frac{\sqrt{2}}{2} \cong -0,56 + j \cdot 0,9, \\ \lambda_2 = & \quad 0,7862 \cdot \left(-\frac{\sqrt{2}}{2}\right) + j \cdot 1,272 \cdot \left(-\frac{\sqrt{2}}{2}\right) \cong -0,56 - j \cdot 0,9. \end{aligned}$$

Their plot is displayed in Fig. 5.9 (left).

The transfer function of the 2nd order Chebyshev filter (type I) is given by

$$\begin{aligned} H(s) = & \quad \frac{\frac{1}{\sqrt{1+\varepsilon^2}}}{\left(1 - \frac{s}{\lambda_1}\right)\left(1 - \frac{s}{\lambda_2}\right)} = \frac{1}{\sqrt{1+\varepsilon^2}} \frac{\lambda_1\lambda_2}{(\lambda_1 - s)(\lambda_2 - s)} \\ H(s) = & \quad \frac{1}{\sqrt{1+\varepsilon^2}} \frac{\lambda_1\lambda_2}{(s^2 - s(\lambda_1 + \lambda_2) + \lambda_1\lambda_2)}. \end{aligned}$$

To get the magnitude response, we can use directly (5.25) as follows:

$$A(\omega) = \frac{1}{\sqrt{1 + \frac{1}{4}C_2^2(\omega)}}, \quad \text{where} \quad C_2(\omega) = 2\omega^2 - 1.$$

The magnitude response of the 2nd order low-pass Chebyshev filter (type I) is

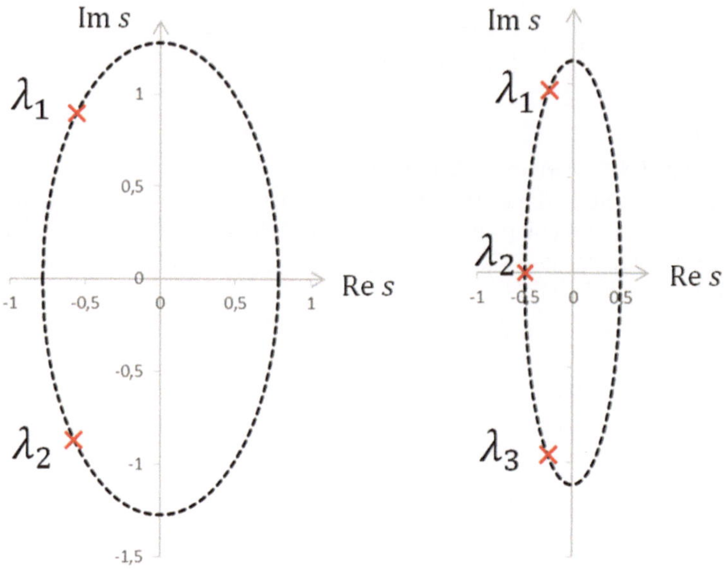

Fig. 5.9 Poles of the transfer functions of Chebyshev filters (type I) with order 2 (left) and 3 (right)

$$A(\omega) = \frac{1}{\sqrt{1 + \frac{1}{4}(4\omega^4 - 4\omega^2 + 1)}} = \frac{1}{\sqrt{\omega^4 - \omega^2 + \frac{5}{4}}}.$$

Now, let us consider the Chebyshev filter (type I) with order $n = 3$. Once again, we can apply (5.23). The poles of the transfer function of the 3rd order Butterworth filter calculated in Problem 5.6 are: $\widehat{\lambda}_1 = -\frac{1}{2} + j\frac{\sqrt{3}}{2}, \widehat{\lambda}_2 = -1, \widehat{\lambda}_3 = -\frac{1}{2} - j\frac{\sqrt{3}}{2}$. Thus, the parameters r and R of an ellipse are

$$\begin{aligned} r = &\quad \sinh\left(\sinh^{-1}(2)/3\right) = 0,5, \\ R = &\quad \cosh\left(\sinh^{-1}(2)/3\right) = 1,118. \end{aligned}$$

Then the corresponding poles $\lambda_k = r \cdot \mathrm{Re}\left(\widehat{\lambda}_k\right) + jR \cdot \mathrm{Im}\left(\widehat{\lambda}_k\right)$ of the Chebyshev filter transfer function are given by

$$\begin{aligned} \lambda_1 = &\quad 0,5 \cdot \left(-\frac{1}{2}\right) + j \cdot 1,118 \cdot \frac{\sqrt{3}}{2} \cong -0,25 + j \cdot 0,97, \lambda_2 \cong -0,5, \\ \lambda_3 = &\quad 0,5 \cdot \left(-\frac{1}{2}\right) + j \cdot 1,118 \cdot \left(-\frac{\sqrt{3}}{2}\right) \cong -0,25 - j \cdot 0,97. \end{aligned}$$

They are shown on a diagram in Fig. 5.9 (right).

The magnitude response of the 3rd order low-pass Chebyshev filter (type I) is

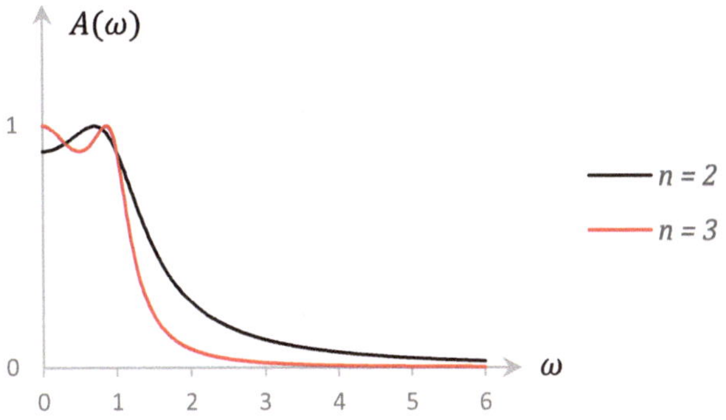

Fig. 5.10 Magnitude responses of the 1st and 2nd order Chebyshev filter (type I)

$$A(\omega) = \frac{1}{\sqrt{1 + \frac{1}{4} C_3^2(\omega)}}, \quad \text{where} \quad C_3(\omega) = 4\omega^3 - 3\omega.$$

Then

$$A(\omega) = \frac{1}{\sqrt{1 + \frac{1}{4}(16\omega^6 - 24\omega^4 + 9\omega^2)}} = \frac{1}{\sqrt{4\omega^6 - 6\omega^4 + \frac{9}{4}\omega^2 + 1}}.$$

In Fig. 5.10, we can observe plots of magnitude responses of the 2nd- and 3rd order Chebyshev filters (type I). We can notice characteristic ripples in the passband. Their amplitude depends on the value of the ripple factor ε and varies between $\frac{1}{\sqrt{1+\varepsilon^2}} \cong 0,89$ and 1. In the transition band, the slope becomes steeper with growing n.

The ripple factor ε is related to the minimal gain A_p in the passband by the formula

$$\varepsilon = \sqrt{10^{0,1A_p} - 1}. \tag{5.28}$$

Since $\frac{1}{\sqrt{1+\varepsilon^2}} = 1 - \delta_p$, then $1 + \varepsilon^2 = (1 - \delta_p)^{-2}$ and

$$\varepsilon = \sqrt{10^{-2\log(1-\delta_p)} - 1} \tag{5.29}$$

and $\log(1 - \delta_p) = -A_p/20$.(see (5.15)).

In the stopband, the maximal attenuation δ_s is

$$\delta_s = \frac{1}{\sqrt{1 + \varepsilon^2 C_n^2\left(\frac{\omega_s}{\omega_p}\right)}}. \tag{5.30}$$

Using (5.27), we get

$$1+\varepsilon^2\cosh^2\left(n\cosh^{-1}\left(\frac{\omega_s}{\omega_p}\right)\right)=\delta_s^{-2}.$$

Let us take a logarithm of both sides of the above equality as follows:

$$\log\left(1+\varepsilon^2\cosh^2\left(n\cosh^{-1}\left(\frac{\omega_s}{\omega_p}\right)\right)\right)=-2\cdot\log(\delta_s).$$

Then, since $\log(\delta_s)=-A_s/20$ (formula (5.16)), we obtain

$$A_s/10=\log\left(1+\varepsilon^2\cosh^2\left(n\cosh^{-1}\left(\frac{\omega_s}{\omega_p}\right)\right)\right).$$

After some calculations, we get the formula defining the order of the Chebyshev filter (type I) in the function of parameters A_s, ε, ω_s and ω_p as follows:

$$n\geq\frac{\cosh^{-1}\left(\sqrt{\frac{10^{0,1A_s}-1}{\varepsilon^2}}\right)}{\cosh^{-1}\left(\frac{\omega_s}{\omega_p}\right)}. \tag{5.31}$$

Problem 5.9 Determine the order n of the Chebyshev filter (type I) if: the cutoff frequency of the passband $f_c=25$ Hz, the minimal amplitude of ripples in the passband is 1 dB, the cutoff frequency of the stopband $f_s=40$ Hz and the maximal attenuation in the stopband $A_s=40$ dB.

Solution Evidently, 1 [dB]$=-20\log x\Leftrightarrow x=10^{-1/20}\cong 0,89125$.

Now, we can calculate the ripple factor ε from the formula $\frac{1}{\sqrt{1+\varepsilon^2}}=0,89125$ that gives $\varepsilon=0,5088$. At last, using (5.31), we can determine the order n as follows:

$$n\geq\frac{\cosh^{-1}\left(\sqrt{\frac{10^4-1}{0,5088^2}}\right)}{\cosh^{-1}\left(\frac{40}{25}\right)}\cong 5,705964$$

The filter n order meeting required assumptions is 6.

5.5.3 *Chebyshev Filters (Type II)*

The transfer function of the nth order Chebyshev filter (type II) is

$$H(s) = H_0 \frac{\prod_{k=1}^{n}\left(s - \frac{1}{\mu_k}\right)}{\prod_{k=1}^{n}\left(s - \frac{1}{\lambda_k}\right)}, \qquad H_0 = \prod_{k=1}^{n} \frac{\mu_k}{\lambda_k}, \tag{5.32}$$

where μ_k, $k = 1, 2, \ldots, n$ are defined by

$$\mu_k = j \sin \phi_k, \phi_k = \frac{\pi}{2} + \frac{\pi}{n}\left(\frac{3}{2} + k\right) \tag{5.33}$$

and λ_k, $k = 1, 2, \ldots, n$ are corresponding poles of the Chebyshev filter (type I) defined by (5.23) and (5.24).

The magnitude response of the nth order Chebyshev filter (type II) is

$$A(\omega) = \frac{1}{\sqrt{1 + \frac{1}{\varepsilon^2 C_n^2\left(\frac{\omega_s}{\omega}\right)}}}, \tag{5.34}$$

where $C_n(\cdot)$ are Chebyshev polynomials from Table 5.4 defined by (5.26) and (5.27) and ε is a ripple factor in the stopband.

In Problem 5.10, we will show how to apply the above formulas for the Chebyshev filter (type II) with order $n = 2$.

Problem 5.10 Assuming the ripple factor $\varepsilon = 0, 1$ and the edge angular frequency in the stopband $\omega_s = 1$ rad/s, determine the transfer function of the Chebyshev filter (type II) with order 2 and sketch its magnitude response.

Solution Using (5.33), we can calculate μ_1 and μ_2 as follows:

$$\mu_1 = j \sin\left[\frac{\pi}{2} + \frac{\pi}{2}\left(\frac{3}{2} + 1\right)\right] = -\frac{j}{\sqrt{2}}, \ \mu_2 = j \sin\left[\frac{\pi}{2} + \frac{\pi}{2}\left(\frac{3}{2} + 2\right)\right] = \frac{j}{\sqrt{2}}.$$

Using (5.23), the poles λ_1, λ_2 of the Chebyshev filter (type I) are

$$\lambda_1 = r \cdot \mathrm{Re}\left(\widehat{\lambda}_1\right) + jR \cdot \mathrm{Im}\left(\widehat{\lambda}_1\right), \ \lambda_2 = r \cdot \mathrm{Re}\left(\widehat{\lambda}_2\right) + jR \cdot \mathrm{Im}\left(\widehat{\lambda}_2\right)$$

where

$$r = \sinh\left(\sinh^{-1}(10)/2\right) \cong 2,127, R = \cosh\left(\sinh^{-1}(10)/2\right) \cong 2,351$$

and

$$\widehat{\lambda}_1 = -\frac{\sqrt{2}}{2} + j\frac{\sqrt{2}}{2},\ \widehat{\lambda}_2 = -\frac{\sqrt{2}}{2} - j\frac{\sqrt{2}}{2}$$

are the poles of the Butterworth filter transfer function calculated in Problem 5.6.
We have

$$\lambda_1 = 2{,}127 \cdot \left(-\frac{\sqrt{2}}{2}\right) + j \cdot 2{,}351 \cdot \left(\frac{\sqrt{2}}{2}\right) \cong -1{,}504 + j \cdot 1{,}662,$$

$$\lambda_2 = 2{,}127 \cdot \left(-\frac{\sqrt{2}}{2}\right) + j \cdot 2{,}351 \cdot \left(-\frac{\sqrt{2}}{2}\right) \cong -1{,}504 - j \cdot 1{,}662.$$

Using (5.32), let us compute the transfer function of the 2nd order Chebyshev filter (type II):

$$H(s) = \frac{\mu_1\mu_2}{\lambda_1\lambda_2} \cdot \frac{\left(s - \frac{1}{\mu_1}\right)\left(s - \frac{1}{\mu_2}\right)}{\left(s - \frac{1}{\lambda_1}\right)\left(s - \frac{1}{\lambda_2}\right)} = \frac{\mu_1\mu_2}{\lambda_1\lambda_2} \cdot \frac{s^2 - \left(\frac{1}{\mu_1} + \frac{1}{\mu_2}\right)s + \frac{1}{\mu_1}\frac{1}{\mu_2}}{s^2 - \left(\frac{1}{\lambda_1} + \frac{1}{\lambda_2}\right)s + \frac{1}{\lambda_1}\frac{1}{\lambda_2}}$$

$$H(s) = 0{,}01 \cdot \frac{s^2 + 2}{s^2 + 0{,}599s + 0{,}199}.$$

To calculate the magnitude response of the filter, we have two possibilities: from $H(s)$ via the frequency response or directly from (5.24). We choose the second method and get

$$A(\omega) = \frac{1}{\sqrt{1 + \frac{1}{0{,}01C_2^2\left(\frac{1}{\omega}\right)}}},$$

where $C_2\left(\frac{1}{\omega}\right) = 2\left(\frac{1}{\omega}\right)^2 - 1$ is the 2nd order Chebyshev polynomial from Table 5.4. Then

$$A(\omega) = \frac{1}{\sqrt{1 + \frac{1}{0{,}01\left(4\left(\frac{1}{\omega}\right)^4 - 4\left(\frac{1}{\omega}\right)^2 + 1\right)}}}.$$

The plot of $A(\omega)$ is shown in Fig. 5.11.

In Problem 5.11, we will determine and plot the magnitude responses of higher order Chebyshev filters (type II).

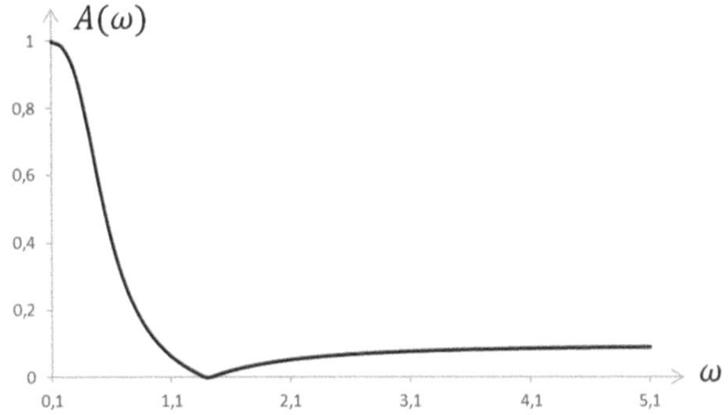

Fig. 5.11 Magnitude response of the Chebyshev filter (type II) with order $n = 2$

Problem 5.11 Assuming the ripple factor $\varepsilon = 0, 1$ and the edge angular frequency in the stopband $\omega_s = 1$ rad/s, determine and sketch magnitude responses of the Chebyshev filter (type II) with order $n = 5$ and 7.

Solution First, let us consider the filter with order $n = 5$. Its magnitude response is:

$$A(\omega) = \frac{1}{\sqrt{1 + \frac{1}{\varepsilon^2 C_5^2\left(\frac{\omega_S}{\omega}\right)}}},$$

where $C_5(x) = 16x^5 - 20x^3 + 5x$ is the 5th order Chebyshev polynomial from Table 5.4 and $x = \frac{\omega_s}{\omega}$. The magnitude response of the 5th order Chebyshev filter (type II) is

$$A(\omega) = \frac{1}{\sqrt{1 + \frac{1}{0,01\left[16\left(\frac{\omega_S}{\omega}\right)^5 - 20\left(\frac{\omega_S}{\omega}\right)^3 + 5\left(\frac{\omega_S}{\omega}\right)\right]^2}}}.$$

In order to determine the magnitude responses of the 7th order filter, we can apply Table 5.4 and the recurrence formula (5.26): $C_n(x) = 2xC_{n-1}(x) - C_{n-2}(x)$, getting

$$C_6(x) = 2x\left(16x^5 - 20x^3 + 5x\right) - \left(8x^4 - 8x^2 + 1\right) = 32x^6 - 48x^4 + 18x^2 - 1,$$
$$C_7(x) = 2x\left(32x^6 - 48x^4 + 18x^2 - 1\right) - \left(16x^5 - 20x^3 + 5x\right) = = 64x^7 - 112x^5 + 56x^3 - 7x.$$

Then the magnitude response of the 7th order Chebyshev filter (type II) is

$$A(\omega) = \frac{1}{\sqrt{1 + \frac{1}{\varepsilon^2 C_7^2\left(\frac{\omega_S}{\omega}\right)}}} = \frac{1}{\sqrt{1 + \frac{1}{\varepsilon^2\left[64\left(\frac{\omega_S}{\omega}\right)^7 - 112\left(\frac{\omega_S}{\omega}\right)^5 + 56\left(\frac{\omega_S}{\omega}\right)^3 - 7\left(\frac{\omega_S}{\omega}\right)\right]^2}}}.$$

If we substitute filter parameters in the above formula, we obtain

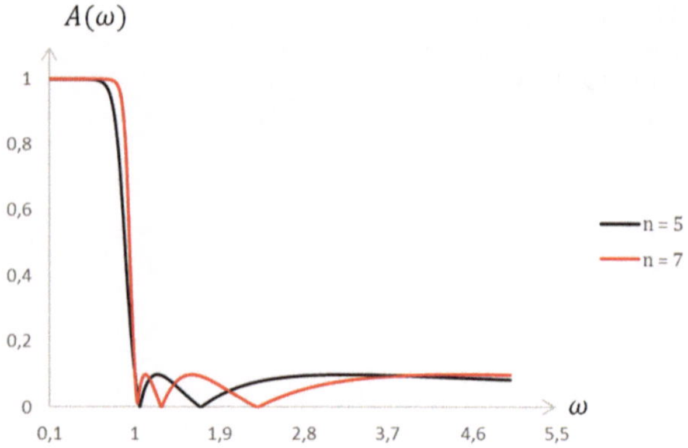

Fig. 5.12 Magnitude response of the Chebyshev filter (type II) with order $n = 5$ and 7

$$A(\omega) = \frac{1}{\sqrt{1 + \frac{1}{0,01\left[64\left(\frac{1}{\omega}\right)^7 - 112\left(\frac{1}{\omega}\right)^5 + 56\left(\frac{1}{\omega}\right)^3 - 7\left(\frac{1}{\omega}\right)\right]^2}}}.$$

The magnitude responses of the Chebyshev filter (type II) with order $n = 5$ and 7 are shown in Fig. 5.12. We can see that for filters of higher order, there are characteristic ripples in the stopband. It has not been observed for the 2nd order filter (see Fig. 5.11). In the passband, the magnitude characteristic is maximally flat.

The ripple factor ε is related to the maximal attenuation A_s in the stopband by the formula

$$\varepsilon = \frac{1}{\sqrt{10^{0,1A_s} - 1}}. \tag{5.35}$$

In the passband, the attenuation for the 3 dB cutoff angular frequency ω_c is

$$1 - \delta_p = \frac{1}{\sqrt{1 + \frac{1}{\varepsilon^2 C_n^2\left(\frac{\omega_s}{\omega_c}\right)}}} \Leftrightarrow 1 + \frac{1}{\varepsilon^2 C_n^2\left(\frac{\omega_s}{\omega_c}\right)} = \left(1 - \delta_p\right)^{-2}.$$

Taking the logarithm of both sides of the above equality results in

$$\log\left(1 + \frac{1}{\varepsilon^2 C_n^2\left(\frac{\omega_s}{\omega_c}\right)}\right) = -2\log\left(1 - \delta_p\right) = 0,1A_p$$

and further, since $C_n\left(\frac{\omega_s}{\omega_c}\right) = \cosh\left(n\cosh^{-1}\left(\frac{\omega_s}{\omega_c}\right)\right)$, we have

$$\cosh\left(n\cosh^{-1}\left(\frac{\omega_s}{\omega_c}\right)\right)=\frac{1}{\varepsilon\sqrt{10^{0,1A_p}-1}}.$$

The order n of the Chebyshev filter (type II) can be calculated as

$$n\geq\frac{\cosh^{-1}\left(\frac{1}{\varepsilon\sqrt{10^{0,1A_p}-1}}\right)}{\cosh^{-1}\left(\frac{\omega_s}{\omega_c}\right)}\tag{5.36}$$

or from (5.35) as

$$n\geq\frac{\cosh^{-1}\left(\sqrt{\frac{10^{0,1A_s}-1}{10^{0,1A_p}-1}}\right)}{\cosh^{-1}\left(\frac{\omega_s}{\omega_c}\right)}.\tag{5.37}$$

Problem 5.12 Determine the order n of the Chebyshev filter (type II) if: the 3 dB cutoff frequency of the passband $f_c = 50$ Hz, the edge frequency of the stopband $f_s = 100$ Hz and the maximal attenuation in the stopband $A_s = 50$ dB.

Solution First, using (5.35), we can calculate the ripple factor

$$\varepsilon=\frac{1}{\sqrt{10^{0,1\cdot 50}-1}}\cong 0,0032.$$

Since $A_p = 3$ dB, substituting the known filter parameters into (5.37) we have

$$n\geq\frac{\cosh^{-1}\left(\sqrt{\frac{10^5-1}{10^{0,3}-1}}\right)}{\cosh^{-1}(2)}\cong 4,8992$$

The filter order n meeting all specifications is 5.

Exercises

5.1. Given the input-output equation of an LTI analog system:

$$\frac{d^2y(t)}{dt^2} + 3\frac{dy(t)}{dt} + 2y(t) = x(t)$$

excited with $x(t) = \mathbf{1}(t - 1)$. Assuming initial conditions: $y(0^+) = 0$, $y'(0^+) = 1$, compute its zero-input and zero-state responses: $y_{0i}(t)$ and $y_{0s}(t)$. Sketch the total system response.

5.2. A LTI continuous-time system is defined by the following input-output equation:

$$\frac{d^2y(t)}{dt^2} - y(t) = x(t), \quad y(0^+) = -1, y'(0^+) = 2.$$

Compute the system response $y(t)$ to the input signal $x(t) = [\sin(t) + 3\cos(2t)] \cdot \mathbf{1}(t)$.

5.3. Given a LTI analog system defined by the input-output relation

$$x(t) = \left(e^{-t} + e^{-3t}\right) \cdot \mathbf{1}(t) \rightarrow y(t) = 2\left(e^{-t} - e^{-4t}\right) \cdot \mathbf{1}(t).$$

Compute the transfer function $H(s)$ of the system and its impulse response $h(t)$. Determine the differential equation of the system and draw its block diagram.

5.4. Given a LTI analog system defined by the input-output relation

$$x(t) = e^{-2t} \cdot \mathbf{1}(t) \rightarrow y(t) = (2t - 1)\exp(-2t + 1) \cdot \mathbf{1}\left(t - \frac{1}{2}\right).$$

Compute the transfer function $H(s)$ of the system and its impulse response $h(t)$. Determine the system response to the input $x(t) = \cos(2t)$, $t \in \mathbb{R}$.

5.5. Given an LTI analog system with the block diagram shown in Fig. 5.13. Determine the input-output differential equation of the system. Compute its transfer function $H(s)$ and the impulse response $h(t)$. Determine the system response to the following input signals: (a) $x(t) = e^{-t} \cdot \mathbf{1}(t)$; (b) $x(t) = \cos(t)$, $t \in \mathbb{R}$.

5.6. Given the transfer function of a LTI analog filter:

$$H(s) = \frac{s^2 + 4}{(s + 2)^2}.$$

Determine the input-output differential equation of the filter and propose its block diagram. Compute and sketch the magnitude response of the filter. Compute the response to the input signal $x(t) = \cos(2t) \cdot \mathbf{1}(t)$.

5.7. Given the transfer function $H(s) = \frac{1}{(s+1)(s+2)}$ of an analog LTI system S that can be realized as: (a) a cascade interconnection; (b) a parallel interconnection; (c) a feedback interconnection of two subsystems S_1 and S_2. Propose three

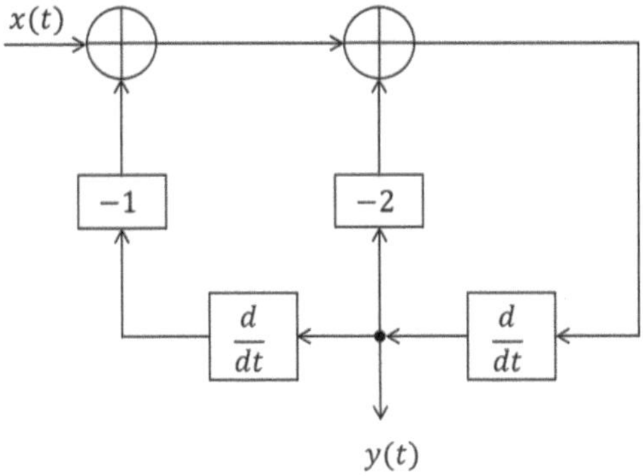

Fig. 5.13 Block diagram of the system from Exercise 5.5

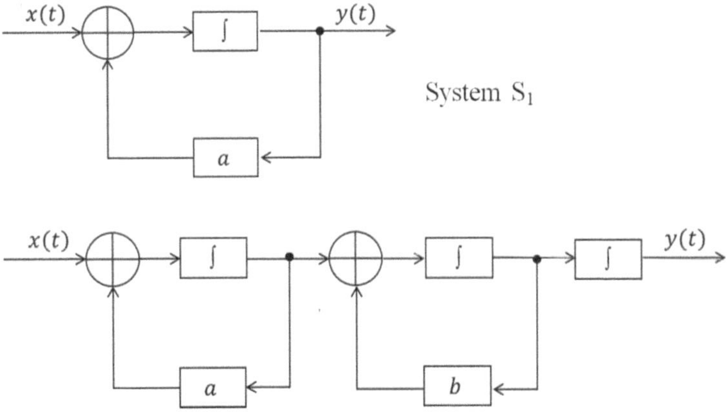

Fig. 5.14 Block diagrams of systems from Exercise 5.8

different block diagrams corresponding to interconnections (a), (b) and (c). What are the transfer functions $H_1(s)$ and $H_2(s)$ of subsystems S_1 and S_2 in each case?

5.8. (a) Given an analog system S_1 shown in Fig. 5.14. Determine its transfer function $H_1(s)$ and sketch its magnitude response. Examine the stability of the system depending on a value of the parameter $a \in \mathbb{R}$.

Figure 5.14 shows a cascade interconnection of two subsystems with a similar structure as S_1 from (a) and of an integrating block, $a, b \in \mathbb{R}$. Compute the transfer function $H(s)$ of the whole system. Calculate its zeros and poles. Examine the stability of the cascade interconnection depending on values of parameters $a, b \in \mathbb{R}$.

5.9. Given a feedback system shown in Fig. 5.15. $H_1(s)$ and $H_2(s)$ are, respectively, transfer functions of subsystems S_1 and S_2. The input-output relations of subsystems S_1 and S_2, respectively, are:

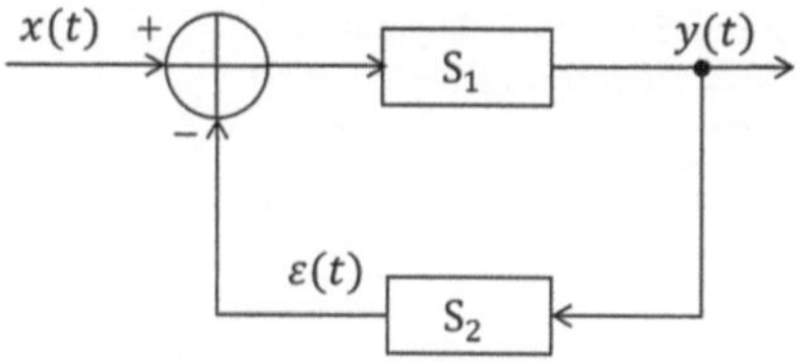

Fig. 5.15 Block diagram of the feedback system from Exercise 5.9

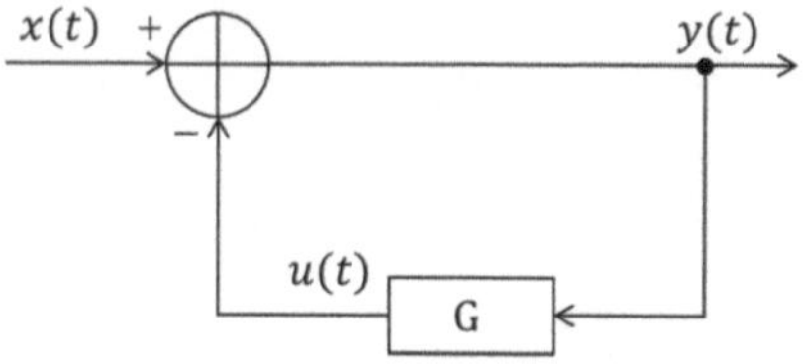

Fig. 5.16 Block diagram of the feedback system from Exercise 5.10

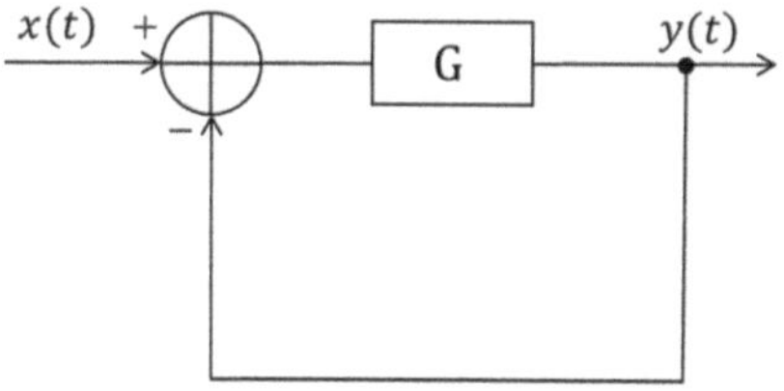

Fig. 5.17 Block diagram of the feedback system from Exercise 5.11

$$y(t) = \int_0^t [x(\tau) - \varepsilon(\tau)]d\tau, \quad \varepsilon(t) = \int_0^t b \cdot e^{-2(t-\tau)} y(\tau)d\tau.$$

Compute the transfer functions $H_1(s)$, $H_2(s)$ and $H(s) = Y(s)/X(s)$. For which value of the parameter b, the impulse response $h(t)$ of the whole system is $h(t) = (1 + t)e^{-t} \cdot \mathbf{1}(t)$?

5.10. The block G in the feedback loop of the system shown in Fig. 5.16 is described by the input-output equation: $\frac{du(t)}{dt} + u(t) = y(t)$. Determine the input-output equation of the whole system. Compute the system response to the following input signals: (a) $x(t) = 1$, $t \in \mathbb{R}$; (b) $x(t) = \mathbf{1}(t)$, $t \in \mathbb{R}$.

5.11. Given a feedback system shown in Fig. 5.17. The step response of the whole system is $k(t) = \frac{1}{2}\delta(t) - \frac{1}{4}e^{-\frac{1}{2}t} \cdot \mathbf{1}(t)$. Compute the transfer function of the whole system and the transfer function $G(s)$ of the open-loop system G. Determine the input-output differential equation of the whole system. What is the system response to the input signal $x(t) = \Pi\left(t - \frac{1}{2}\right)$?

5.12. Given a feedback system S with the block diagram shown in Fig. 5.15. The transfer functions of blocks S_1 and S_2, respectively, are: $H_1(s) = \frac{s}{(s+1)(s+a)}$ and $H_2(s) = \frac{b}{s}$, where $a, b \in \mathbb{R}$. Determine the transfer function $H(s)$ of the whole system. Mark the region of stability of the system S on the (a, b)–plane.

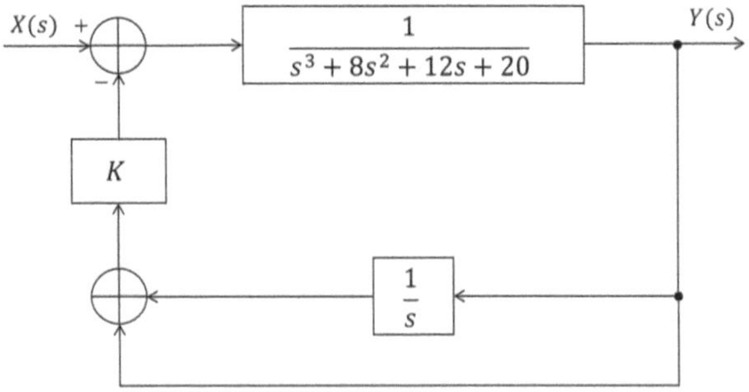

Fig. 5.18 Block diagram of the system from Exercise 5.13

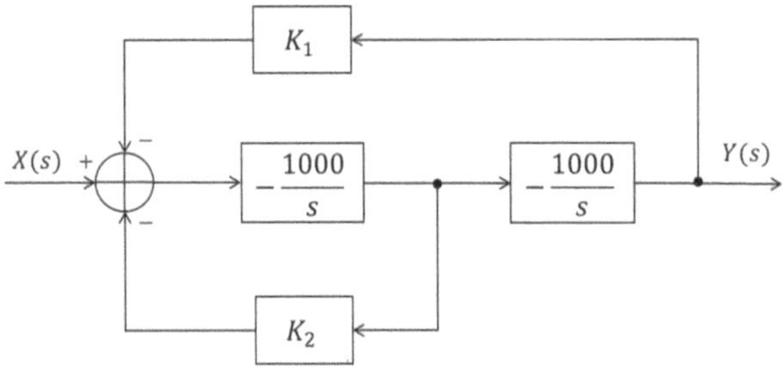

Fig. 5.19 Block diagram of the system from Exercise 5.14

5.13. Given an analog system with the block diagram shown in Fig. 5.18. For which values of $K > 0$ is the system stable?

5.14. Given an analog system with the block diagram shown in Fig. 5.19. Mark the region of stability on the (K_1, K_2)–plane. Then assuming $K_1 = 2$, $K_2 = -3$, compute the step response of the whole system.

5.15. Using the Hurwitz criterion and the Routh-Hurwitz criterion, study stability of two different systems with the following transfer functions:

(a) $H(s) = (s + 2)/(s^4 + 7s^3 + 12s^2 + 23s + 10)$;
(b) $H(s) = (s^2 + 2)/(s^5 + 2s^4 + s^3 + 2s^2 + 23s + 1)$.

Chapter 6
Sampling and Frequency Representation of Discrete-Time Signals

In this chapter, we present the theory of signal sampling, the operation permitting to transform a continuous-time signal into a discrete-time signal in the form of an infinite set of signal samples. We will introduce the notion of the *sampling frequency* and formulate the *Nyquist-Shannon sampling* theorem. Focused on a periodic spectrum of the sampled signal, we will show the consequences of the *aliasing effect,* which occurs if the sampling frequency is too small, that is, smaller than the *Nyquist frequency,* defined as twice the maximal frequency of an analog signal. Normalization of the angular frequency in a spectrum of the sampled signal will allow us to define the Discrete-time Fourier Transform. On the other hand, we will show that the Discrete Fourier Transform is a result of discretization of the frequency domain.

6.1 Ideal Sampling

The *ideal sampling* is based on a multiplication of the analog signal $x(t)$, $t \in \mathbb{R}$ by the comb distribution $\delta_{T_s}(t), t \in \mathbb{R}$, called the *sampling signal* (in the form of a periodic sequence of Dirac delta pulses with period T_s (see Chap. 1).

Definition 6.1 The signal $\delta_{T_s}(t)$ is called the *sampling signal* and its fundamental period T_s - the *sampling period* or the *sampling interval*. The inverse of T_s, that is, $f_s = 1/T_s$ is the *sampling frequency*, while $\omega_s = 2\pi f_s$ is the *sampling angular frequency*.

If we multiply $x(t)$ by $\delta_{T_s}(t)$, we can get $x_s(t)$ called the *sampled signal*, that is,

$$x_s(t) = x(t) \cdot \delta_{T_s}(t) = x(t) \cdot \sum_{n \in \mathbb{Z}} \delta(t - nT_s) = \sum_{n \in \mathbb{Z}} x(t)\delta(t - nT_s).$$

K. M. Snopek, *Hands-on Signals and Systems Theory*,
https://doi.org/10.1007/978-3-031-56081-1_6

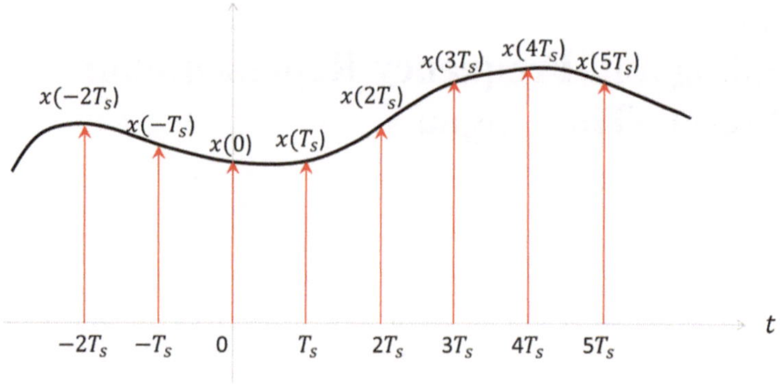

Fig. 6.1 Uniformly sampled signal in the form of a set of Dirac delta pulses

The sampling property of the Dirac delta (see Table 1.1) lets us express $x_s(t)$ in the following form:

$$x_s(t) = \sum\nolimits_{n\in\mathbb{Z}} x(nT_s)\delta(t - nT_s), \tag{6.1}$$

where $x(nT_s)$, $n \in \mathbb{Z}$ are called the *samples* of $x(t)$. Notice that the ideal sampling is an *uniform sampling* since the sampling period T_s is constant. Fig. 6.1 shows a sampled signal (marked in red) represented by Dirac pulses at $t = nT_s$, $n \in \mathbb{Z}$.

6.1.1 Spectrum of the Ideally Sampled Signal

The sampled signal $x_s(t) = x(t) \cdot \delta_{T_s}(t)$ is a product of the analog signal $x(t)$ and the comb distribution $\delta_{T_s}(t)$. As the FT of a product of two signals is a convolution of their spectra, that is, $x_1(t) \cdot x_2(t) \overset{\mathcal{F}}{\leftrightarrow} \frac{1}{2\pi}[X_1(\omega) * X_2(\omega)]$ (Table 2.2) and $\delta_{T_s}(t) \overset{\mathcal{F}}{\leftrightarrow} \frac{2\pi}{T_s}\sum\limits_{n\in\mathbb{Z}} \delta\left(\omega - n\frac{2\pi}{T_s}\right) = \frac{2\pi}{T_s}\sum\nolimits_{n\in\mathbb{Z}} \delta(\omega - n\omega_s)$ (see Problem 2.15), we can express $X_s(\omega)$ as a convolution:

$$X_s(\omega) = \frac{1}{2\pi}\left[X(\omega) * \frac{2\pi}{T_s}\sum\nolimits_{n\in\mathbb{Z}} \delta(\omega - n\omega_s)\right].$$

Using distributiveness over the addition of a convolution (Table 1.4), we get $X_s(\omega) = \frac{1}{T_s}\sum\nolimits_{n\in\mathbb{Z}} X(\omega) * \delta(\omega - n\omega_s)$ and, finally,

$$\text{(ideal sampling) } X_s(\omega) = \frac{1}{T_s} \sum_{n \in \mathbb{Z}} X(\omega - n\omega_s). \tag{6.2}$$

Important
The sampled signal spectrum is periodic in the domain ω and consists of copies of $X(\omega)$ repeating periodically with ω_s and scaled in amplitude by $1/T_s$.

In the domain f, spectrum of the sampled signal is given by

$$\text{(ideal sampling) } X_s(f) = f_s \sum_{n \in \mathbb{Z}} X(f - nf_s). \tag{6.3}$$

6.1.2 Nyquist-Shannon Theorem

The Nyquist-Shannon sampling theorem gives an answer to the question of what value of ω_s (or f_s) should be chosen to ensure a perfect recovering of an analog signal $x(t)$ from the set of its samples $x(nT_s)$. Let us formulate this important theorem in the domain f.

Theorem 6.1 If the signal $x(t)$ is low-pass, that is, it exists the edge frequency $f_{\max}$ such that the spectrum $X(f) = 0$ for $|f| > f_{\max}$, then the samples $x(nT_s)$ taken with frequency $f_s > 2f_{\max}$ completely determine the signal $x(t)$.

Definition 6.2 The frequency $2f_{\max}$ is called the *Nyquist rate* denoted with f_N and the inequality $f_s > 2f_{\max}$ is known as the *Nyquist criterion*.

Important If the Nyquist criterion is fulfilled, we are able to recover the analog signal $x(t)$ from its samples $x(nT_s)$ by use of an ideal low-pass filter with gain $1/f_s$ and cutoff frequency $f_{\max} \leq f_c < f_s - f_{\max}$.

The next two problems will show two different situations: when the Nyquist criterion is fulfilled and not. In the second case, when the sampling frequency f_s is smaller than the Nyquist rate f_N (*undersampling*), we will observe overlapping of spectrum components known as the *aliasing effect*.

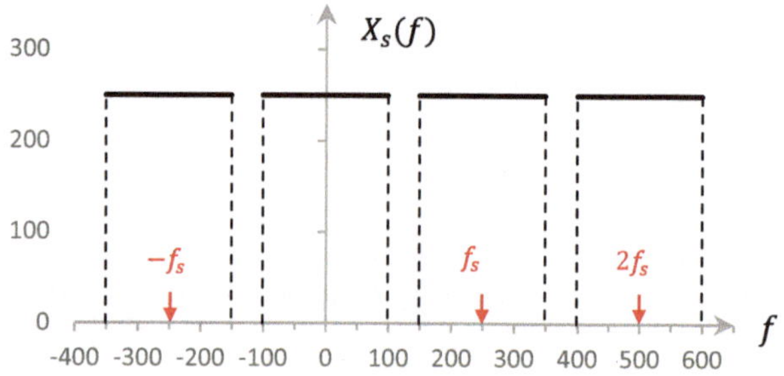

Fig. 6.2 Spectrum of the sampled signal from Problem 6.1, $f_s = 250$ Hz

Problem 6.1 The analog signal $x(t) = 200 \cdot \mathrm{Sa}(200\pi t)$ is ideally sampled with frequency: (a) $f_s = 250$ Hz; (b) $f_s = 180$ Hz and then filtered with an ideal LP filter with gain $1/f_s$ and cutoff frequency $f_c = 100$ Hz. Determine and sketch the spectrum of the sampled signal $x_s(t)$, and compute the filter response $y(t)$ for two sampling scenarios (a) and (b).

Solution The first step is to determine f_{max} (see Theorem 6.1) of the signal $x(t) = 200\mathrm{Sa}(200\pi t)$. Its spectrum $X(f) = \Pi\left(\frac{f}{200}\right)$; thus, $f_{\mathrm{max}} = 100$ Hz. The Nyquist rate $f_{\mathrm{N}} = 200$ Hz.

(a) Consider the case: $f_s = 250$ Hz, that is, the Nyquist criterion is fulfilled, since $f_s > f_{\mathrm{N}}$.

Using (6.3), we can compute the spectrum $X_s(f)$ of the sampled signal in the form

$$X_s(f) = 250 \sum_{n \in \mathbb{Z}} \Pi\left(\frac{f - 250n}{200}\right).$$

We can see that it consists of copies of $X(f) = \Pi\left(\frac{f}{200}\right)$ repeating every 250 Hz and scaled in amplitude by 250 as shown in Fig. 6.2 (only copies for $n = -1, 0, 1,$ and 2 are shown). There is no aliasing effect.

Now, we are passing the sampled signal through the ideal LP filter with gain 1/250 and cutoff frequency $f_c = 100$ Hz. Using (4.1) from Chap. 4, we can compute the spectrum of the filer response $y(t)$ as $Y(f) = X_s(f)H(f)$, where $H(f)$ is the frequency response of the LP filter given by

$$H(f) = \begin{cases} 1/250, & |f| \leq 100 \\ 0 & \text{elsewhere.} \end{cases}$$

As a result, we get $Y(f) = \Pi\left(\frac{f}{200}\right)$. Using Table 2.1, we obtain the output signal in the following form: $y(t) = 200\mathrm{Sa}(200\pi t) \equiv x(t)$. In this case, we can notice that the analog signal $x(t)$ has been completely recovered.

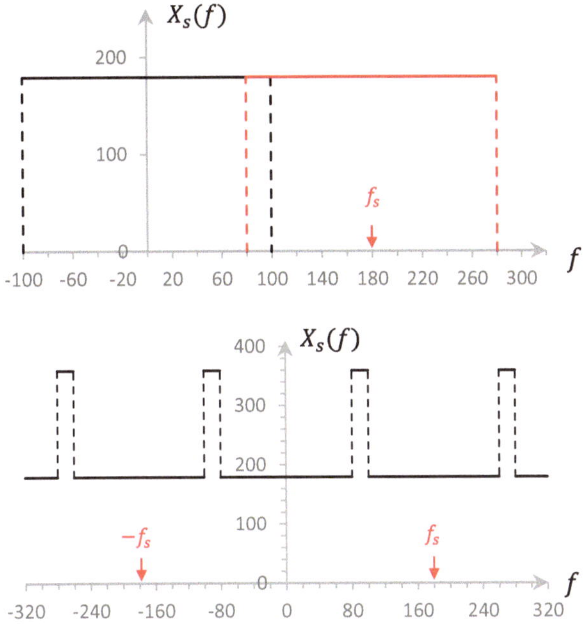

Fig. 6.3 Spectrum of the sampled signal from Problem 6.1, $f_s = 180$ Hz

(b) Consider now the case $f_s = 180$ Hz, in which the Nyquist criterion is not fulfilled, since $f_s < f_N$.

The spectrum of the sampled signal, that is, $X_s(f) = 180\sum_{n\in\mathbb{Z}}\Pi\left(\frac{f-180n}{200}\right)$, consists of copies of $X(f) = \Pi\left(\frac{f}{200}\right)$ repeating every 180 Hz and scaled in amplitude by 180. Fig. 6.3 (up) shows two copies $180X(f)$ and $180X(f - 180)$ for $n = 0$ and 1. We can notice that they overlap for $f \in [80, 100]$ (Hz). Similar situations will be observed in all frequency bands: $f \in [80 + 180n, 100 + 180n]$, $n = 1, 2, \ldots$ and $f \in [100 - 180n, 80 - 180n,]$, $n = 1, 2, \ldots$ In consequence, all copies of $X(f)$ in overlapping intervals are added up, as shown in Fig. 6.3 (down). The deformation of the spectrum in overlapping intervals due to undersampling is called the *aliasing effect*.

The sampled signal with the spectrum $X_s(f)$ shown in Fig. 6.3 is then filtering by the LP filter with the frequency response

$$H(f) = \begin{cases} 1/180, & |f| \le 100 \\ 0 & \text{elsewhere.} \end{cases}$$

The spectrum $Y(f)$ of the output signal $y(t)$ is shown in Fig. 6.4 (up). We can decompose $Y(f)$ into two parts $Y_1(f)$ and $Y_2(f)$ (this is not the only possibility!) (see Fig. 6.4 (down)). Then, using linearity of the FT, we can compute $y(t)$ as the sum of $y_1(t)$ and $y_2(t)$ (resp. inverse FTs of $Y_1(f)$ and $Y_2(f)$). We have

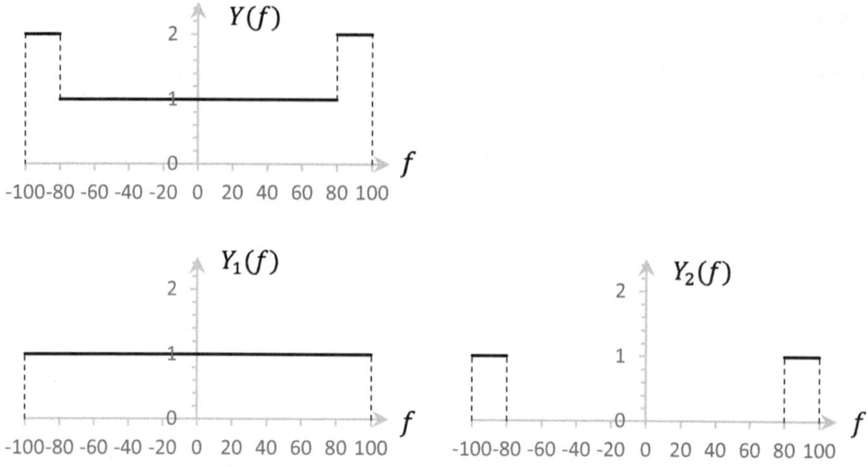

Fig. 6.4 Spectrum $Y(f)$ of the filter response from Problem 6.1

$$Y(f) = \underbrace{\Pi\left(\frac{f}{200}\right)}_{Y_1(f)\equiv X(f)} + \underbrace{\Pi\left(\frac{f-90}{20}\right) + \Pi\left(\frac{f+90}{20}\right)}_{Y_2(f)},$$

from where it follows that

$$y(t) = \underbrace{200 \cdot \mathrm{Sa}(200\pi t)}_{x(t)} + 40 \cdot \mathrm{Sa}(20\pi t)\cos(180\pi t),$$

where the second component is a consequence of the aliasing effect.

Problem 6.2 The analog signal $x(t) = 2 + 2\cos(2\pi t) - \cos(4\pi t)$ is ideally sampled with frequency: (a) $f_s = 5$ Hz; (b) $f_s = 3$ Hz and then filtered by an ideal LP filter with gain $1/f_s$ and cutoff frequency $f_c = 2$ Hz. Determine and sketch the spectrum of the sampled signal $x_s(t)$ and compute the filter response $y(t)$ for two sampling scenarios (a) and (b).

Solution Let us calculate the Nyquist rate. The spectrum of $x(t)$ is given by (see Table 2.1):

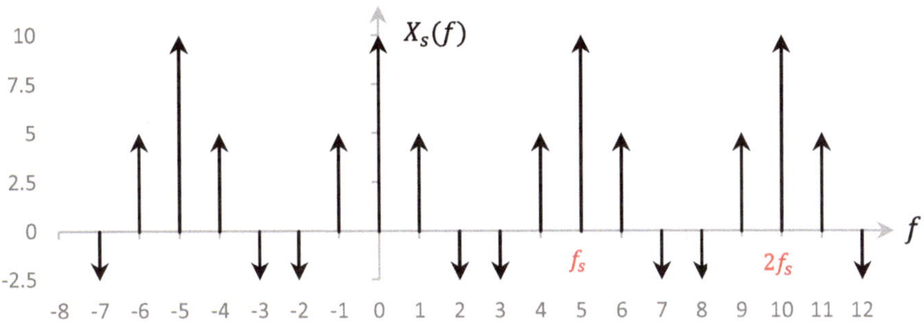

Fig. 6.5 Spectrum of the sampled signal from Problem 6.2; $f_s = 5$ Hz

$$X(f) = 2\delta(f) + \delta(f-1) + \delta(f+1) - \frac{1}{2}\delta(f-2) - \frac{1}{2}\delta(f+2).$$

Thus, $f_{\max} = 2$ Hz and the Nyquist rate $f_N = 4$ Hz.

(a) Let us consider the case $f_s = 5$ Hz, that is, $f_s > f_N$. The spectrum of the sampled signal (Fig. 6.5) is given by

$$\begin{aligned} X_s(f) &= 5\sum\nolimits_{n\in\mathbb{Z}} X(f-5n) \\ &= 5\sum\nolimits_{n\in\mathbb{Z}} \left[2\delta(f-5n) + \delta(f-1-5n) + \delta(f+1-5n) - \frac{1}{2}\delta(f-2-5n) - \frac{1}{2}\delta(f+2-5n)\right]. \end{aligned}$$

We can see that there is no aliasing.

If we pass the sampled signal $x_s(t)$ through the ideal LP filter with the frequency response

$$H(f) = \begin{cases} 1/5, & |f| \leq 2 \\ 0 & \text{elsewhere}, \end{cases}$$

at the output we obtain the signal $y(t) \equiv x(t)$, that is, the reconstruction is perfect.

(b) Let us now consider the case $f_s = 3$ Hz, that is, $f_s < f_N$. The Nyquist criterion is not fulfilled, and we expect to observe an aliasing effect. The spectrum of the sampled signal is

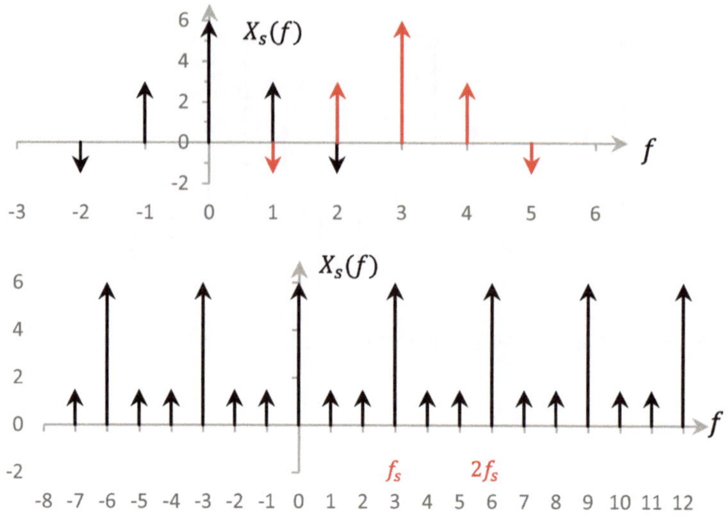

Fig. 6.6 Spectrum of the sampled signal from Problem 6.2; $f_s = 3$ Hz

$$
\begin{aligned}
X_s(f) &= 3\sum_{n\in\mathbb{Z}} X(f-3n) \\
&= 3\sum_{n\in\mathbb{Z}} \left[2\delta(f-3n) + \delta(f-1-3n) + \delta(f+1-3n) - \frac{1}{2}\delta(f-2-3n) - \frac{1}{2}\delta(f+2-3n)\right].
\end{aligned}
$$

Let us sketch two copies: $3X(f)$ and $3X(f-3)$, as shown in Fig. 6.6 (up). We can observe overlapping of peaks at frequencies 1 and 2 [Hz]. These components add up and as a result, we get two delta peaks of "height": 3 – 1, 5 = 1, 5. Identical overlaps will be observed for spectrum components at frequencies $1 \pm 3n$ [Hz] and $2 \pm 3n$ [Hz], and the resulting spectrum $X_s(f)$ is shown in Fig. 6.6 (down)

If we pass the sampled signal through the ideal LP filter with the frequency response

$$
H(f) = \begin{cases} 1/3, & |f| \leq 2 \\ 0 & \text{elsewhere,} \end{cases}
$$

at the output we can get the signal $y(t)$ with the spectrum given by

$$Y(f) = X_s(f)H(f) = 2\delta(f) + \frac{1}{2}\delta(f-1) + \frac{1}{2}\delta(f+1) + \frac{1}{2}\delta(f-2) + \frac{1}{2}\delta(f+2).$$

The output signal is $y(t) = 2 + \cos(2\pi t) + \cos(4\pi t)$.

6.1.3 The Stroboscopic Effect

The *stroboscopic effect* is a very interesting phenomenon observed during the sampling of periodic signals with fundamental frequency f_0. If the sampling frequency f_s is close to f_0 (in consequence, the Nyquist criterion is not fulfilled), the spectrum components overlap in such a way that the resulting spectrum corresponds to a periodic signal with a lower frequency. Problem 6.3 will illustrate this effect.

Problem 6.3 The analog signal $x(t) = 2\cos(2\pi t) + \cos(4\pi t) + 0,5\cos(6\pi t)$ is ideally sampled with frequency $f_s = 0,9$ Hz and then filtered with an ideal LP filter with gain $1/f_s$ and cutoff frequency $f_c = 0,5$ Hz. Sketch the spectrum of the sampled signal $x_s(t)$ for $f \in [-1,3; 1,3]$ [Hz], and compute the filter response $y(t)$. Sketch plots of signals $x(t)$ and $y(t)$.

Solution The fundamental frequency of $x(t)$ is $f_0 = 1$ Hz, and we can see that the sampling frequency is close to f_0. The spectrum of $x(t)$ is

$$X(f) = \delta(f-1) + \delta(f+1) + \frac{1}{2}\delta(f-2) + \frac{1}{2}\delta(f+2) + \frac{1}{4}\delta(f-3) + \frac{1}{4}\delta(f+3).$$

We know that the spectrum of the sampled signal is a periodic repetition of $X(f)$ with "period" $f_s = 0,9$ Hz, that is,

$$X_s(f) = 0,9\sum_{n\in\mathbb{Z}} X(f - 0,9n).$$

In the table below, we present the components of $X_s(f)$ for $n = -4, \ldots, 4$ indicating frequency and amplitude of Dirac delta peaks. The "grey" cells correspond to peaks in the frequency band $f \in [-1,3; 1,3]$ Hz. For example, at the frequency $-0,6$ [Hz], we have a peak with amplitude 0,25 as a result of moving $\delta(f-3)$ to the left by $4f_s = 3,6$ [Hz], and so on. A row corresponding to $X(f)$ is marked in red.

Peak amplitudes $X_s(f)/f_s$		0,25 ↓	0,5 ↓	1 ↓	1 ↓	0,5 ↓	0,25 ↓
n	$X(f-0{,}9n)$	Peak frequencies [Hz]					
−4	$X(f+3{,}6)$	−6,6	−5,6	−4,6	−2,6	−1,6	−0,6
−3	$X(f+2{,}7)$	−5,7	−4,7	−3,7	−1,7	−0,7	0,3
−2	$X(f+1{,}8)$	−4,8	−3,8	−2,6	−0,8	0,2	1,2
−1	$X(f+0{,}9)$	−3,9	−2,9	−1,9	0,1	1,1	2,1
0	$X(f)$	−3	−2	−1	1	2	3
1	$X(f-0{,}9)$	−2,1	−1,1	−0,1	1,9	2,9	3,9
2	$X(f-1{,}8)$	−1,2	−0,2	0,8	2,8	3,8	4,8
3	$X(f-2{,}7)$	−0,3	0,7	1,7	3,7	4,7	5,7
4	$X(f-3{,}6)$	0,6	1,6	2,6	4,6	5,6	6,6

Basing on the above table, let us sketch the plot of $X_s(f)/f_s$ for $f \in [-1,3;\, 1,3]$ Hz, as shown in Fig. 6.7. We can observe that the "shape" of the spectrum is the same as $X(f)$ but all peaks are more concentrated. In other words, we deal with *scaling in the frequency domain.*

Let us now determine the signal $y(t)$ at the output of the filter with the frequency response

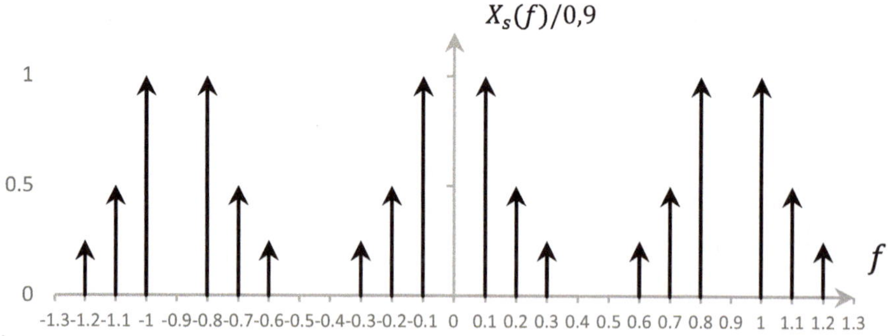

Fig. 6.7 Spectrum of the sampled signal from Problem 6.3; $f_s = 0,9$ Hz

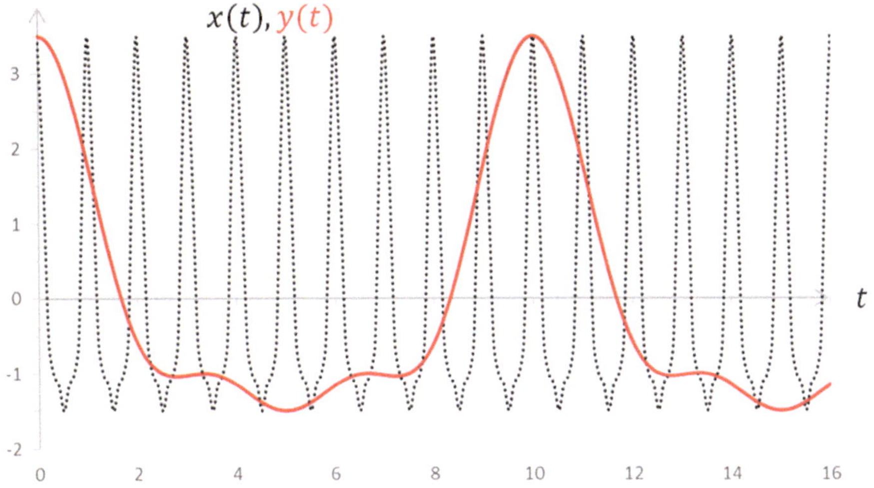

Fig. 6.8 Signals $x(t)$ and $y(t)$ from Problem 6.3

$$H(f) = \begin{cases} 10/9, |f| \leq 0,5 \\ 0 \text{ elsewhere.} \end{cases}$$

The spectrum $Y(f) = X_s(f)H(f)$ is given by

$$Y(f) = \delta(f - 0,1) + \delta(f + 0,1) + \frac{1}{2}\delta(f - 0,2) + \frac{1}{2}\delta(f + 0,2) + \frac{1}{4}\delta(f - 0,3) + \frac{1}{4}\delta(f + 0,3).$$

Using the inverse FT, we obtain

$$y(t) = 2\cos(0,2\pi t) + \cos(0,4\pi t) + \frac{1}{2}\cos(0,6\pi t).$$

Both signals $x(t)$ (dotted black line) and $y(t)$ (red line) are shown in Fig. 6.8. We can observe that $y(t)$ is a periodic signal of the same shape as $x(t)$ but its fundamental frequency is 10 times smaller than f_0. In other words, we deal with *scaling in the time domain.*

6.2 Flat–Top Sampling

The *flat–top sampling* is a technique applied in sample-and-hold circuits [21]. It is based on the multiplication of the analog signal $x(t)$, $t \in \mathbb{R}$ by the unit-amplitude rectangle pulse train $p(t)$ with period T_s (right shifted by $\tau/2$) and on holding the value $x(nT_s)$ for the time τ (equal the duration time of a single rectangular pulse), called the *aperture time*. The signal $p(t)$ plays a role of a sampling sequence. The result of the flat–top sampling is a periodic sequence of rectangular pulses with amplitudes defined by $x(nT_s)$, as shown in Fig. 6.9.

Important The flat–top sampling differs from the *natural sampling*, which is based on the multiplication of the analog signal $x(t)$ by the unit-amplitude rectangle pulse train $p(t) = \sum_{n\in\mathbb{Z}} \Pi\left(\frac{t-nT_s}{\tau}\right)$ with period T_s. The product is the modulated rectangle pulse train $\tilde{x}(t)$, composed of "slices" of $x(t)$ repeating with T_s having the following spectrum:

$$\text{(natural sampling) } \tilde{X}_s(\omega) = \frac{\tau}{T_s}\sum_{n\in\mathbb{Z}} \mathrm{Sa}\left(\frac{n\omega_s\tau}{2}\right) X(\omega - n\omega_s). \tag{6.4}$$

6.2.1 Spectrum of the Flat–Top Sampled Signal

Let us denote with $\widehat{x}(t)$ the flat–top sampled signal from Fig. 6.9. We can write it as

$$\widehat{x}(t) = \sum_{n\in\mathbb{Z}} x(nT_s) z(t - nT_s), \tag{6.5}$$

where $z(t) = \Pi\left(\frac{t - \frac{\tau}{2}}{\tau}\right)$ and $x(nT_s)$, $n \in \mathbb{Z}$ are samples of $x(t)$, $t \in \mathbb{R}$. As

$$z(t - nT_s) = z(t) * \delta(t - nT_s)$$

(convolution property of Dirac delta from Table 4.1), we can rewrite (6.5) in the following form:

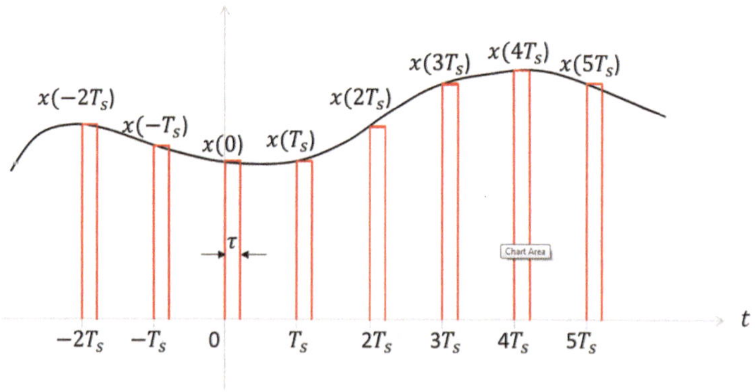

Fig. 6.9 Flat–top sampling

$$\widehat{x}(t) = \sum_{n\in\mathbb{Z}} x(nT_s)(z(t) * \delta(t - nT_s)) \tag{6.6}$$

and then, using distributiveness over the addition of a convolution, we have

$$\widehat{x}(t) = z(t) * \sum_{n\in\mathbb{Z}} x(nT_s)\delta(t - nT_s). \tag{6.7}$$

In (6.7), we can recognize the ideally sampled signal $x_s(t)$ from (6.1), thus $\widehat{x}(t) = z(t) * x_s(t)$. Since the FT of a convolution of two signals is a product of their spectra and $x_s(t) \overset{\mathcal{F}}{\leftrightarrow} \frac{1}{T_s}\sum_{n\in\mathbb{Z}} X(\omega - n\omega_s)$, $z(t) \overset{\mathcal{F}}{\leftrightarrow} Z(\omega) = \tau \mathrm{Sa}\left(\frac{\omega\tau}{2}\right)e^{-j\omega\frac{\tau}{2}}$, we can get the spectrum $\widehat{X}(\omega)$ of the flat–top sampled signal in the form

$$\text{(flat–top sampling) } \widehat{X}(\omega) = \frac{\tau}{T_s}\mathrm{Sa}\left(\frac{\omega\tau}{2}\right)e^{-j\omega\frac{\tau}{2}}\sum_{n\in\mathbb{Z}} X(\omega - n\omega_s). \tag{6.8}$$

We can notice that the spectrum (6.8) consists of copies $X(\omega - n\omega_s)$ repeating periodically with ω_s (similarly as for ideal sampling) and, additionally, attenuated with $\mathrm{Sa}\left(\frac{\omega\tau}{2}\right)$. Since the sampling signal has been right-shifted by $\tau/2$, in (6.8) appears the factor $e^{-j\omega\frac{\tau}{2}}$.

Now, if $\tau = T_s$, $\widehat{x}(t)$ gets the form of a *staircase signal* with the spectrum

$$\widehat{X}(\omega) = \mathrm{Sa}\left(\frac{\omega T_s}{2}\right)e^{-j\omega\frac{T_s}{2}}\sum_{n\in\mathbb{Z}} X(\omega - n\omega_s) \tag{6.9}$$

expressed in the f–domain as

$$\widehat{X}(f) = \mathrm{Sa}(\pi f T_s)e^{-j\pi f T_s}\sum_{n\in\mathbb{Z}} X(f - nf_s). \tag{6.10}$$

Problem 6.4 illustrates the flat–top sampling with $\tau = T_s$.

Problem 6.4 The analog signal $x(t) = \cos(2\pi t)$ is flat–top sampled with frequency $f_s = 10$ Hz. Assuming the time-duration of a rectangular pulse $\tau = T_s$, plot the staircase signal $\widehat{x}(t)$. Compute and sketch the amplitude spectrum of the signal $y(t) = \widehat{x}\left(t + \frac{T_s}{2}\right)$.

Solution The fundamental period of $x(t)$ is $T_0 = 1$ [s]. Let us calculate the samples $x(nT_s)$, $n \in \{0, 1, \ldots, 9\}$, $T_s = 0, 1$ [s] within one period of $x(t)$:

$x(0) = 1$; $x(0, 1) = \cos(0, 2\pi) \cong 0, 809$; $x(0, 2) = \cos(0, 4\pi) \cong 0, 309$;
$x(0, 3) = \cos(0, 6\pi) \cong -0, 309$; $x(0, 4) = \cos(0, 8\pi) \cong -0, 809$; $x(0, 5) = -1$;
$x(0, 6) = \cos(1, 2\pi) \cong -0, 809$; $x(0, 7) = \cos(1, 4\pi) \cong -0, 309$;
$x(0, 8) = \cos(1, 6\pi) \cong 0, 309$; $x(0, 9) = \cos(1, 8\pi) \cong 0, 809$.

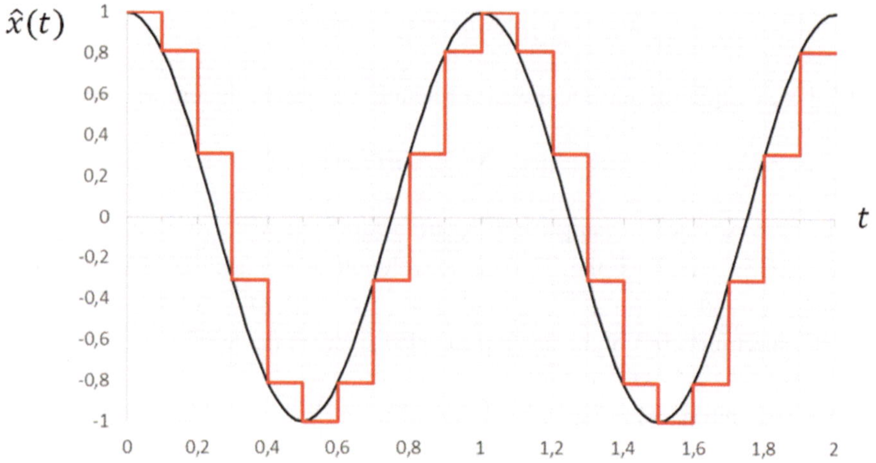

Fig. 6.10 Flat–top sampled signal $\widehat{x}(t)$ from Problem 6.4, $f_s = 10$ Hz

For $\tau = T_s$, the flat–top sampled signal $\widehat{x}(t)$ has the form of a staircase "cosine" signal shown in Fig. 6.10 (in red). For comparison, we also show the graph of $x(t) = \cos(2\pi t)$.

As we want to determine the spectrum $Y(f)$ of $y(t) = \widehat{x}\left(t + \frac{T_s}{2}\right)$, we should apply (6.10) in which we omit the factor $e^{-j\pi f T_s}$, that is,

$$Y(f) = \mathrm{Sa}(\pi f T_s) \sum_{n \in \mathbb{Z}} X(f - n f_s) = \mathrm{Sa}(0,1\pi f) \sum_{n \in \mathbb{Z}} X(f - 10n),$$

where $X(f) = \frac{1}{2}\delta(f-1) + \frac{1}{2}\delta(f+1)$ (spectrum of $x(t)$). We have

$$\begin{aligned} Y(f) = & \quad \frac{1}{2}\mathrm{Sa}(0,1\pi f) \sum_{n \in \mathbb{Z}} (\delta(f - 1 - 10n) + \delta(f + 1 - 10n)) \\ = & \quad \frac{1}{2} \sum_{n \in \mathbb{Z}} \mathrm{Sa}(0,1\pi f)(\delta(f - 1 - 10n) + \delta(f + 1 - 10n)) \\ = & \quad \frac{1}{2} \sum_{n \in \mathbb{Z}} \mathrm{Sa}(0,1\pi f)\delta(f - 1 - 10n) + \frac{1}{2} \sum_{n \in \mathbb{Z}} \mathrm{Sa}(0,1\pi f)\delta(f + 1 - 10n). \end{aligned}$$

Further, using the sampling property of the Dirac delta, we get

$$Y(f) = \frac{1}{2} \sum_{n \in \mathbb{Z}} \mathrm{Sa}(\pi(n + 0,1))\delta(f - 1 - 10n) + \frac{1}{2} \sum_{n \in \mathbb{Z}} \mathrm{Sa}(\pi(n - 0,1))\delta(f + 1 - 10n).$$

We can see that the spectrum $Y(f)$ consists of pairs of delta peaks repeating periodically with 10 Hz and attenuated by $\mathrm{Sa}(\cdot)$. Values of amplitudes and

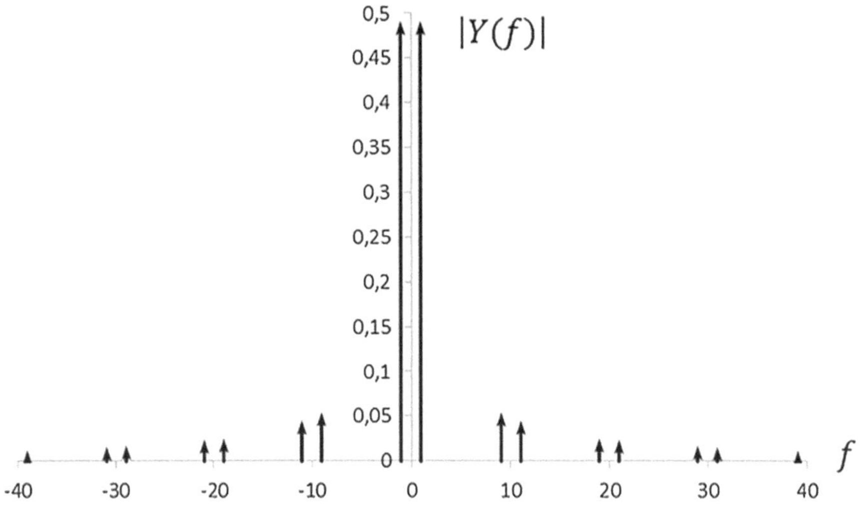

Fig. 6.11 Amplitude spectrum of the signal $y(t)$ from Problem 6.4

frequencies of delta peaks of $Y(f)$ are presented in the table below for $n \in -5, \ldots, 5$. Figure 6.11 shows the amplitude spectrum $|Y(f)|$ plotted based on values from this table.

n	$\frac{1}{2}\mathrm{Sa}(\pi(n+0{,}1))$	f [Hz]	$\frac{1}{2}\mathrm{Sa}(\pi(n-0{,}1))$	f [Hz]
−5	0,010	−49	−0,010	−51
−4	−0,013	−39	0,012	−41
−3	0,017	−29	−0,016	−31
−2	−0,026	−19	0,023	−21
−1	0,055	−9	−0,045	−11
0	0,492	1	0,492	−1
1	−0,045	11	0,055	9
2	0,023	21	−0,026	19
3	−0,016	31	0,017	29
4	0,012	41	−0,013	39
5	−0,010	51	0,010	49

6.3 Discrete-Time Fourier Transform

Let us come back to the formula (6.1) defining the ideally sampled signal. Our task is computing its Fourier transform, that is, $X_s(\omega) = \mathcal{F}\left\{\sum_{n\in\mathbb{Z}} x(nT_s)\delta(t - nT_s)\right\}$. Using the linearity of the FT, we can express $X_s(\omega)$ in the equivalent form to (6.2) as

$$X_s(\omega) = \sum_{n\in\mathbb{Z}} x(nT_s)\mathcal{F}\{\delta(t - nT_s)\} = \sum_{n\in\mathbb{Z}} x(nT_s)e^{-j\omega nT_s}. \tag{6.11}$$

The formula (6.11) is *the complex (exponential) Fourier series* (in the domain ω) of the periodic function $X_s(\omega)$ (with fundamental period ω_s). The coefficients $x(nT_s)$ of the series (6.11) are given by

$$x(nT_s) = \frac{1}{\omega_s}\int_{-\omega_s/2}^{\omega_s/2} X_s(\omega)e^{j\omega nT_s}d\omega. \tag{6.12}$$

Formulas (6.11) and (6.12) provide the basis for the definition of the Discrete-Time Fourier Transform (DTFT) of a discrete-time signal that is also widely used in frequency analysis of discrete-time systems, as it will be shown in Chap. 7. Its definition is based on the *normalization* of the variable ω as follows.

Definition 6.3 The *normalized angular frequency* Ω is defined as $\Omega = \omega T_s$ expressed in radians. The *normalized frequency* F is given by $\mathrm{F} = fT_s = f/f_s$ (no units).

Now, let us substitute $\Omega = \omega T_s$ into (6.11) and (6.12) getting

$$X_s(\omega) = \sum_{n\in\mathbb{Z}} x(nT_s)e^{-jn\Omega} \tag{6.13}$$

Ans, as $d\omega = \frac{1}{T_s}d\Omega$ and integration limits change from $(-\omega_s/2, \omega_s/2)$ to $(-\omega_sT_s/2, \omega_sT_s/2) \equiv (-\pi, \pi)$, we obtain

$$x(nT_s) = \frac{1}{\omega_sT_s}\int_{-\pi}^{\pi} X_s(\omega)e^{jn\Omega}d\Omega = \frac{1}{2\pi}\int_{-\pi}^{\pi} X_s(\omega)e^{jn\Omega}d\Omega. \tag{6.14}$$

Normalization of ω means that the spectrum of the sampled signal depends no more on sampling parameters T_s, ω_s or f_s. Therefore, we can simply denote $X_s(\omega)$ with $X(\Omega)$ and $x(nT_s)$ with $x(n)$.

Definition 6.4 The *Discrete-Time Fourier Transform* (DTFT) of a discrete-time signal $x(n)$, $n \in \mathbb{Z}$ is defined in the domain Ω as

$$X(\Omega) = \sum_{n\in\mathbb{Z}} x(n)e^{-jn\Omega} \tag{6.15}$$

or, in the domain F, as

$$X(\mathrm{F})=\sum\nolimits_{n\in\mathbb{Z}}x(n)e^{-j2\pi n\mathrm{F}}. \tag{6.16}$$

The DTFT $X(\Omega)$, called the *spectrum* of $x(n)$, is a periodic function of variable Ω with period 2π. As a consequence, the spectrum $X(\mathrm{F})$ is periodic with period 1.

Definition 6.5 The *inverse Discrete-Time Fourier Transform* is defined as

$$x(n)=\frac{1}{2\pi}\int_{-\pi}^{\pi}X(\Omega)e^{jn\Omega}d\Omega \tag{6.17}$$

or, equivalently, by change of variables: $\Omega=2\pi\mathrm{F}$, $d\Omega=2\pi d\mathrm{F}$, as

$$x(n)=\int_{-1/2}^{1/2}X(\mathrm{F})e^{j2\pi n\mathrm{F}}d\mathrm{F}. \tag{6.18}$$

A pair of Discrete-Time Fourier transforms will be denoted as $x(n)\overset{\mathcal{F}_d}{\leftrightarrow}X(\Omega)$. In Table 6.1, we present some chosen DTFT pairs, and the properties of the DTFT are collected in Table 6.2.

Problems 6.5, 6.6 and 6.7 illustrate the practical usage of Tables 6.1 and 6.2.

Problem 6.5 Given the spectrum $X(\Omega)=2\cos\Omega+\cos 2\Omega$ of a discrete-time signal $x(n)$. Determine $x(n)$.

Solution First, using Euler's equality $\cos x=\frac{1}{2}e^{jx}+\frac{1}{2}e^{-jx}$, let us rewrite $X(\Omega)$ as

Table 6.1 DTFT pairs

Signal $x(n)$, $n\in\mathbb{Z}$	$X(\Omega)$, $\Omega\in\mathbb{R}$
$\delta(n)$	1
$\delta(n-n_0)$	$e^{-j\Omega n_0}$
$\Pi_N(n)$	$\frac{\sin\left[\Omega\left(N+\frac{1}{2}\right)\right]}{\sin\left(\frac{\Omega}{2}\right)}$
$\frac{\sin(\Omega_0 n)}{\pi n}$, $\lvert\Omega_0\rvert<\pi$	$\begin{cases}1, \lvert\Omega\rvert\le\Omega_0\\ 0, \Omega_0<\lvert\Omega\rvert\le\pi\end{cases}$
$a^n\cdot\mathbf{1}(n)$, $\lvert a\rvert<1$	$\frac{1}{1-ae^{-j\Omega}}$
$(n+1)a^n\cdot\mathbf{1}(n)$, $\lvert a\rvert<1$	$\frac{1}{(1-ae^{-j\Omega})^2}$
$a^{\lvert n\rvert}$, $\lvert a\rvert<1$	$\frac{1-a^2}{1-2a\cos\Omega+a^2}$
1	$2\pi\delta(\Omega)$
$e^{j\Omega_0 n}$, $\lvert\Omega_0\rvert<\pi$	$2\pi\delta(\Omega-\Omega_0)$
$\cos(\Omega_0 n)$, $\lvert\Omega_0\rvert<\pi$	$\pi\delta(\Omega-\Omega_0)+\pi\delta(\Omega+\Omega_0)$
$\sin(\Omega_0 n)$, $\lvert\Omega_0\rvert<\pi$	$-j\pi\delta(\Omega+\Omega_0)+j\pi\delta(\Omega-\Omega_0)+$
$\mathbf{1}(n)$	$\frac{1}{1-e^{-j\Omega}}+\pi\delta(\Omega)$

Table 6.2 Properties of DTFT

Property	Signal $x(n)$, $n \in \mathbb{Z}$	$X(\Omega)$, $\Omega \in \mathbb{R}$
Linearity	$\alpha x_1(n) + \beta x_2(n)$	$\alpha X_1(\Omega) + \beta X_2(\Omega)$
Time reversal	$x(-n)$	$X(-\Omega)$
Complex conjugate	$x^*(n)$	$X^*(-\Omega)$
Time-shift	$x(n - n_0)$	$X(\Omega)e^{-j\Omega n}$
Frequency-shift	$x(n)e^{j\Omega_0 n}$, $\vert\Omega_0\vert < \pi$	$X(\Omega - \Omega_0)$
Modulation	$x(n) \cos(\Omega_0 n)$, $\vert\Omega_0\vert < \pi$	$\frac{1}{2}X(\Omega+\Omega_0)+\frac{1}{2}X(\Omega-\Omega_0)$
	$x(n) \sin(\Omega_0 n)$, $\vert\Omega_0\vert < \pi$	$-\frac{1}{2j}X(\Omega+\Omega_0)+\frac{1}{2j}X(\Omega-\Omega_0)$
Derivation in frequency	$nx(n)$	$j\frac{dX(\Omega)}{d\Omega}$
Accumulation	$\sum\limits_{l=-\infty}^{n} x(l)$	$\frac{1}{1-e^{-j\Omega}}X(\Omega)+\pi X(0)\delta(\Omega)$
Time-convolution	$x_1(n) * x_2(n)$	$X_1(\Omega) \cdot X_2(\Omega)$
Multiplication	$x_1(n) \cdot x_2(n)$	$\frac{1}{2\pi}\int\limits_{-\pi}^{\pi} X_1(\lambda)X_2(\Omega-\lambda)d\lambda$
Plancherel's equality (for energy signals)	$\sum\limits_{n\in\mathbb{Z}} \vert x(n)\vert^2 = \frac{1}{2\pi}\int\limits_{-\pi}^{\pi} \vert X(\Omega)\vert^2 d\Omega = \int\limits_{-1/2}^{1/2} \vert X(F)\vert^2 dF$	

$$X(\Omega) = e^{j\Omega} + e^{-j\Omega} + \frac{1}{2}e^{j2\Omega} + \frac{1}{2}e^{-j2\Omega}.$$

Using the linearity of DTFT and the time-shift theorem from Table 2.2, we can get the signal $x(n)$ in the form

$$x(n) = \delta(n+1) + \delta(n-1) + \frac{1}{2}\delta(n+2) + \frac{1}{2}\delta(n-2),$$

where $\delta(n) \overset{\mathcal{F}_d}{\leftrightarrow} 1$. At last, using (1.26), we can calculate the energy of $x(n)$ as follows:

$$E_x = 2 \cdot 1 + 2 \cdot \frac{1}{4} = 2,5.$$

Problem 6.6 Compute the convolution $x_1(n) * x_2(n)$ of two signals $x_1(n)$ and $x_2(n)$, where: (a) $x_1(n) = \delta(n) + 2\delta(n-1) + 3\delta(n-2)$ and $x_2(n) = \delta(n+1) + \delta(n-1)$; (b) $x_1(n) = \left(\frac{1}{2}\right)^n \cdot \mathbf{1}(n)$ and $x_2(n) = \left(\frac{1}{3}\right)^n \cdot \mathbf{1}(n)$.

Solution (a) The convolution of $x_1(n) = \delta(n) + 2\delta(n-1) + 3\delta(n-2)$ and $x_2(n) = \delta(n+1) + \delta(n-1)$ has already been calculated in Problem 1.11 using the graphical procedure. Here, we can apply an alternative method basing on the convolution property of DTFT. At first, we have to compute the spectra $X_1(\Omega)$ and $X_2(\Omega)$ of $x_1(n)$ and $x_2(n)$ as follows:

$$X_1(\Omega)=1+2e^{-j\Omega}+3e^{-j2\Omega},\quad X_2(\Omega)=e^{j\Omega}+e^{-j\Omega}.$$

Then we can calculate their product, that is,

$$X_1(\Omega)\cdot X_2(\Omega)=\left(1+2e^{-j\Omega}+3e^{-j2\Omega}\right)\left(e^{j\Omega}+e^{-j\Omega}\right)=e^{j\Omega}+2+4e^{-j\Omega}+2e^{-j2\Omega}+3e^{-j3\Omega}.$$

The inverse DTFT of $X_1(\Omega)\cdot X_2(\Omega)$ gives the convolution $x_1(n)*x_2(n)$ in the form

$$x_1(n)*x_2(n)=\delta(n+1)+2\delta(n)+4\delta(n-1)+2\delta(n-2)+3\delta(n-3).$$

We can notice that we got exactly the same result as in Problem 1.11. The plot of $x_1(n)*x_2(n)$ has been shown in Fig. 1.43.

(b) Using Table 6.1, we get the spectra $X_1(\Omega)$ and $X_2(\Omega)$ of $x_1(n)=\left(\frac{1}{2}\right)^n\cdot\mathbf{1}(n)$ and $x_2(n)=\left(\frac{1}{3}\right)^n\cdot\mathbf{1}(n)$. We have

$$X_1(\Omega)=\frac{1}{1-\frac{1}{2}e^{-j\Omega}},X_2(\Omega)=\frac{1}{1-\frac{1}{3}e^{-j\Omega}}.$$

Their product $X_1(\Omega)\cdot X_2(\Omega)$ is

$$X_1(\Omega)\cdot X_2(\Omega)=\frac{1}{\left(1-\frac{1}{2}e^{-j\Omega}\right)\left(1-\frac{1}{3}e^{-j\Omega}\right)}.$$

To compute the inverse DTFT of the product $X_1(\Omega)\cdot X_2(\Omega)$, we first need to apply the following partial fraction decomposition:

$$\frac{1}{\left(1-\frac{1}{2}e^{-j\Omega}\right)\left(1-\frac{1}{3}e^{-j\Omega}\right)}=\frac{A}{1-\frac{1}{2}e^{-j\Omega}}+\frac{B}{1-\frac{1}{3}e^{-j\Omega}},$$

$$\frac{1}{\left(1-\frac{1}{2}e^{-j\Omega}\right)\left(1-\frac{1}{3}e^{-j\Omega}\right)}=\frac{A\left(1-\frac{1}{3}e^{-j\Omega}\right)+B\left(1-\frac{1}{2}e^{-j\Omega}\right)}{\left(1-\frac{1}{2}e^{-j\Omega}\right)\left(1-\frac{1}{3}e^{-j\Omega}\right)}.$$

We have to solve the set of two equations: $A+B=1$; $-\frac{1}{3}A-\frac{1}{2}B=0$. Its solution is $A=3$, $B=-2$. Thus, we have

$$X_1(\Omega)\cdot X_2(\Omega)=\frac{3}{1-\frac{1}{2}e^{-j\Omega}}-\frac{2}{1-\frac{1}{3}e^{-j\Omega}}.$$

The inverse DTFT of the above product gives the final result in the form

$$x_1(n)*x_2(n)=\left[3\left(\frac{1}{2}\right)^n-2\left(\frac{1}{3}\right)^n\right]\cdot\mathbf{1}(n).$$

Problem 6.7 The DTFT of a discrete-time signal $y(n)$ is

$$Y(\Omega)=\frac{1}{1-e^{-j\Omega}}\frac{\sin(\frac{5}{2}\Omega)}{\sin(\frac{\Omega}{2})}+3\pi\delta(\Omega).$$

Compute $y(n)$.

Solution Let us notice that, according to the accumulation property from Table 6.2,

$$\sum_{l=-\infty}^{n}x(l)\overset{\mathcal{F}_d}{\leftrightarrow}\frac{1}{1-e^{-j\Omega}}X(\Omega)+\pi X(0)\delta(\Omega),$$

we can write

$$X(\Omega)=\frac{\sin\left(\frac{5}{2}\Omega\right)}{\sin\left(\frac{\Omega}{2}\right)}=\frac{\frac{5}{2}\Omega\,\mathrm{Sa}\left(\frac{5}{2}\Omega\right)}{\frac{\Omega}{2}\,\mathrm{Sa}\left(\frac{\Omega}{2}\right)}=5\frac{\mathrm{Sa}\left(\frac{5}{2}\Omega\right)}{\mathrm{Sa}\left(\frac{5}{2}\Omega\right)}.$$

Since $X(0)=5$, then $Y(\Omega)$ should be rewritten as the difference of two DTFTs $Y_1(\Omega)$ and $Y_2(\Omega)$ as follows:

$$Y(\Omega)=\underbrace{\frac{1}{1-e^{-j\Omega}}\frac{\sin(\frac{5}{2}\Omega)}{\sin(\frac{\Omega}{2})}+5\pi\delta(\Omega)}_{Y_1(\Omega)}-\underbrace{2\pi\delta(\Omega)}_{Y_2(\Omega)}.$$

Using Table 6.1, we get the following DTFT pairs:

$$\Pi_2(n)\overset{\mathcal{F}_d}{\leftrightarrow}\frac{\sin\left(\frac{5}{2}\Omega\right)}{\sin\left(\frac{\Omega}{2}\right)}\text{ and }1\overset{\mathcal{F}_d}{\leftrightarrow}2\pi\delta(\Omega).$$

Then, the signal $y(n)$ equals the difference $y(n)=y_1(n)-y_2(n)$ of two inverse DTFTs of $Y_1(\Omega)$ and $Y_2(\Omega)$. We get

$$y_1(n) = \sum_{l=-\infty}^{n} \Pi_2(l) \text{ and } y_2(n) = 1.$$

Using (1.8), we have

$$y_1(n) = \begin{cases} 0 \text{ for } n \leq -1, \\ n+3 \text{ for } -2 \leq n \leq 2, \\ 5 \text{ for } n \geq 3. \end{cases}$$

So finally,

$$y(n) = \begin{cases} -1 \text{ for } n \leq -1, \\ n+2 \text{ for } -2 \leq n \leq 2, \\ 4 \text{ for } n \geq 3. \end{cases}$$

6.3.1 DTFT of Discrete-Time Periodic Signals

Let $x(n)$, $n \in \mathbb{Z}$ be a discrete-time periodic signal with fundamental period N_0. The Discrete-time Fourier Transform $X(\Omega)$ of $x(n)$ is given by

$$X(\Omega) = 2\pi \sum_{k \in \mathbb{Z}} c_k \delta\left(\Omega - 2\pi \frac{k}{N_0}\right) \tag{6.19}$$

or in the F-domain

$$X(\mathrm{F}) = \sum_{k \in \mathbb{Z}} c_k \delta\left(\mathrm{F} - \frac{k}{N_0}\right) \tag{6.20}$$

with coefficients c_k defined by a sum

$$c_k = \frac{1}{N_0} \sum\nolimits_{n=0}^{N_0-1} x(n) e^{-j2\pi \frac{kn}{N_0}}. \tag{6.21}$$

Important
We can notice that (6.21) is periodic with period N_0, that is, $c_k = c_{k+N_0}$. It means that a discrete-time periodic signal has a periodic spectrum $X(\Omega)$ (with

(continued)

the period N_0) that is discrete in the Ω–domain (respectively X(F) is periodic with N_0 in the F–domain).

6.3.2 Amplitude and Phase Spectra of Discrete-Time Signals

Definition 6.6 The *amplitude spectrum* of a discrete-time signal $x(n)$, $n \in \mathbb{Z}$ is defined as absolute value of its spectrum, that is, $|X(\Omega)|$. The *phase spectrum* is defined as argument of the spectrum $X(\Omega)$, that is, arg $X(\Omega)$. We have

$$X(\Omega) = |X(\Omega)|e^{j\arg X(\Omega)}. \tag{6.22}$$

Both $|X(\Omega)|$ and arg $X(\Omega)$ are periodic functions of variable Ω (normalized angular frequency) with period 2π. Therefore, we usually sketch their plots for $|\Omega| \leq \pi$. The amplitude spectrum is an even function of Ω (or of F) while the phase spectrum is an odd one.

Definition 6.7 The *real spectrum* Re $X(\Omega)$ of a discrete-time signal $x(n)$ is defined as

$$\mathrm{Re}\, X(\Omega) = |X(\Omega)|\cos(\arg X(\Omega)). \tag{6.23}$$

The *imaginary spectrum* Im $X(\Omega)$ is defined as

$$\mathrm{Im}\, X(\Omega) = |X(\Omega)|\sin(\arg X(\Omega)). \tag{6.24}$$

Re $X(\Omega)$ is always an even function of Ω (or of F) while Im $X(\Omega)$ is an odd one. There are the following relations between real-, imaginary-, amplitude- and phase spectra:

$$|X(\Omega)| = \sqrt{[\mathrm{Re}X(\Omega)]^2 + [\mathrm{Im}\, X(\Omega)]^2}, \tag{6.25}$$

$$\arg X(\Omega) = \mathrm{atan2}(\mathrm{Im}\, X(\Omega), \mathrm{Re}\, X(\Omega)). \tag{6.26}$$

Problem 6.8 Determine the amplitude- and phase spectra of the signal x-$(n) = a^n \cdot \mathbf{1}(n)$, $|a| < 1$. Assuming $a = \pm 1/2$, sketch their graphs as functions of F (normalized frequency). Calculate the energy of $x(n)$.

Solution From Table 6.1, we can get

$$x(n) = a^n \cdot \mathbf{1}(n) \overset{\mathcal{F}_d}{\leftrightarrow} X(\Omega) = \frac{1}{1 - ae^{-j\Omega}}.$$

Then

$$X(\Omega) = \frac{1}{1 - ae^{-j\Omega}} = \frac{1}{1 - a\cos\Omega + ja\sin\Omega}$$

The amplitude spectrum $|X(\Omega)|$ is given by the formula

$$|X(\Omega)| = \frac{1}{|1 - a\cos\Omega + ja\sin\Omega|} = \frac{1}{\sqrt{(1 - a\cos\Omega)^2 + a^2\sin^2\Omega}} = \frac{1}{\sqrt{1 + a^2 - 2a\cos\Omega}}$$

Or, in the F-domain,

$$|X(\mathrm{F})| = \frac{1}{\sqrt{1 + a^2 - 2a\cos(2\pi\mathrm{F})}}$$

Then, for $a = \pm 1/2$, we have

$$|X(\mathrm{F})| = \begin{cases} \dfrac{1}{\sqrt{\frac{5}{4} - \cos(2\pi\mathrm{F})}}, a = \dfrac{1}{2} \\ \dfrac{1}{\sqrt{\frac{5}{4} + \cos(2\pi\mathrm{F})}}, a = -\dfrac{1}{2} \end{cases}$$

The phase spectrum $\arg X(\Omega)$ is expressed as

$$\arg X(\Omega) = \arg\left(\frac{1}{1 - a\cos\Omega + ja\sin\Omega}\right) = -\arg(1 - a\cos\Omega + ja\sin\Omega) = -\operatorname{atan}\left(\frac{a\sin\Omega}{1 - a\cos\Omega}\right).$$

Then, for $a = \pm 1/2$, we get the formula

$$\arg X(\Omega) = \begin{cases} -\operatorname{atan}\left(\dfrac{\frac{1}{2}\sin\Omega}{1 - \frac{1}{2}\cos\Omega}\right), a = \dfrac{1}{2} \\ \operatorname{atan}\left(\dfrac{\frac{1}{2}\sin\Omega}{1 + \frac{1}{2}\cos\Omega}\right), a = -\dfrac{1}{2} \end{cases}$$

or

$$\arg X(\mathrm{F}) = \begin{cases} -\operatorname{atan}\left(\dfrac{\frac{1}{2}\sin(2\pi \mathrm{F})}{1-\frac{1}{2}\cos(2\pi \mathrm{F})}\right), a = \dfrac{1}{2} \\ \operatorname{atan}\left(\dfrac{\frac{1}{2}\sin(2\pi \mathrm{F})}{1+\frac{1}{2}\cos(2\pi \mathrm{F})}\right), a = -\dfrac{1}{2} \end{cases}$$

Figure 6.12 shows the amplitude- and phase spectra of $x(n) = \left(\frac{1}{2}\right)^n \cdot \mathbf{1}(n)$. Basing on the amplitude characteristic, we can conclude that this is a low-frequency signal. In Fig. 6.13, we presented the amplitude- and phase spectra of the signal $x(n) = \left(-\frac{1}{2}\right)^n \cdot \mathbf{1}(n)$. In this case, we deal with a high-frequency signal.

Problem 6.9 Given a discrete-time periodic signal $x(n)$ with fundamental period $N_0 = 6$ shown in Fig. 6.14. Calculate its spectrum $X(\mathrm{F})$ and sketch amplitude and phase spectra.

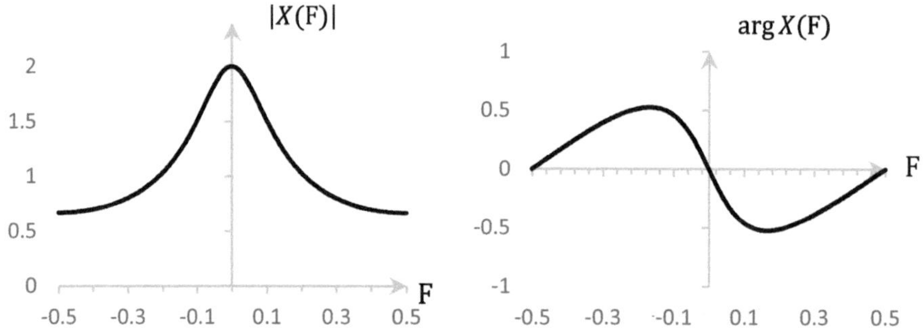

Fig. 6.12 Amplitude- and phase spectra of the signal $x(n) = \left(\frac{1}{2}\right)^n \cdot \mathbf{1}(n)$ from Problem 6.8

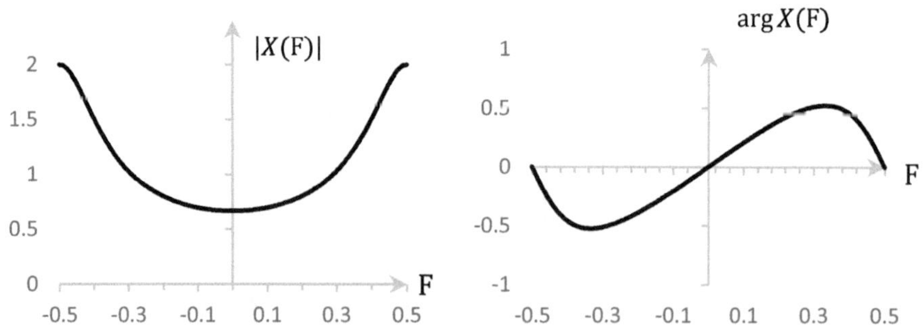

Fig. 6.13 Amplitude- and phase spectra of the signal $x(n) = \left(-\frac{1}{2}\right)^n \cdot \mathbf{1}(n)$ from Problem 6.8

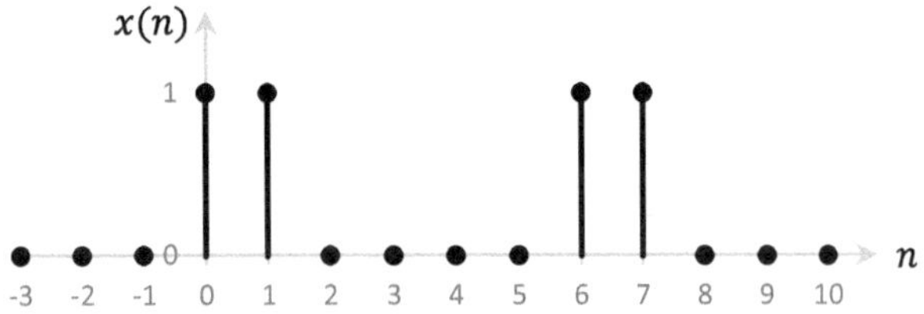

Fig. 6.14 Discrete-time periodic signal $x(n)$ from Problem 6.9

Solution For computing the spectrum of $x(n)$, we apply formulas (6.20) and (6.21) with $N_0 = 6$, that is,

$$X(\mathrm{F}) = \sum_{k \in \mathbb{Z}} c_k \delta\left(\mathrm{F} - \frac{k}{6}\right)$$

with coefficients c_k defined by

$$c_k = \frac{1}{6}\sum_{n=0}^{5} x(n) e^{-j\pi\frac{kn}{3}} = \frac{1}{6}\left(1 + e^{-j\pi\frac{k}{3}}\right).$$

All coefficients c_k are periodic with period $N_0 = 6$, so we can calculate them only for $k = 0, 1, 2, 3, 4, 5$ as follows:

$$c_0 = \frac{1}{3};$$
$$c_1 = \frac{1}{6}\left(1 + e^{-j\frac{\pi}{3}}\right) = \frac{1}{6}\left(1 + \cos\frac{\pi}{3} - j\sin\frac{\pi}{3}\right) = \frac{1}{4} - \frac{\sqrt{3}}{12}j; c_2 = \frac{1}{6}\left(1 + e^{-j\frac{2\pi}{3}}\right)$$
$$= \frac{1}{6}\left(1 + \cos\frac{2\pi}{3} - j\sin\frac{2\pi}{3}\right) = \frac{1}{12} - \frac{\sqrt{3}}{12}j; c_3 = \frac{1}{6}\left(1 + e^{-j\pi}\right) = 0; c_4$$
$$= \frac{1}{6}\left(1 + e^{-j\frac{4\pi}{3}}\right) = \frac{1}{6}\left(1 + \cos\frac{4\pi}{3} - j\sin\frac{4\pi}{3}\right) = \frac{1}{12} + \frac{\sqrt{3}}{12}j; c_5 = \frac{1}{6}\left(1 + e^{-j\frac{5\pi}{3}}\right)$$
$$= \frac{1}{6}\left(1 + \cos\frac{5\pi}{3} - j\sin\frac{5\pi}{3}\right) = \frac{1}{4} + \frac{\sqrt{3}}{12}j.$$

The spectrum $X(\mathrm{F})$ for $F \in [0, 0.5)$ is given by the formula

$$X(\mathrm{F}) = c_0\delta(\mathrm{F}) + c_1\delta\left(\mathrm{F}-\frac{1}{6}\right) + c_2\delta\left(\mathrm{F}-\frac{1}{3}\right) + c_3\delta\left(\mathrm{F}-\frac{1}{2}\right) + c_4\delta\left(\mathrm{F}-\frac{2}{3}\right) + c_5\delta\left(\mathrm{F}-\frac{5}{6}\right);$$

$$X(\mathrm{F}) = \frac{1}{3}\delta(\mathrm{F}) + \left(\frac{3}{12}-\frac{\sqrt{3}}{12}j\right)\delta\left(\mathrm{F}-\frac{1}{6}\right) + \left(\frac{1}{12}-\frac{\sqrt{3}}{12}j\right)\delta\left(\mathrm{F}-\frac{1}{3}\right)$$
$$+\left(\frac{1}{12}+\frac{\sqrt{3}}{12}j\right)\delta\left(\mathrm{F}-\frac{2}{3}\right) + \left(\frac{3}{12}+\frac{\sqrt{3}}{12}j\right)\delta\left(\mathrm{F}-\frac{5}{6}\right).$$

The amplitude spectrum $|X(\mathrm{F})|$ for $F \in [0, 0.5)$ in the F–domain is defined by the sequence $\{|c_k|\}$, $k = 0, 1, \ldots, 5$, that is,

$$|c_0| = \frac{1}{3}; |c_1| = \sqrt{\frac{1}{16}+\frac{3}{4}} = \frac{\sqrt{13}}{4} = |c_5|; |c_2| = \sqrt{\frac{1}{144}+\frac{3}{4}} = \frac{\sqrt{109}}{12} = |c_4|; |c_3| = 0.$$

Its plot is shown in Fig. 6.15.

The phase spectrum $\arg X(\mathrm{F})$ for $F \in [0, 0.5)$ in the F–domain is defined by the following sequence $\{\arg c_k\}$, $k = 0, 1, \ldots, 5$:

$$\arg c_0 = 0; \arg c_1 = \operatorname{atan}\left(-\frac{\sqrt{3}}{3}\right) = -\pi/6; \arg c_2 = \operatorname{atan}\left(-\sqrt{3}\right) = -\pi/3;$$
$$\arg c_3 = 0; \arg c_4 = \operatorname{atan}\left(\sqrt{3}\right) = \pi/3; \arg c_5 = \operatorname{atan}\left(\frac{\sqrt{3}}{3}\right) = \pi/6.$$

The plot of the phase spectrum is shown in Fig. 6.16.

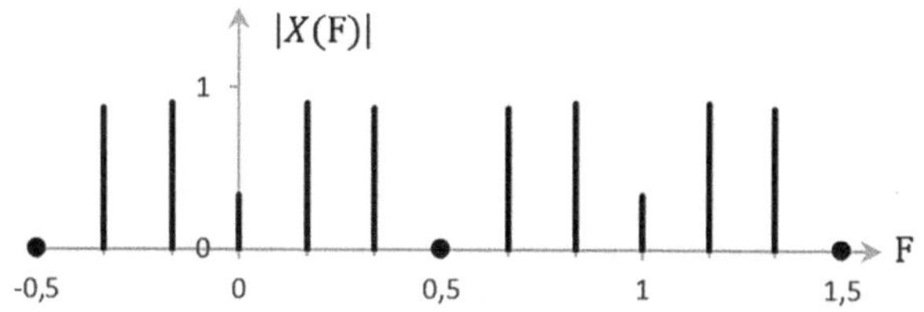

Fig. 6.15 Amplitude spectrum of the periodic signal $x(n)$ from Problem 6.9

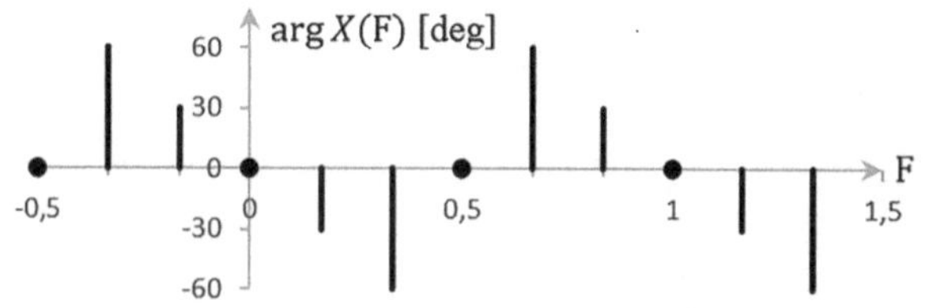

Fig. 6.16 Phase spectrum of the periodic signal $x(n)$ from Problem 6.9

6.4 Discrete Fourier Transform

The Discrete Fourier Transform (DFT) is widely applied in spectral analysis of discrete-time signals. In this section, we present its definition and basic properties illustrated with a few examples. It should be noted that DFT is the *discretized spectrum* $X_s(f)$ of an ideally sampled signal given by

$$X_s(f) = \sum_{n\in\mathbb{Z}} x(nT_s)e^{-j2\pi fnT_s}. \tag{6.27}$$

Discretization of the variable f means that there are N frequency samples in one period f_s of $X_s(f)$. Let us introduce the discretization step Δf called the *frequency resolution* of DFT in the form

$$\Delta f = f_s/N. \tag{6.28}$$

The result of discretization $f \rightarrow k\Delta f$ ($k = 0, 1, \ldots, N - 1$) of the function exp($-j2\pi fnT_s$) in (6.27) is the function

$$\exp(-j2\pi k\Delta fnT_s) = \exp\left(-j2\pi k\frac{f_s}{N}nT_s\right) = \exp\left(-j2\pi k\frac{n}{N}\right), k \in \mathbb{Z}.$$

We can easily prove that $\exp\left(-j2\pi k\frac{n}{N}\right)$ is periodic w.r.t. n or k and its period is N. The frequency discretization permits to define the *N-point* DFT.

Definition 6.8 The *N-point Discrete Fourier Transform* $X_N(k)$ of a discrete-time signal $x(n)$, $n = 0, 1, \ldots, N - 1$ is defined as

$$X_N(k) = \sum_{n=0}^{N-1} x(n)\exp\left(-j2\pi k\frac{n}{N}\right), k = 0, 1, \ldots, N - 1. \tag{6.29}$$

The N-point DFT (6.29) is called the *spectrum* of $x(n)$. Let us note that in situations when it is evident, we can omit the subscript N and simply write $X(k)$ (as in Table 6.3).

Definition 6.9 The *N-point Inverse Discrete Fourier Transform (IDFT)* of $X_N(k)$, $k = 0, 1, \ldots, N$ is defined as

$$x(n) = \frac{1}{N}\sum_{k=0}^{N-1} X_N(k)\exp\left(j2\pi k\frac{n}{N}\right), n = 0, 1, \ldots, N - 1. \tag{6.30}$$

Important In literature, there are definitions of DFT and IDFT differing with the factor $\frac{1}{N}$. Here, we use the convention applied, for example, in a Matlab environment.

Table 6.3 presents some properties of the DFT. Some of them will be applied in solutions of Problems 6.9, 6.10, 6.11 and 6.12.

Problem 6.10 Given two sequences: $x(n) = \{1, 2, 3, 0\}$, $y(n) = \{0, 1, 0, 1\}$, $n = 0, 1, 2, 3$. Calculate their circular convolution $z(n) = \sum_{n=0}^{N-1} x(k)y(n-k), n = 0, 1, 2, 3$; Then, performing appropriate calculations, check the validity of the time-convolution property of the DFT (see Table 6.3).

Solution First, let us calculate the time-convolution $z(n)$ defined as

$$z(n) = \sum_{n=0}^{3} x(k)y(n-k) = y(n) + 2y(n-1) + 3y(n-2).$$

For $n = 0, 1, 2, 3$, we obtain

$$z(0) = y(0) + 2y(-1) + 3y(-2) = 2.$$

Since $y(-1) = y(3) = 1$ and $y(-2) = y(2) = 0$ (because of periodicity with $N = 4$), we have

Table 6.3 Properties of DFT

Property	Signal $x(n)$, $n = 0, 1, \ldots, N-1$	$X_N(k) \equiv X(k)$, $k = 0, 1, \ldots, N-1$
Linearity	$\alpha x(n) + \beta y(n)$	$\alpha X(k) + \beta Y(k)$
Periodicity	$\forall n \in \mathbb{Z}$, $x(n) = x(n + N)$	$\forall k \in \mathbb{Z}$, $X(k) = X(k + N)$
Duality	$\frac{1}{N} X(n)$	$x(-k)$
Reversal	$x(N - n)$	$X(N - k)$
Complex conjugate		$X^*\left(\frac{N}{2} - k\right) = X\left(\frac{N}{2} + k\right)$, $k = 0, 1, \ldots, \frac{N}{2}$
Time-shift	$x(n - m)$	$X(k)\exp\left(-j2\pi k\frac{m}{N}\right)$
Frequency-shift	$x(n)\exp\left(j2\pi n\frac{l}{N}\right)$	$X(k - l)$
Time-convolution (circular convolution)	$\sum_{n=0}^{N-1} x(k)y(n-k)$	$X(k) \cdot Y(k)$
Multiplication	$x(n) \cdot y(n)$	$\frac{1}{N}\sum_{l=0}^{N-1} X(l)Y(k-l)$
Parseval's equality	$\sum_{n=0}^{N-1} \lvert x(n)\rvert^2 = \frac{1}{N}\sum_{k=0}^{N-1} \lvert X(k)\rvert^2$	

$z(1) = y(1) + 2y(0) + 3y(-1) = 4; z(2) = y(2) + 2y(1) + 3y(0) = 2;$
$z(3) = y(3) + 2y(2) + 3y(1) = 4.$

To check the validity of the time-convolution property of the DFT, we have to calculate two 4-point DFTs of signals $x(n)$, $y(n)$, denoted, respectively, with $X(k)$, $Y(k)$. Using (6.29), we have

$$X(k) = \sum_{n=0}^{3} x(n) \exp\left(-j\pi k\frac{n}{2}\right) = 1 + 2e^{-j\pi k/2} + 3e^{-j\pi k}$$

and

$$Y(k) = \sum\nolimits_{n=0}^{3} y(n) \exp\left(-j\pi k\frac{n}{2}\right) = e^{-j\pi k/2} + e^{-j3\pi k/2}.$$

Let us calculate values of $X(k)$, $Y(k)$ for $k = 0, 1, 2, 3$ as follows:

$$\begin{aligned} X(0) &= 6; X(1) = 1 + 2e^{-j\pi/2} + 3e^{-j\pi} = -2 - 2j; \\ X(2) &= 1 + 2e^{-j\pi} + 3e^{-j\pi 2} = 2; \\ X(3) &= X^*(1) = -2 + 2j \text{ (see complex conjugate property of DFT)} \end{aligned}$$

and

$$Y(0) = 2; Y(1) = e^{-j\pi/2} + e^{-\frac{j3\pi}{2}} = 0; Y(2) = e^{-j\pi} + e^{-j3\pi} = -2; Y(3) = 0.$$

Now, we have to calculate the product $Z(k) = X(k) \cdot Y(k)$ of two DFTs. It is defined by the sequence $\{12, 0, -4, 0\}$ since $Z(0) = X(0) \cdot Y(0) = 12$, etc.

The last step is calculation of the inverse DFT $z(n)$ given by

$$z(n) = \frac{1}{4}\sum_{k=0}^{3} Z(k) \exp\left(j\pi k\frac{n}{2}\right)$$

(see formula (6.30)). We have

$$z(n) = \frac{1}{4}\sum_{k=0}^{3} Z(k) \exp\left(j\pi k\frac{n}{2}\right) = \frac{1}{4}\left(12 - 4e^{j\pi n}\right) = 3 - e^{j\pi n}.$$

For $n = 0, 1, 2, 3$, we obtain: $z(0) = 2$; $z(1) = 4$; $z(2) = 2$; $z(3) = 4$, that is exactly the same as previously calculated circular convolution of $x(n)$ and $y(n)$.

6.4.1 *Amplitude- and Phase Spectra of* x(n)

Definition 6.10 The *amplitude spectrum* of a discrete-time signal $x(n)$, $n = 0, 1, \ldots, N - 1$ is defined as absolute value $|X_N(k)|$ of the DFT defined by (6.29) for $k = 0, 1, \ldots, N - 1$.

Definition 6.11 The *phase spectrum* of a discrete-time signal $x(n)$, $n = 0, 1, \ldots, N - 1$ is defined as argument $\arg X_N(k)$, of the DFT defined by (6.29) for $k = 0, 1, \ldots, N - 1$.

Definition 6.12 The *power spectrum* of a discrete-time signal $x(n)$, $n = 0, 1, \ldots, N - 1$ is defined as square $|X_N(k)|^2$ of the amplitude spectrum for $k = 0, 1, \ldots, N - 1$.

Let us note that all spectra from Definitions 6.10, 6.11 and 6.12 are periodic with period N.

Let us compare the DTFT with the DFT of a given signal solving Problem 6.10.

Problem 6.11 Given the discrete-time signal $x(n) = \delta(n + 1) - \delta(n - 1)$. Compute and compare its Discrete-Time Fourier Transform $X(\mathrm{F})$ with the 8-point DFT $X_8(k)$. Plot the corresponding amplitude spectra: $|X(\mathrm{F})|$, $|\mathrm{F}| \le 1/2$ and $|X_8(k)|$, $k = 0, 1, \ldots, 7$.

Solution From Table 6.1, we get the following pair of DTFTs:

$$x(n) = \delta(n+1) - \delta(n-1) \overset{\mathcal{F}_d}{\leftrightarrow} X(\Omega) = e^{j\Omega} - e^{-j\Omega} = 2j \sin \Omega.$$

Thus, $X(\mathrm{F}) = 2j \sin(2\pi \mathrm{F})$. The amplitude spectrum, given by $|X(\mathrm{F})| = 2|\sin(2\pi \mathrm{F})|$, is displayed in Fig. 6.18.

To calculate the 8-point DFT using (6.29), we have to introduce the periodic signal $x_8(n)$ with period $N = 8$, defined by the sequence $\{0, -1, 0, 0, 0, 0, 0, 1\}$ for $n = 0, 1, \ldots, 7$. We can see that it is a result of the periodic repetition of $x(n)$ (marked in red in Fig. 6.17) with period $N = 8$. It is a common way to replace a discrete-time

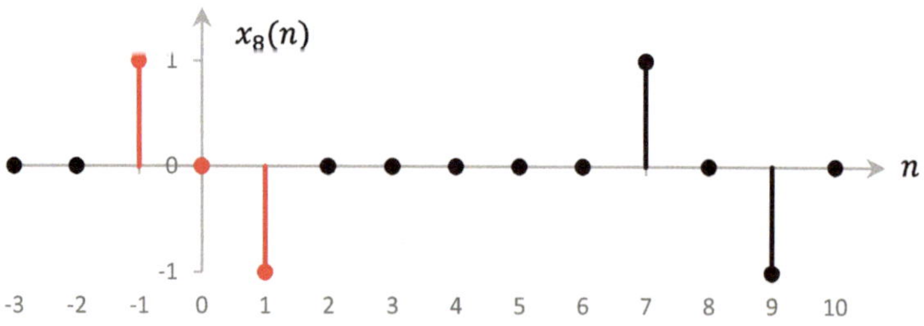

Fig. 6.17 Periodic sequence $x_8(n)$, in which $x(n)$ from Problem 6.11 is marked in red

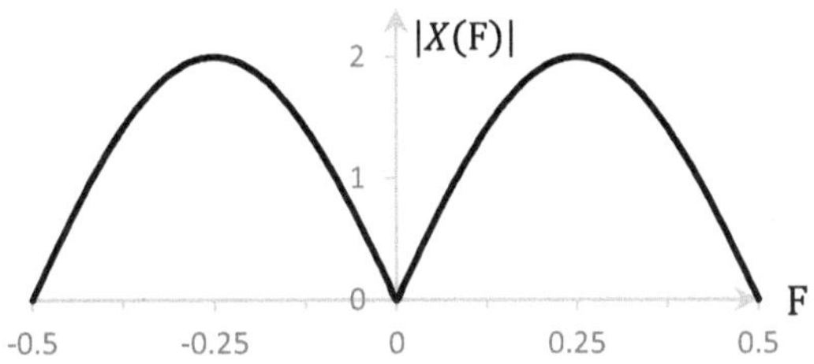

Fig. 6.18 Amplitude spectrum |X(F)| of the signal $x(n)$ from Problem 6.11

signal $x(n)$ with a periodic sequence $x_N(n)$ (with the period N) and to replace all "missing" samples with zeros.

From (6.29), the 8-point DFT of $x_8(n)$ is

$$X_8(k) = \sum_{n=0}^{7} x_8(n)\exp\left(-j2\pi k\frac{n}{8}\right) = \sum_{n=0}^{7} x_8(n)\exp\left(-j\pi k\frac{n}{4}\right).$$

Knowing that $x_8(1) = -1$ and $x_8(7) = 1$, we have

$$\begin{aligned} X_8(k) &= -\exp\left(-j\frac{\pi}{4}k\right) + \exp\left(-j\frac{7\pi}{4}k\right) \\ &= -\exp\left(-j\frac{\pi}{4}k\right) + \exp\left(-j\left(2\pi - \frac{\pi}{4}\right)k\right) \\ &= -\exp\left(-j\frac{\pi}{4}k\right) + \exp\left(j\frac{\pi}{4}k\right) = 2j\sin\left(\frac{\pi}{4}k\right). \end{aligned}$$

For $k = 0, 1, \ldots, 7$, we get

$$\begin{aligned} &X_8(0) = 0; X_8(1) = 2j\sin\left(\frac{\pi}{4}\right) = \sqrt{2}j; X_8(2) = 2j\sin\left(\frac{\pi}{2}\right) = 2j; \\ &X_8(3) = 2j\sin\left(\frac{3\pi}{4}\right) = \sqrt{2}j; X_8(4) = 0; X_8(5) = X_8^*(3) = -\sqrt{2}j; \\ &X_8(6) = X_8^*(2) = -2j; X_8(7) = X_8^*(1) = -\sqrt{2}j. \end{aligned}$$

The amplitude spectrum $|X_8(k)|$ is defined by $\{0, \sqrt{2}, 2, \sqrt{2}, 0, \sqrt{2}, 2, \sqrt{2}\}$ for $k = 0, 1, \ldots, 7$, as shown in Fig. 6.19.

Both spectra presented in Figs. 6.18 and 6.19 are periodic: |X(F)| with period 1 in the domain of $\mathrm{F} = fT_s = f/f_s$ and $|X_8(k)|$—with period $N = 8$. The frequency resolution of the 8-point DFT is $\Delta f = f_s/8$. We can notice, for instance, $|X_8(1)| = \sqrt{2}$ and $k = 1$ corresponds to frequency $f_s/8$. On the other hand, $f = \mathrm{F}\cdot f_s = f_s/8 \Leftrightarrow \mathrm{F} = 1/8$ and $|X(1/8)| = |X_8(1)| = \sqrt{2}$. We have shown that the DFT is the DTFT discretized in frequency.

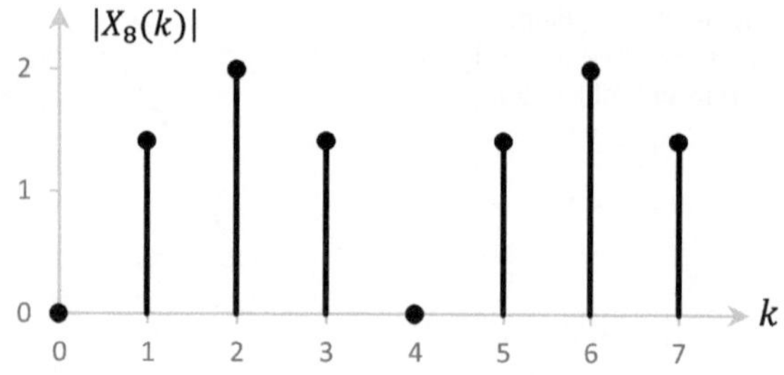

Fig. 6.19 Amplitude spectrum $|X_8(k)|$ of the signal $x(n)$ from Problem 6.11

Problem 6.12 Given the sequence $x(n) = \{0, 1, 0, 2\}$, $n = 0, 1, 2, 3$. Compute its 4-point DFT $X(k)$. Then calculate the DFT of the signal $y(n) = x(n - 1)$ (*Hint*: treat $x(n)$ as a periodic signal with period $N = 4$). Determine amplitude and phase spectra of $x(n)$ and $y(n)$.

Solution From (6.29), we can calculate the 4-point DFT of $x(n)$ as follows:

$$X(k) = \sum_{n=0}^{3} x(n) \exp\left(-j\pi k \frac{n}{2}\right) = \exp\left(-j\frac{\pi}{2}k\right) + 2\exp\left(-j\frac{3\pi}{2}k\right).$$

For $k = 0, 1, 2, 3$, we get $X(0) = 3$; $X(1) = \exp\left(-j\frac{\pi}{2}\right) + 2\exp\left(-j\frac{3\pi}{2}\right) = j$;

$$X(2) = \exp(-j\pi) + 2\exp(-j3\pi) = -3; X(3) = X^*(1) = -j.$$

The amplitude spectrum of $x(n)$ is given by the sequence $|X(k)| = \{3, 1, 3, 1\}$. The phase spectrum of $x(n)$ is defined by $\arg X(k) = \left\{0, \frac{\pi}{2}, \pi, -\frac{\pi}{2}\right\}$.

Now, let us calculate the 4-point DFT of $y(n) = x(n - 1)$. First, let us represent $x(n)$ as a periodic signal, as shown in Fig. 6.20. Then we shift it right by 1, which is illustrated in Fig. 6.21. We can notice that right-shifting of a “periodic” $x(n)$ from Fig. 6.20 corresponds to *circular right-shifting* by 1 of four samples of $x(n) = \{0, 1, 0, 2\}$ for $n = 0, 1, 2, 3$ marked in red in Fig. 6.20.

In Fig. 6.21, we marked in red the signal $y(n)$, $n = 0, 1, 2, 3$, for which we can calculate the 4-point DFT as follows:

$$Y(k) = \sum_{n=0}^{3} y(n) \exp\left(-j\pi k \frac{n}{2}\right) = 2 + \exp(-j\pi k).$$

For $k = 0, 1, 2, 3$, we get $Y(0) = 3$; $Y(1) = 1$; $Y(2) = 3$; $Y(3) = 1$. Thus, the amplitude spectrum of $y(n)$ is also defined by $\{3, 1, 3, 1\}$ and the phase spectrum is zero.

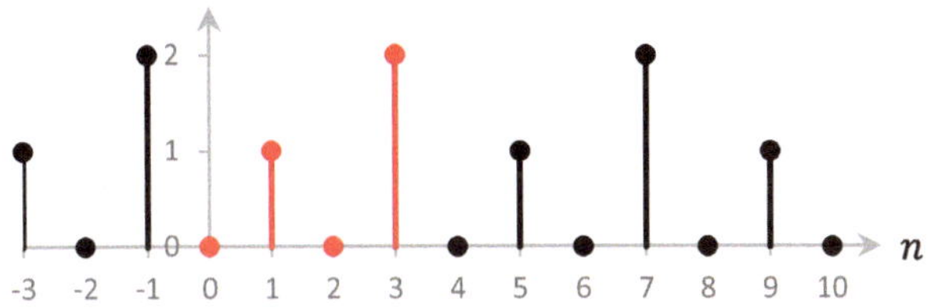

Fig. 6.20 Signal $x(n)$ from Problem 6.12 (marked in red) repeated with period $N = 4$

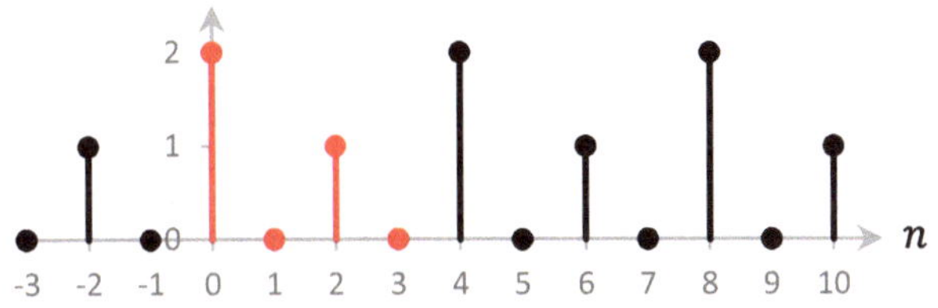

Fig. 6.21 Signal from Fig. 6.20 right-shifted by 1. The signal $y(n)$ is marked in red

Important We can notice that *time-shifting does not change the amplitude spectrum of a signal* and only *changes the phase spectrum*. In what way? Let us look at the time-shifting property of the DFT (Table 6.3). Here,

$$x(n-1) \overset{\text{DFT}}{\leftrightarrow} X(k)\exp\left(-j\frac{\pi}{2}k\right).$$

We can see that the phase spectrum $\arg Y(k) = \arg X(k) - \frac{\pi}{2}k$, $k = 0, \ldots, 3$, that is, $\arg Y(0) = 0$; $\arg Y(1) = \frac{\pi}{2} - \frac{\pi}{2} = 0$; $\arg Y(2) = \pi - \pi = 0$ and $\arg Y(3) = -\frac{\pi}{2} - \frac{3\pi}{2} = -2\pi \equiv 0$.

6.4.2 *Zero Padding*

Zero padding is a technique typically employed to make the size of a signal the power of 2 as it is applied, for example, in Fast Fourier Transform (FFT) algorithms. It is based on adding zero samples to the end of a signal. The result is a "denser sampling" of the spectrum in the frequency domain. In Problem 6.13, we demonstrate this technique with a simple example.

Problem 6.13 Given the discrete-time sequence $x(n) = \{1, 1, 0, 0\}$, $n = 0, 1, 2, 3$. We are adding four zero samples to $x(n)$ to get $y(n) = \{1, 1, 0, 0, 0, 0, 0, 0\}$. Compute the DFTs of both signals and plot their amplitude spectra for $k = -N/2, \ldots, N/2$, where: $N = 4$ for $x(n)$ and $N = 8$ for $y(n)$.

Solution Using (6.29), we can compute the 4-point DFT of $x(n)$ as follows:

$$X(k)=\sum_{n=0}^{3}x(n)\exp\left(-j\pi k\frac{n}{2}\right)=1+\exp\left(-j\frac{\pi}{2}k\right).$$

We have

$$X(0)=2;X(1)=1+e^{-j\frac{\pi}{2}}=1-j;X(2)=1+e^{-j\pi}=0;X(3)=X^*(1)=1+j.$$

The amplitude spectrum of $x(n)$ is defined by $|X(k)|=\{2,\sqrt{2},0,\sqrt{2}\}, k=0,1,2,3$. Using its periodicity, we easily get $|X(-2)| = |X(2)| = 0$ and $|X(-1)|=|X(3)|=\sqrt{2}$. It is shown in Fig. 6.22 for $k = -2, \ldots, 2$.

Now, let us calculate the 8-point DFT of $y(n)$ using the formula

$$Y(k)=\sum_{n=0}^{7}y(n)\exp\left(-j\pi k\frac{n}{4}\right)=1+\exp\left(-j\frac{\pi}{4}k\right).$$

We get the following values:

$$Y(0)=2;\ Y(1)=1+e^{-j\frac{\pi}{4}}=1+\frac{\sqrt{2}}{2}-j\frac{\sqrt{2}}{2};Y(2)=1+e^{-j\frac{\pi}{2}}=1-j;Y(3)=$$
$$+e^{-j\frac{3\pi}{4}}=1-\frac{\sqrt{2}}{2}-j\frac{\sqrt{2}}{2};Y(4)=0;Y(5)=Y^*(3)=1-\frac{\sqrt{2}}{2}+j\frac{\sqrt{2}}{2};Y(6)$$
$$=Y^*(2)=1+j;Y(7)=Y^*(1)=1+\frac{\sqrt{2}}{2}+j\frac{\sqrt{2}}{2}.$$

For $k = 0, \ldots, 7$, the amplitude spectrum $|Y(k)|$ of $y(n)$ is

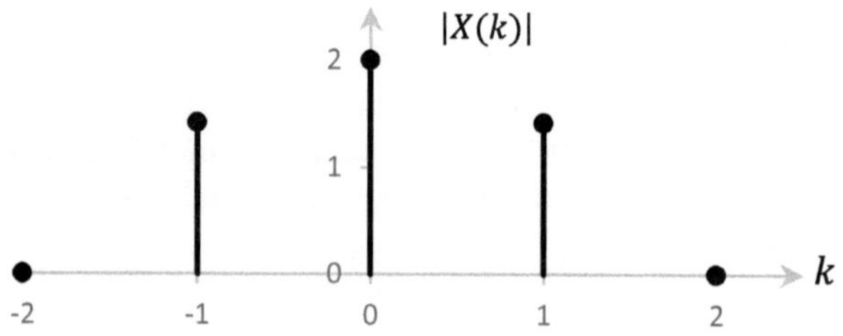

Fig. 6.22 Amplitude spectrum of the signal $x(n)$ from Problem 6.13

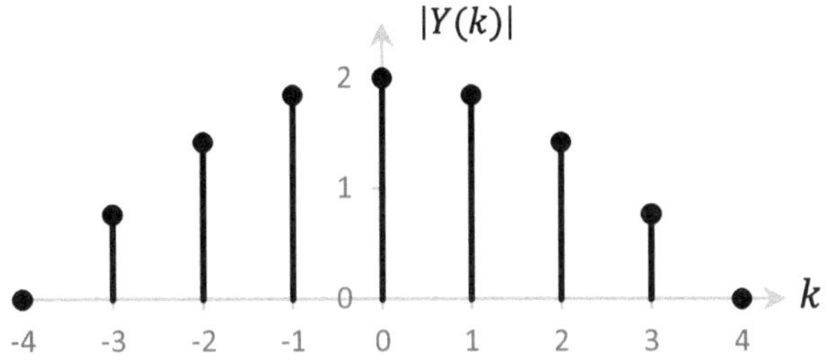

Fig. 6.23 Amplitude spectrum of the signal $y(n)$ from Problem 6.13

$$|Y(k)| = \left\{2, \sqrt{2+\sqrt{2}}, \sqrt{2}, \sqrt{2-\sqrt{2}}, 0, \sqrt{2-\sqrt{2}}, \sqrt{2}, \sqrt{2+\sqrt{2}}\right\}$$
$$\cong \left\{2, 1.8478, \sqrt{2}, 0.7654, 0, 0.7654, \sqrt{2}, 1.8478\right\}.$$

Using periodicity of $|Y(k)|$ with period $N = 8$, we can get $|Y(-4)| = |Y(4)| = 0$, $|Y(-3)| = |Y(5)| \cong 0.7654$, $|Y(-2)| = |Y(6)| \cong 1.4142$ and $|Y(-1)| = |Y(7)| \cong 1.8478$. The amplitude spectrum for $k = -4, \ldots, 4$ is shown in Fig. 6.23.

6.4.3 Spectrum Leakage

Spectrum leakage is a phenomenon connected with sampling a periodic signal $x(t)$ with the sampling rate f_s such that the DFT of the sampled signal $x(n) = x(nT_s)$ "produces" additional frequency components not present in $x(t)$. Let us illustrate it in Problem 6.14.

Problem 6.14 Given two continuous-time signals: (a) $x(t) = \cos(2\pi t)$; (b) $y(t) = \cos(3\pi t)$, both sampled with frequency $f_s = 8$ Hz. Using the 8-point DFT compute and plot spectra $X(k)$ and $Y(k)$, $k = 0, \ldots, 7$ of sampled signals: $x(n) = x(nT_s)$ and $y(n) = y(nT_s)$. Then, using Parseval's theorem, calculate and compare energies contained in 8 samples of $X(k)$ and $Y(k)$.

Solution First, let us notice that the frequency resolution of the 8-point DFT equals $\Delta f = f_s/N = 1$ Hz.

(a) Let us recall that the signal $x(t) = \cos(2\pi t)$ has the fundamental period 1 [s] and its frequency is 1 [Hz]. In the frequency domain, it is represented by two peaks at frequencies ± 1 Hz. The signal $x(t)$ is sampled 8 times within one period and its samples are: $x(n) = x(nT_s) = \cos(2\pi n T_s) = \cos\left(\frac{\pi}{4}n\right), n = 0, 1, \ldots, 7$, that is, $x(0) = 1$; $x(1) = 1/\sqrt{2}$; $x(2) = 0$; $x(3) = -1/\sqrt{2}$; $x(4) = -1$; $x(5) = -1/\sqrt{2}$; $x(6) = 0$ and $x(7) = 1/\sqrt{2}$.

Let us now calculate its 8-point DFT using the following formula:

$$
\begin{aligned}
X(k) &= \sum_{n=0}^{7} x(n)\exp\left(-j\pi k\frac{n}{4}\right) \\
&= 1 + \frac{1}{\sqrt{2}}e^{-j\frac{\pi}{4}k} - \frac{1}{\sqrt{2}}e^{-j\frac{3\pi}{4}k} - e^{-j\pi k} - \frac{1}{\sqrt{2}}e^{-j\frac{5\pi}{4}k} + \frac{1}{\sqrt{2}}e^{-j\frac{7\pi}{4}k}.
\end{aligned}
$$

Then, since $e^{-j\frac{7\pi}{4}k} = e^{-j\left(2\pi - \frac{\pi}{4}\right)k} = e^{j\frac{\pi}{4}k}$, $e^{-j\frac{3\pi}{4}k} = e^{-j\left(\pi - \frac{\pi}{4}\right)k} = e^{-j\pi k}\, e^{j\frac{\pi}{4}k}$, $e^{-j\frac{5\pi}{4}k} = e^{-j\left(\pi + \frac{\pi}{4}\right)k} = e^{-j\pi k}\, e^{-j\frac{\pi}{4}k}$, we can group the terms of $X(k)$ as follows:

$$
\begin{aligned}
X(k) &= 1 + \frac{1}{\sqrt{2}}\left(e^{-j\frac{\pi}{4}k} + e^{j\frac{\pi}{4}k}\right) - \frac{1}{\sqrt{2}}e^{-j\pi k}\left(e^{j\frac{\pi}{4}k} + e^{-j\frac{\pi}{4}k}\right) - e^{-j\pi k} \\
&= 1 - e^{-j\pi k} + \frac{1}{\sqrt{2}}\left(e^{-j\frac{\pi}{4}k} + e^{j\frac{\pi}{4}k}\right)\left(1 - e^{-j\pi k}\right) \\
&= \left(1 - e^{-j\pi k}\right)\left(1 + \sqrt{2}\cos\left(\frac{\pi}{4}k\right)\right).
\end{aligned}
$$

We can see that $X(k) = 0$ for all k except $k = 1$ and $k = 7$, for which $X(1) = X(7) = 4$. Periodicity of $X(k)$ permits also to notice that $X(-1) = X(7) = 4$. It means that for $k = \pm 1$ (corresponding to frequencies $\pm$ 1 Hz) we have two peaks of amplitude 4, which is exactly what we expected to obtain. There is *no spectrum leakage*.

Let us calculate energy contained in 8 samples of $X(k)$ as follows:

$$
E_x = \frac{1}{8}\sum\nolimits_{k=0}^{7}|X(k)|^2 = \frac{1}{8}(16 + 16) = 4.
$$

(b) Let us now consider the second signal $y(t) = \cos(3\pi t)$. Its fundamental frequency is 1,5 [Hz] and the fundamental period 2/3 [s]. The spectrum constitutes two peaks at frequencies ±1, 5 Hz. We can see that $y(t)$ does not contain components at frequencies being multiples of $\Delta f = 1$ [Hz].

The signal $y(t)$ is sampled with sampling rate 8 Hz and its samples are: $y(n) = y(nT_s) = \cos(3\pi n T_s) = \cos\left(\frac{3\pi}{8}n\right), n = 0, 1, \ldots, 7$, that is,

$$
\begin{aligned}
&y(0) = 1; y(1) = \cos\left(\frac{3\pi}{8}\right) \cong 0{,}3827; y(2) = \cos\left(\frac{3\pi}{4}\right) \cong -0{,}7071; \\
&y(3) = \cos\left(\frac{9\pi}{8}\right) \cong -0{,}9239; y(4) = 0; y(5) = \cos\left(\frac{15\pi}{8}\right) \cong 0{,}9239; \\
&y(6) = \cos\left(\frac{9\pi}{4}\right) \cong 0{,}7071 \text{ and } y(7) = \cos\left(\frac{21\pi}{8}\right) \cong -0{,}3827.
\end{aligned}
$$

Let us now calculate its 8-point DFT as follows:

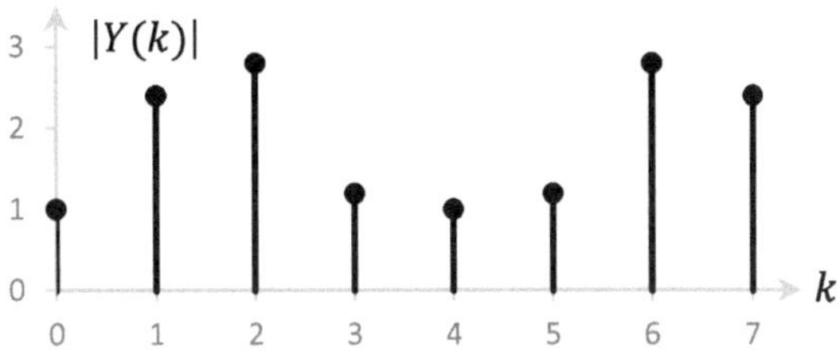

Fig. 6.24 Amplitude spectrum of the signal $y(n)$ from Problem 6.14

$$\begin{aligned}Y(k) &= \sum_{n=0}^{7} y(n)\exp\left(-j\pi k\frac{n}{4}\right)\\ &= 1 + 0{,}3827e^{-j\frac{\pi}{4}k} - 0{,}7071e^{-j\frac{\pi}{2}k} - 0{,}9239e^{-j\frac{3\pi}{4}k}\\ &\quad +0{,}9239e^{-j\frac{5\pi}{4}k} + 0{,}7071e^{-j\frac{3\pi}{2}k} - 0{,}3827e^{-j\frac{7\pi}{4}k}.\end{aligned}$$

Then $Y(0) = 1$; $Y(1) \cong 1 + 2{,}\ 1796j$; $Y(2) \cong 1 - 2{,}\ 6131j$; $Y(3) \cong 1 - 0{,}\ 6488j$; $Y(4) = 1$; $Y(5) \cong 1 + 0{,}\ 6488j$; $Y(6) \cong 1 + 2{,}\ 6131j$; $Y(7) \cong 1 - 2{,}\ 1796j$.

The amplitude spectrum of $y(n)$ (displayed in Fig. 6.24) is defined by $\{1, 2.398, 2.7979, 1.1921, 1, 1.1921, 2.7979, 2.398\}$. We can see that all frequency components are present in the spectrum. However, energy calculated from 8 samples of $Y(k)$ is exactly the same as E_x and given by

$$E_y = \frac{1}{8}\sum\nolimits_{k=0}^{7}|Y(k)|^2 = \frac{32}{8} = 4.$$

We can say that energy contained in two peaks of the signal was “leaking” to all components of its DFT. This phenomenon is known as the *spectrum leakage*.

Having in mind that $y(n)$ is periodic with period $N_0 = 16$ (it is easy to check by the reader), the 16-point DFT would not produce a “leakage”.

Exercises

6.1. Determine the Nyquist rate of the following signals: (a) $x_1(t) = \frac{\sin(20\pi t)}{\pi t}$; (b) $x_2(t) = \left(\frac{\sin(20\pi t)}{\pi t}\right)^2$; (c) $x_1(t) = \sin(20\pi t) + \sin^2(20\pi t)$.

6.2. The analog signal $x(t) = 3/2 + \cos(\pi t) - \cos^2(2\pi t)$ is ideally sampled with frequency: (a) $f_s = 5$ Hz; (b) $f_s = 3$ Hz and then filtered with an ideal LP filter with gain $1/f_s$ and cutoff frequency $f_c = 1{,}\ 5$ Hz. Calculate the Nyquist rate of $x(t)$. Determine and sketch the spectrum $X_s(f)$ of the sampled signal $x_s(t)$, and compute the filter response $y(t)$ for two sampling scenarios (a) and (b).

6.3. Given the spectrum $X(f) = \delta(f + 70) + \delta(f + 50) + \delta(f - 50) + \delta(f - 70)$ of an analog signal $x(t)$. The signal $x(t)$ is ideally sampled with frequency: (a) $f_s = 40$ Hz; (b) $f_s = 90$ Hz; (c) $f_s = 200$ Hz and then filtered with an ideal LP filter with gain $1/f_s$ and cutoff frequency $f_c = 50$ Hz. Plot the spectrum $X_s(f)$ of the sampled signal and compute the filter response for three sampling scenarios (a), (b), and (c).

6.4. The analog signal $x(t) = 2\cos(20\pi t) - \cos(40\pi t)$ is ideally sampled with frequency $f_s = 9$ Hz and then filtered with an ideal LP filter with gain $1/f_s$ and cutoff frequency $f_c = 3$ Hz. Plot the spectrum $X_s(f)$ of the sampled signal and compute the filter response. How do you name the phenomenon you are observing?

6.5. Given an analog signal $x(t)$ with the spectrum $X(f) = \Pi\left(\frac{f}{4}\right)$. The signal is ideally sampled with frequency: (i) $f_s = 3$ Hz; (b) $f_s = 6$ Hz. Sketch the spectrum $X_s(f)$ of the sampled signal. Repeat all calculations for the analog signal $x(t)$ with the spectrum $X(f) = \Lambda\left(\frac{f}{2}\right)$.

6.6. The signal $x(t) = 1000 \cdot \mathrm{Sa}(1000\pi t)$ is passing through a filter with the frequency response

$$H(f) = \begin{cases} 1 - \dfrac{|f|}{500} & \text{for } |f| \leq 500 \\ 0 \text{ elsewhere.} \end{cases}$$

The filter output $y(t)$ is then ideally sampled with sampling rate $f_s = 1000$ Hz to get the sampled signal $y_s(t)$. Plot the spectrum $Y_s(f)$ of $y_s(t)$ and propose the frequency response of a filter that could recover $x(t)$ from samples of $y(t)$.

6.7. The analog signal $x(t) = \mathrm{Sa}(\pi t)$ is ideally sampled with frequency $f_s = 0,75$ Hz and then filtered with an ideal LP filter with gain $1/f_s$ and cutoff frequency: (a) $f_c = 0,5$ Hz; (b) $f_c = 0,25$ Hz. Sketch the spectrum $X_s(f)$ of the sampled signal. Sketch the spectrum $Y(f)$ of the filter output $y(t)$ and compute $y(t)$ for cases (a) and (b).

6.8. The analog signal $x(t) = \mathrm{Sa}(\pi t)\cos(2\pi t)$ is ideally sampled with frequency $f_s = 1,5$ Hz and then filtered with an ideal LP filter with gain $1/f_s$ and cutoff frequency $f_c = 1$ Hz. Calculate the Nyquist rate for $x(t)$. Sketch the spectrum $X_s(f)$ of the sampled signal. Sketch the spectrum $Y(f)$ of the filter output $y(t)$ and compute $y(t)$.

6.9. Figure 6.25 displays the spectrum $X(f)$ of an analog signal $x(t)$. The signal $x(t)$ is sampled with frequency f_s and then filtered with an ideal LP filter with gain $1/f_s$ and cutoff frequency $f_c = 2f_0$. Determine the signal $x(t)$. Then plot the spectrum $X_s(f)$ of the sampled signal and compute the filter response $y(t)$ if: (a) $f_s = 4f_0$; (b) $f_s = 3f_0$; (c) $f_s = \frac{7}{2}f_0$; (d) $f_s = 2f_0$.

6.10. The signal $x(t) = 9\mathrm{Sa}^2(3\pi t)$ is passing through an ideal LP filter I with gain 1 and cutoff frequency $f_c = 2$ Hz. The output $w(t)$ of this filter is then ideally sampled with frequency $f_s = 2$ Hz. At the end, the sampled signal $w_s(t)$ is filtered with the second ideal LP filter II to get the output signal $y(t) = 4\,\mathrm{Sa}(4\pi t)$. Determine spectra $X(f)$, $W(f)$, and $W_s(f)$ of signals $x(t)$, $w(t)$, and $w_s(t)$, respectively. Define the frequency response of the filter II.

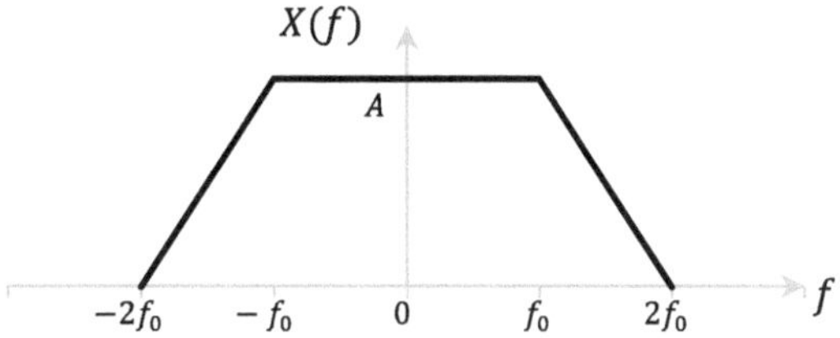

Fig. 6.25 Spectrum $X(f)$ of the signal $x(t)$ from Exercise 6.9

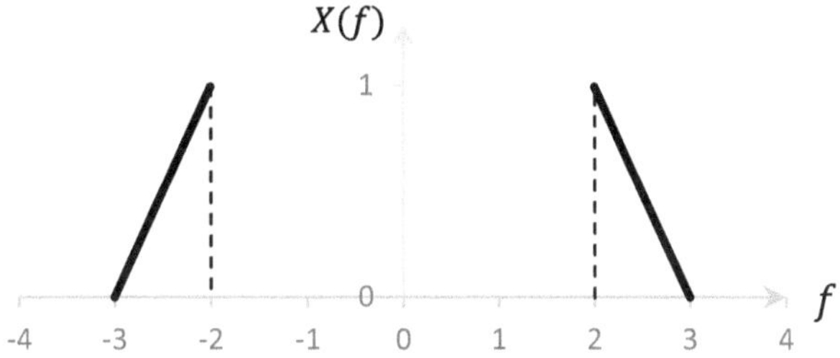

Fig. 6.26 Spectrum $X(f)$ of the signal $x(t)$ from Exercise 6.11

6.11. Figure 6.26 displays the spectrum $X(f)$ of an analog signal $x(t)$. The signal $x(t)$ is sampled with frequency $f_s = 4$ Hz and then filtered with an ideal LP filter with gain $1/f_s$ and cutoff frequency $f_c = 3$ Hz. Sketch the spectrum $X_s(f)$ of the sampled signal and compute the filter response $y(t)$.

6.12. The analog signal $x(t) = \cos(\pi t)$ is flat–top sampled with frequency $f_s = 4$ Hz. Assuming the time-duration of the rectangular pulse $\tau = 0, 25$ [s], sketch the staircase signal $\widehat{x}(t)$. Compute and plot the amplitude spectrum of the signal $y(t) = \widehat{x}\left(t + \frac{T_s}{2}\right)$.

6.13. Given the discrete-time rectangular pulse $\Pi_{10}(n)$. Determine its DTFT, plot amplitude and phase spectra and calculate energy of $x(n)$.

6.14. Given the discrete-time signal $x(n) = a^{|n|}$, $|a| < 1$. Plot the amplitude and phase spectra for $a = \pm\frac{1}{2}$ and calculate energy of $x(n)$.

6.15. Calculate the DTFT of the following signals: (a) $x_1(n) = \delta(n - 2) - \delta(n + 2)$; (b) $x_1(n) = \delta(n + 2) + \delta(n - 3)$. Plot their amplitude spectra.

6.16. Compute the discrete-time convolution $y(n) = x_1(n) * x_2(n)$ of the following signals:

$$x_1(n) = \begin{cases} 1 \text{ for } n \in [0, 4] \\ 0 \text{ elsewhere} \end{cases}, \quad x_2(n) = \begin{cases} \left(\frac{1}{2}\right)^n \text{ for } n \in [0, 3] \\ 0 \text{ elsewhere} \end{cases}.$$

Compare the answer with solution of Exercise 1.7 (Chap. 1).

6.17. The DTFT of a discrete-time signal $y(n)$ is given by:

$$Y(\Omega) = \frac{1}{1 - e^{-j\Omega}} \frac{\sin\left(\frac{3}{2}\Omega\right)}{\sin\left(\frac{\Omega}{2}\right)} + 7\pi\delta(\Omega).$$

Calculate $x(n)$.

6.18. Given two sequences: $x(n) = \{2, 0, 1, 0\}$, $y(n) = \{3, 2, 1, 0\}$, $n = 0, 1, 2, 3$. Calculate their circular convolution $z(n) = \sum_{n=0}^{N-1} x(k)y(n-k), n = 0, 1, 2, 3;$ Then, performing appropriate calculations, check the validity of the time-convolution property of DFT: $z(n) \overset{\text{DFT}}{\leftrightarrow} X(k) \cdot Y(k)$, where $X(k)$ and $Y(k)$ are, respectively, 4-point DFTs of $x(n)$ and $y(n)$.

6.19. Given two sequences: $x(n) = \{2, 0, 1, 0\}$, $y(n) = \{3, 2, 1, 0\}$, $n = 0, 1, 2, 3$. Using results of Exercise 6.20 and performing appropriate calculations, check validity of multiplication property of DFT: $x(n) \cdot y(n) \overset{\text{DFT}}{\leftrightarrow} \frac{1}{4} \times \sum_{l=0}^{3} X(l)Y(k-l)$, where $X(k)$ and $Y(k)$ are, respectively, 4-point DFTs of $x(n)$ and $y(n)$.

6.20. Compute and plot the 8-point DFTs of the following signals:

(*a*) $x_1(n) = 1, n = 0, 1, \ldots, 7$; (b) $x_2(n) = \begin{cases} 1, n = 0 \\ 0, n = 1, 2, \ldots, 7. \end{cases}$

6.21. Compute the 8-point DFT of the signal $x(n) = \begin{cases} 1, n = 0, 4 \\ 0, n = 1, 2, 3, 5, 6, 7 \end{cases}$

Then, using the obtained results and chosen properties of DFT, determine the 8-point DFT of the sequence $y(n) = \left\{1, 0, \frac{1}{2}, 0, 1, 0, \frac{1}{2}, 0\right\}$.

Chapter 7
Analysis of Discrete-Time Systems in Frequency Domain

In this chapter, we apply the theory of discrete-time Fourier transformation (DTFT) presented in Chap. 6 to study the properties of discrete-time LTI systems (digital filters) in the frequency domain. The careful reader will be able to notice a lot of similarities to the frequency description of analog filters presented in Chap. 4. Here also, the notion of the frequency response function is introduced, as well as magnitude and phase characteristics. We will present ideal and real filters of different types (low-pass, high-pass, band-pass, and band-stop) and show how to calculate the response of a system to the known input signal, including constant and harmonic excitations. A special section will be devoted to the frequency transformation method used in digital filter design.

7.1 Frequency Response Function

Given a discrete-time ($n \in \mathbb{Z}$) signal $x(n) \overset{\mathcal{F}_d}{\leftrightarrow} X(\Omega)$ at the input of a discrete-time LTI system. At its output, we observe the discrete-time response $y(n) \overset{\mathcal{F}_d}{\leftrightarrow} Y(\Omega)$. From (3.5), we know that $y(n) = x(n) * h(n)$ where $h(n)$ is the system impulse response. Let us apply the convolution property of the DTFT to (3.5) and get

$$Y(\Omega) = X(\Omega)H(\Omega), \tag{7.1}$$

where $H(\Omega)$ is the DTFT of the impulse response $h(n)$, that is, $h(n) \overset{\mathcal{F}_d}{\leftrightarrow} H(\Omega)$.

Definition 7.1 The Discrete-Time Fourier Transform of the impulse response $h(n)$ of a discrete-time LTI system is called its *frequency response function*.

Evidently, we can also define the frequency response function as a function of the normalized frequency F. Let us illustrate the above relations solving next Problems 7.1 and 7.2.

K. M. Snopek, *Hands-on Signals and Systems Theory*,
https://doi.org/10.1007/978-3-031-56081-1_7

Problem 7.1 Given a discrete-time system with the frequency response function $H(\Omega)$ satisfying: $H\left(\frac{\pi}{2}\right) = 1$; $H(\Omega) = H(\Omega - \pi)$. Assuming the input-output relation $x(n) = \left(\frac{1}{2}\right)^n \cdot \mathbf{1}(n) \rightarrow y(n)$, where $y(n) = 0$ for $n < 0$ and for $n \geq 2$, calculate the impulse response $h(n)$ of this system.

Solution Using Table 6.1, we can get the DTFT of $x(n)$ in the form

$$X(\Omega) = \frac{1}{1 - \frac{1}{2}e^{-j\Omega}}.$$

The system output $y(n) = 0$ for $n < 0$ and for $n \geq 2$, so it can be written in a general form as $y(n) = A\delta(n) + B\delta(n - 1)$, where A, B are real constants. The DTFT transform of $y(n)$ is $Y(\Omega) = A + Be^{-j\Omega}$.

Using (7.1), we can calculate the frequency response function of the filter by the formula

$$H(\Omega) = \frac{Y(\Omega)}{X(\Omega)} = \frac{A + Be^{-j\Omega}}{\frac{1}{1-\frac{1}{2}e^{-j\Omega}}} = \left(A + Be^{-j\Omega}\right)\left(1 - \frac{1}{2}e^{-j\Omega}\right).$$

Knowing that $H\left(\frac{\pi}{2}\right) = 1$ and $H(\Omega) = H(\Omega - \pi)$, we can write the following set of equations:

$$\begin{cases} \left(A + Be^{-j\frac{\pi}{2}}\right)\left(1 - \frac{1}{2}e^{-j\frac{\pi}{2}}\right) = 1 \\ \left(A + Be^{-j\Omega}\right)\left(1 - \frac{1}{2}e^{-j\Omega}\right) = \left(A + Be^{-j(\Omega-\pi)}\right)\left(1 - \frac{1}{2}e^{-j(\Omega-\pi)}\right). \end{cases}$$

Then

$$\begin{cases} A - j\left(B - \frac{1}{2}A\right) + \frac{1}{2}B = 1 \\ A + \left(B - \frac{1}{2}A\right)e^{-j\Omega} - \frac{1}{2}Be^{-j2\Omega} = A + \left(\frac{1}{2}A - B\right)e^{-j\Omega} - \frac{1}{2}Be^{-j2\Omega} \end{cases},$$

$$\begin{cases} A - j\left(B - \frac{1}{2}A\right) + \frac{1}{2}B = 1 \\ \left(B - \frac{1}{2}A\right)e^{-j\Omega} = \left(\frac{1}{2}A - B\right)e^{-j\Omega} \end{cases},$$

$$\begin{cases} A - j\left(B - \frac{1}{2}A\right) + \frac{1}{2}B = 1 \\ 2B - A = 0 \end{cases}.$$

Substituting $A = 2B$ into the first equation, we easily get $2B - j(B - B) + \frac{1}{2}B = 1 \Leftrightarrow B = \frac{2}{5}$ and thus $A = \frac{4}{5}$. The frequency response function of the system is

$$H(\Omega) = \left(\frac{4}{5} + \frac{2}{5}e^{-j\Omega}\right)\left(1 - \frac{1}{2}e^{-j\Omega}\right) = \frac{4}{5} - \frac{1}{5}e^{-j2\Omega}$$

and the impulse response (equal to the inverse DTFT of $H(\Omega)$) is $h(n) = \frac{4}{5}\delta(n) - \frac{1}{5}\delta(n-2)$.

Problem 7.2 Given a discrete-time LTI system with the following input-output relation: $x(n) = \sum_{k\in\mathbb{Z}} \delta(n-4k) \rightarrow y(n) = \cos\left(\frac{\pi}{2}n + \frac{\pi}{4}\right)$. Determine values of the frequency response function $H(\Omega)$ for $\Omega = 0, \frac{\pi}{2}, \pi, \frac{3\pi}{2}$.

Solution The signal $x(n)$ is periodic with fundamental period $N_0 = 4$. Its spectrum (periodic) is given by (6.19), that is,

$$X(\Omega) = 2\pi \sum_{k\in\mathbb{Z}} c_k \delta\left(\Omega - \frac{\pi}{2}k\right)$$

where the coefficients c_k are calculated using (6.21) as follows:

$$c_k = \frac{1}{4}\sum_{n=0}^{3} x(n)e^{-j\frac{\pi}{2}kn} = \frac{1}{4}.$$

Then

$$X(\Omega) = \frac{\pi}{2}\sum_{k\in\mathbb{Z}} \delta\left(\Omega - \frac{\pi}{2}k\right).$$

Now, let us consider the system output $y(n)$ written in the equivalent form as

$$y(n) = \cos\left(\frac{\pi}{2}n + \frac{\pi}{4}\right) = \cos\left(\frac{\pi}{4}\right)\cos\left(\frac{\pi}{2}n\right) - \sin\left(\frac{\pi}{4}\right)\sin\left(\frac{\pi}{2}n\right) = \frac{1}{\sqrt{2}}\left[\cos\left(\frac{\pi}{2}n\right) - \sin\left(\frac{\pi}{2}n\right)\right].$$

Its spectrum $Y(\Omega)$ for $|\Omega| \leq \pi$ (see Table 6.1) is

$$\begin{aligned} Y(\Omega) &= \frac{\pi}{\sqrt{2}}\left[\delta\left(\Omega - \frac{\pi}{2}\right) + \delta\left(\Omega + \frac{\pi}{2}\right) + j\delta\left(\Omega - \frac{\pi}{2}\right) - j\delta\left(\Omega + \frac{\pi}{2}\right)\right] \\ &= \frac{\pi}{\sqrt{2}}\left[(1+j)\delta\left(\Omega - \frac{\pi}{2}\right) + (1-j)\delta\left(\Omega + \frac{\pi}{2}\right)\right]. \end{aligned}$$

From (7.1), we have $H(\Omega) = Y(\Omega)/X(\Omega)$. Let us calculate values of $H(\Omega)$ for $\Omega = 0, \frac{\pi}{2}, \pi, \frac{3\pi}{2}$. We have.

$$H(0) = \frac{Y(0)}{X(0)} = \frac{0}{\pi/2} = 0; H\left(\frac{\pi}{2}\right) = \frac{Y(\pi/2)}{X(\pi/2)} = \frac{\frac{\pi(1+j)}{\sqrt{2}}}{\pi/2} = \sqrt{2}(1+j);$$

$$H(\pi) = \frac{Y(\pi)}{X(\pi)} = \frac{0}{\pi/2} = 0; H\left(\frac{3\pi}{2}\right) = \frac{Y(3\pi/2)}{X(3\pi/2)} = \frac{Y(-\pi/2)}{\pi/2} = \frac{\frac{\pi(1-j)}{\sqrt{2}}}{\pi/2} = \sqrt{2}(1-j).$$

In general, $H(\Omega)$ is a periodic (with period 2π) complex-valued function. It can be expressed as

$$H(\Omega) = |H(\Omega)| e^{j\arg H(\Omega)} = A(\Omega) e^{j\varphi(\Omega)}, \tag{7.2}$$

where $A(\Omega) = |H(\Omega)|$ is the *magnitude response* and $\varphi(\Omega) = \arg H(\Omega)$—the *phase response* of a system. Both characteristics are periodic with period 2π. That is why we usually plot them for $\Omega \in [-\pi, \pi]$. The magnitude response is always an even function of Ω: $A(\Omega) = A(-\Omega)$ while the phase response is an odd function, that is, $\varphi(-\Omega) = -\varphi(\Omega)$.

Definition 7.2 The *real frequency response* $P(\Omega)$ is the real part of the complex-valued frequency response function $H(\Omega)$. The imaginary part $Q(\Omega)$ is called the *imaginary frequency response*. We have

$$H(\Omega) = P(\Omega) + jQ(\Omega). \tag{7.3}$$

From (7.2) and (7.3), we get the following relations:

$$A(\Omega) = \sqrt{[P(\Omega)]^2 + [Q(\Omega)]^2}, \tag{7.4}$$

$$\varphi(\Omega) = \text{atan2}(Q(\Omega), P(\Omega)). \tag{7.5}$$

The real frequency response is always an even function of Ω: $P(\Omega) = P(-\Omega)$ and the imaginary frequency response is an odd one, that is, $Q(-\Omega) = -Q(\Omega)$. Evidently, they are also periodic with period 2π.

In case, when the frequency response of a discrete-time system is a function of F (normalized frequency), all characteristics (amplitude-, phase-, real-, and imaginary-) denoted, respectively, with $A(F)$, $\varphi(F)$, $P(F)$, and $Q(F)$, are periodic with period 1. Their plots are most often sketched for $|F| \leq 1/2$. In Problem 7.3, we will show how to plot different frequency characteristics of a 1st order FIR filter.

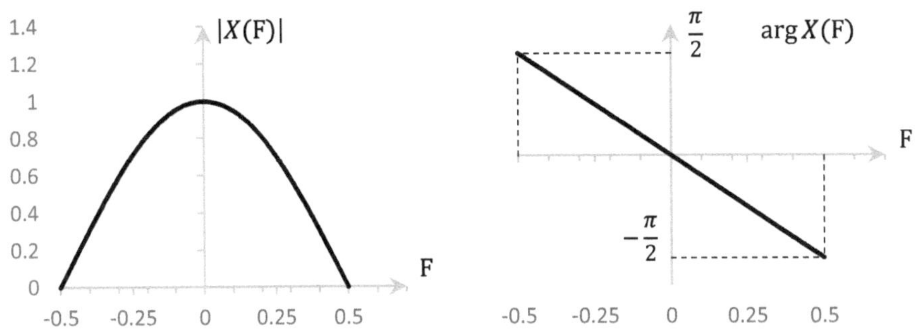

Fig. 7.1 Amplitude and phase responses of the filter from Problem 7.3

Problem 7.3 Given the FIR filter defined by the input-output equation $y(n) = \frac{1}{2}[x(n) + x(n-1)]$. Determine and sketch its amplitude and phase responses as functions of the normalized frequency F.

Solution The impulse response of the filter is $h(n) = \frac{1}{2}[\delta(n) + \delta(n-1)]$. The frequency response $H(\Omega)$ is calculated by means of Tables 6.1 and 6.2 ($\delta(n) \overset{\mathcal{F}_d}{\leftrightarrow} 1$, time shifting property of DTFT). We get

$$H(\Omega) = \frac{1}{2}\left(1 + e^{-j\Omega}\right) = \frac{1}{2}(1 + \cos\Omega - j\sin\Omega).$$

The amplitude response $A(\Omega)$ is calculated using (7.4) as

$$A(\Omega) = \sqrt{\left[\frac{1}{2}(1 + \cos\Omega)\right]^2 + \left[-\frac{1}{2}\sin(\Omega)\right]^2} = \sqrt{\frac{1}{2} + \frac{1}{2}\cos\Omega} = \sqrt{\cos^2\left(\frac{\Omega}{2}\right)}$$
$$= \left|\cos\left(\frac{\Omega}{2}\right)\right|.$$

Since $\Omega = 2\pi\text{F}$, then $A(\text{F}) = |\cos(\pi\text{F})|$. Its plot is shown in Fig. 7.1 (left). We can see that the considered system is a low-pass filter. Let us now calculate its phase response $\varphi(\Omega)$ as follows:

$$\varphi(\Omega) = \arg\left[\frac{1}{2} + \frac{1}{2}\cos\Omega - j\frac{1}{2}\sin\Omega\right] = -\operatorname{atan}\left(\frac{\sin\Omega}{1 + \cos\Omega}\right)$$
$$= -\operatorname{atan}\left(\frac{2\sin\left(\frac{\Omega}{2}\right)\cos\left(\frac{\Omega}{2}\right)}{2\cos^2\left(\frac{\Omega}{2}\right)}\right) = -\operatorname{atan}\left(\tan\left(\frac{\Omega}{2}\right)\right) = -\frac{\Omega}{2},$$

then $\varphi(\text{F}) = -\pi\text{F}$. Its plot is shown in Fig. 7.1 (right). We can notice that the phase response of the filter is a linear function.

7.2 Spectrum Input-Output Relations

Using (7.1) and (7.2), we can derive important relations between amplitude- and phase spectra of input and output signals. The amplitude spectrum $|Y(\Omega)|$ of the output signal $y(n)$ is

$$|Y(\Omega)| = |X(\Omega)| \cdot |H(\Omega)| = |X(\Omega)| \cdot A(\Omega) \tag{7.6}$$

while the phase spectrum arg $Y(\Omega)$ is

$$\arg Y(\Omega) = \arg X(\Omega) + \arg H(\Omega) = \arg X(\Omega) + \varphi(\Omega). \tag{7.7}$$

In analogy to the theory presented in Sect. 4.2 (Chap. 4), let us present the response of a discrete-time LTI system to the input signal in the form of: a complex harmonic signal $e^{j\Omega_0 n}$, a cosine $\cos(\Omega_0 n)$ and a sine $\sin(\Omega_0 n)$.

7.2.1 Response to a Complex Harmonic Excitation

Using an analogous method as in Sect. 4.2, we can derive the following input-output relation:

$$e^{j\Omega_0 n} \rightarrow e^{j\Omega_0 n} H(\Omega_0) \tag{7.8}$$

or its special case, if $\Omega_0 = 0$, that is,

$$K \rightarrow K \cdot H(0), \tag{7.9}$$

where K is a constant signal, $n \in \mathbb{Z}$.

7.2.2 Response to a Harmonic Excitation

Let us consider signals $\cos(\Omega_0 n)$ or $\sin(\Omega_0 n)$, $n \subset \mathbb{Z}$, at the input of a discrete time filter. In this case, the input-output relations are the following:

$$\cos(\Omega_0 n) \rightarrow A(\Omega_0)\cos(\Omega_0 n + \varphi(\Omega_0)) = \mathrm{Re}\{e^{j\Omega_0 n} H(\Omega_0)\}, \tag{7.10}$$

$$\sin(\Omega_0 n) \rightarrow A(\Omega_0)\sin(\Omega_0 n + \varphi(\Omega_0)) = \mathrm{Im}\{e^{j\Omega_0 n} H(\Omega_0)\}. \tag{7.11}$$

Let us show how to use the above relations solving Problem 7.4.

Problem 7.4 Given the IIR filter of the input-output equation $y(n) - \frac{1}{2}y(n-1) = x(n)$. Determine its amplitude and phase responses. Then assuming the input $x(n) = \frac{1}{2} + \cos\left(\frac{\pi}{3}n\right), n \in \mathbb{Z}$, calculate the signal $y(n)$ at filter output.

Solution Let us write the filter equation $y(n) - \frac{1}{2}y(n-1) = x(n)$ in the Ω–domain, that is, $Y(\Omega) - \frac{1}{2}Y(\Omega)e^{-j\Omega} = X(\Omega)$ (see Table 6.2—time shift property). From here, we can get the frequency response function in the form

$$H(\Omega) = \frac{Y(\Omega)}{X(\Omega)} = \frac{1}{1 - \frac{1}{2}e^{-j\Omega}}.$$

In Problem 6.8, we have already dealt with the same Discrete-time Fourier transform. We easily get that the magnitude of $H(\Omega)$ (amplitude response) is

$$A(\Omega) = \frac{1}{\sqrt{5/4 - \cos\Omega}}$$

and the argument of $H(\Omega)$ (phase response) is

$$\varphi(\Omega) = -\operatorname{atan}\left(\frac{\frac{1}{2}\sin(\Omega)}{1 - \frac{1}{2}\cos(\Omega)}\right).$$

The system is linear, so the output $y(n)$ is the sum of two responses: to the constant signal $\frac{1}{2}$ and to the cosine excitation $\cos\left(\frac{\pi}{3}n\right)$. Using (7.9) and (7.10), we get

$$y(n) = \frac{1}{2} \cdot H(0) + A\left(\frac{\pi}{3}\right)\cos\left(\frac{\pi}{3}n + \varphi\left(\frac{\pi}{3}\right)\right)$$

or equivalently,

$$y(n) = \frac{1}{2} \cdot H(0) + \operatorname{Re}\left\{e^{j\frac{\pi}{3}n}H\left(\frac{\pi}{3}\right)\right\}.$$

Let us show that both options give the same result. We have: $H(0) = \frac{1}{1 - \frac{1}{2}e^{-j0}} = 2$,

$$A\left(\frac{\pi}{3}\right)=\frac{1}{\sqrt{\frac{5}{4}-\cos\left(\frac{\pi}{3}\right)}}=\frac{2\sqrt{3}}{3},$$

$$\varphi\left(\frac{\pi}{3}\right)=-\operatorname{atan}\left(\frac{\frac{1}{2}\sin\left(\frac{\pi}{3}\right)}{1-\frac{1}{2}\cos\left(\frac{\pi}{3}\right)}\right)=-\operatorname{atan}\left(\frac{\frac{\sqrt{3}}{4}}{1-\frac{1}{4}}\right)=-\frac{\pi}{6}.$$

The output signal

$$y(n)=1+\frac{2\sqrt{3}}{3}\cos\left(\frac{\pi}{3}n-\frac{\pi}{6}\right)=1+\frac{2\sqrt{3}}{3}\left[\cos\left(\frac{\pi}{3}n\right)\cos\left(\frac{\pi}{6}\right)+\sin\left(\frac{\pi}{3}n\right)\sin\left(\frac{\pi}{6}\right)\right]$$
$$=1+\cos\left(\frac{\pi}{3}n\right)+\frac{\sqrt{3}}{3}\sin\left(\frac{\pi}{3}n\right).$$

or equivalently,

$$y(n)=1+\operatorname{Re}\left\{e^{j\frac{\pi}{3}n}\left(\frac{1}{1-\frac{1}{2}e^{-j\frac{\pi}{3}}}\right)\right\}$$
$$=1+\operatorname{Re}\left\{e^{j\frac{\pi}{3}n}\left(\frac{1}{1-\frac{1}{2}\left(\cos\frac{\pi}{3}-j\sin\frac{\pi}{3}\right)}\right)\right\}$$
$$=1+\operatorname{Re}\left\{e^{j\frac{\pi}{3}n}\left(\frac{1}{\frac{3}{4}+\frac{\sqrt{3}}{4}j}\right)\right\}=1+4\operatorname{Re}\left\{e^{j\frac{\pi}{3}n}\left(\frac{3-\sqrt{3}j}{12}\right)\right\}$$
$$=\quad 1+\operatorname{Re}\left\{e^{j\frac{\pi}{3}n}\left(\frac{3-\sqrt{3}j}{3}\right)\right\}=\frac{1}{2}+\cos\left(\frac{\pi}{3}n\right)+\frac{\sqrt{3}}{3}\sin\left(\frac{\pi}{3}n\right).$$

We have shown that using two different methods, we came to the same result.

7.3 Discrete-Time Ideal Filters

In Sect. 4.3, we presented the theory of analog ideal filters of four types: low-pass (LP), high-pass (HP), band-pass (BP), and band-stop (BS). In full analogy to this, in the Ω–domain, we also deal with ideal filters. The only difference is that their frequency response functions are periodic with period 2π. We will present them in brief.

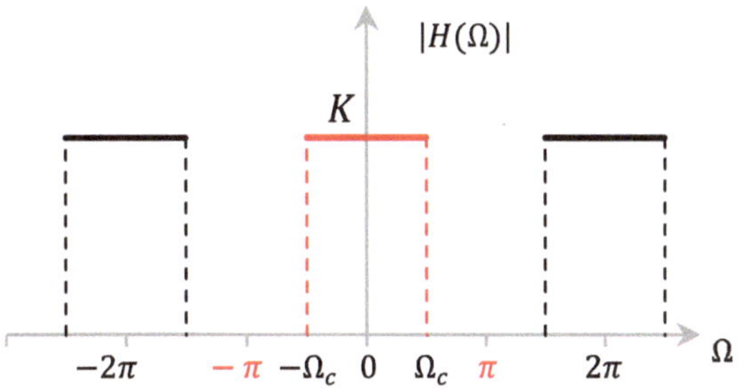

Fig. 7.2 Amplitude response $|H(\Omega)|$ of a discrete-time ideal LP filter. In red: $|H(\Omega)|$ for $|\Omega| \leq \pi$

7.3.1 Discrete-Time Ideal LP Filter

The frequency response function $H(\Omega)$ of a discrete-time ideal LP filter for $|\Omega| \leq \pi$ is given by

$$H(\Omega) = \begin{cases} Ke^{-j\Omega n_0}, & |\Omega| \leq \Omega_c, \\ 0 & \text{elsewhere} \end{cases} \tag{7.12}$$

where $\Omega_c \in (0, \pi)$ is the cutoff angular frequency. In Fig. 7.2, we present the amplitude response of this filter for $|\Omega| \leq 2\pi + \Omega_c$, drawing the reader's attention to the periodicity of this characteristic.

7.3.2 Discrete-Time Ideal HP Filter

The frequency response function $H(\Omega)$ of a discrete-time ideal HP filter for $|\Omega| \leq \pi$ is given by

$$H(\Omega) = \begin{cases} Ke^{-j\Omega n_0}, & \Omega_c \leq |\Omega| \leq \pi, \\ 0 & \text{elsewhere} \end{cases} \tag{7.13}$$

where $\Omega_c \in (0, \pi)$ is the cutoff angular frequency. In Fig. 7.3, we show the periodic amplitude response of this filter for $|\Omega| \leq 2\pi$.

7.3.3 Discrete-Time Ideal BP Filter

The frequency response function $H(\Omega)$ of the discrete-time ideal BP filter for $|\Omega| \leq \pi$ is given by

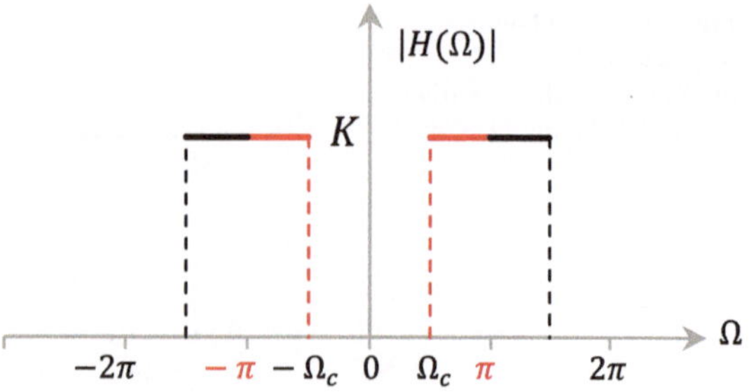

Fig. 7.3 Amplitude response $|H(\Omega)|$ of the discrete-time ideal HP filter. In red: $|H(\Omega)|$ for $|\Omega| \leq \pi$

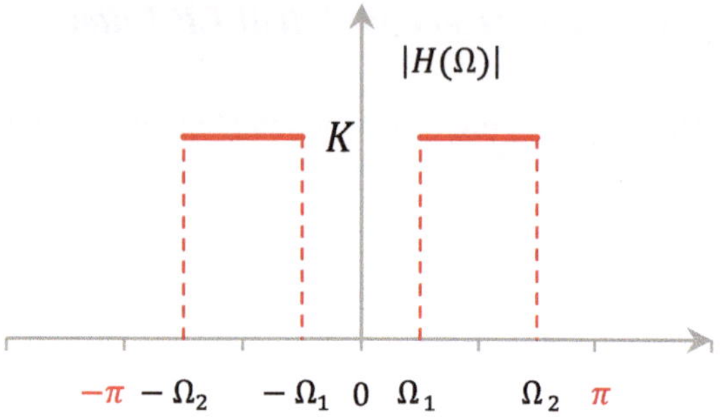

Fig. 7.4 Amplitude response $|H(\Omega)|$ of the discrete-time ideal BP filter

$$H(\Omega) = \begin{cases} Ke^{-j\Omega n_0}, & \Omega_1 \leq |\Omega| \leq \Omega_2, \\ 0 & \text{elsewhere} \end{cases} \tag{7.14}$$

where $\Omega_1, \Omega_2 \in (0, \pi)$ limit the passband of the filter. Let us look at Fig. 7.4, showing the periodic amplitude response of this filter, presented only for $|\Omega| \leq \pi$.

7.3.4 Discrete-Time Ideal BS Filter

The frequency response function $H(\Omega)$ of a discrete-time ideal BS filter for $|\Omega| \leq \pi$ is given by

$$H(\Omega) = \begin{cases} Ke^{-j\Omega n_0}, & \Omega \in [-\pi, -\Omega_2] \cup [-\Omega_1, \Omega_1] \cup [\Omega_2, \pi], \\ 0 & \text{elsewhere} \end{cases} \tag{7.15}$$

where $\Omega_1, \Omega_2 \in (0, \pi)$ limit the stopband of the filter. Figure 7.5 shows the periodic amplitude response of this filter for $|\Omega| \leq \pi + \Omega_2$.

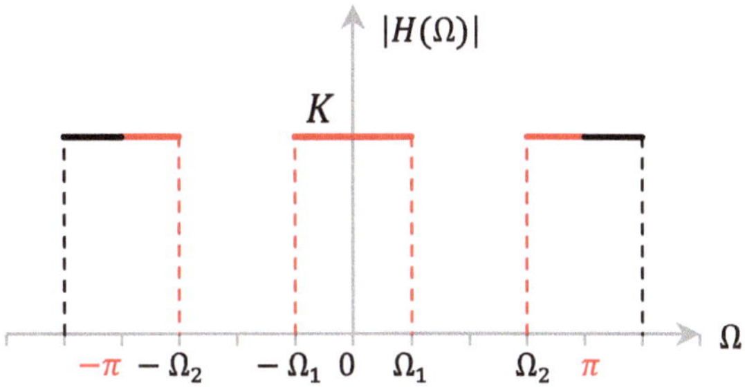

Fig. 7.5 Amplitude response $|H(\Omega)|$ of the discrete-time ideal BS filter

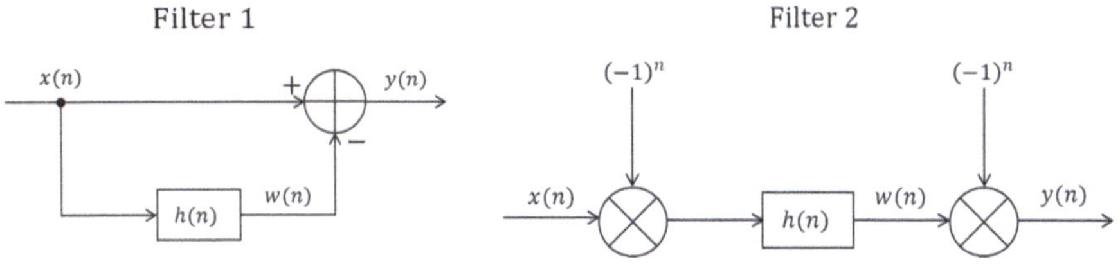

Fig. 7.6 Block diagrams of two discrete-time ideal filters from Problem 7.5

Problem 7.5 Given two discrete-time ideal filters, 1 and 2, with block diagrams shown in Fig. 7.6, where $h(n)$ is the impulse response of an ideal low-pass filter with unit gain, cutoff angular frequency $\Omega_c = \frac{\pi}{4}$, and zero phase response. Compute the frequency response functions of these filters and determine the type (LP/HP/BP/BS) of filtration.

Solution According to (7.12), the frequency response function of the considered ideal LP filter is given by

$$H_{\mathrm{LP}}(\Omega) = \begin{cases} 1, & |\Omega| \leq \dfrac{\pi}{4}, \\ 0 & \text{elsewhere} \end{cases}$$

(see Fig. 7.7).

First, let us consider Filter 1. Basing on its block diagram, we can write the following equalities:

$$\begin{cases} w(n) = x(n) * h(n) \\ y(n) = x(n) - w(n). \end{cases}$$

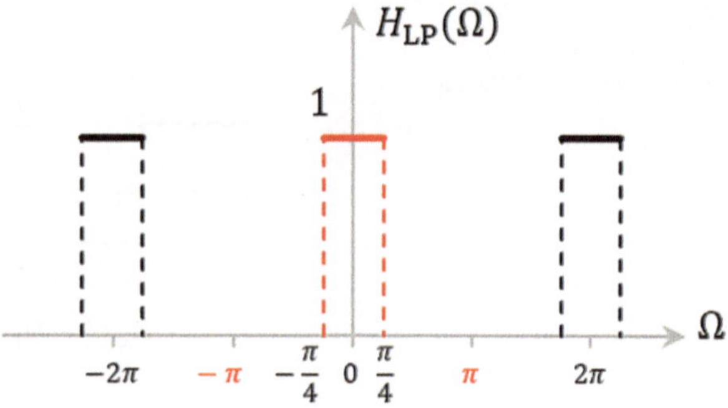

Fig. 7.7 Frequency response function of the LP filter from Problem 7.5

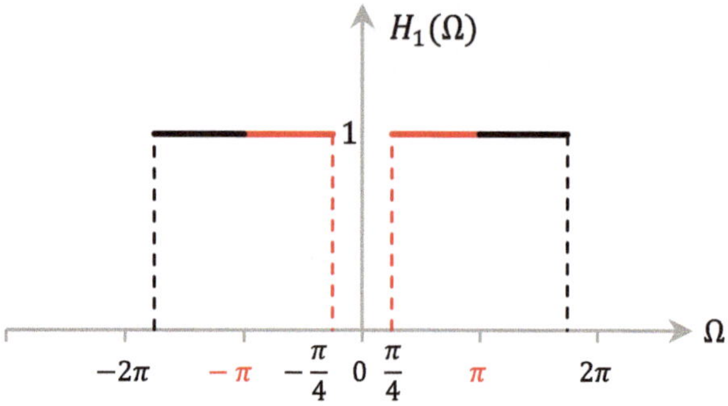

Fig. 7.8 Frequency response function of Filter 1 from Fig. 7.6

Assuming $x(n) \overset{\mathcal{F}_d}{\leftrightarrow} X(\Omega)$, $y(n) \overset{\mathcal{F}_d}{\leftrightarrow} Y(\Omega)$, $w(n) \overset{\mathcal{F}_d}{\leftrightarrow} W(\Omega)$ and using the linearity of DTFT and the convolution theorem, we can write the above set of equations in the domain Ω as follows:

$$\begin{cases} W(\Omega) = X(\Omega) \cdot H_{\text{LP}}(\Omega) \\ Y(\Omega) = X(\Omega) - W(\Omega). \end{cases}$$

Substituting the first equation into the second one, we easily obtain

$$Y(\Omega) = X(\Omega) - X(\Omega) \cdot H_{\text{LP}}(\Omega) = X(\Omega)(1 - H_{\text{LP}}(\Omega)),$$

from where we get the frequency response function $H_1(\Omega)$ of Filter 1 in the form

$$H_1(\Omega) = \frac{Y(\Omega)}{X(\Omega)} = 1 - H_{\text{LP}}(\Omega).$$

The plot of $H_1(\Omega)$ is displayed in Fig. 7.8. We can see that Filter 1 is a discrete-time ideal HP filter with cutoff angular frequency $\Omega_c = \frac{\pi}{4}$ [rad].

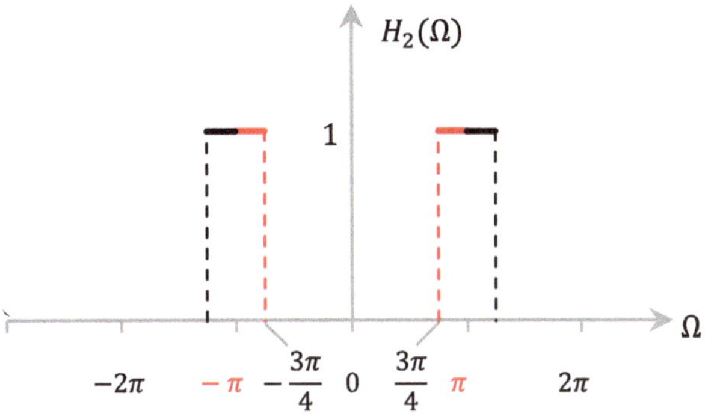

Fig. 7.9 Frequency response function of Filter 2 from Problem 7.5

Now, let us consider the Filter 2. Notice that $(-1)^n = e^{j\pi n}$. Basing on the block diagram of Filter 2, we can write the following set of equations:

$$\begin{cases} w(n) = x(n)e^{j\pi n} * h(n) \\ y(n) = w(n)e^{j\pi n}. \end{cases}$$

Applying to the above set the frequency shift property of the DTFT (Table 6.2) and the convolution theorem, we get

$$\begin{cases} W(\Omega) = X(\Omega - \pi) \cdot H_{\mathrm{LP}}(\Omega) \\ Y(\Omega) = W(\Omega - \pi). \end{cases}$$

Substituting the first equality into the second one, we obtain

$$Y(\Omega) = X(\Omega - 2\pi) \cdot H_{\mathrm{LP}}(\Omega - \pi)$$

but $X(\Omega - 2\pi) = X(\Omega)$ (2π–periodicity), then

$$Y(\Omega) = X(\Omega) \cdot H_{\mathrm{LP}}(\Omega - \pi).$$

The frequency response function $H_2(\Omega)$ of Filter 2 is expressed as

$$H_2(\Omega) = \frac{Y(\Omega)}{X(\Omega)} = H_{\mathrm{LP}}(\Omega - \pi).$$

To sketch its plot, we should shift the frequency response function from Fig. 7.7. right by π. The resulting plot is shown in Fig. 7.9. Once again, we deal with the discrete-time ideal HP filter with cutoff angular frequency $\Omega_c = \frac{3\pi}{4}$ [rad].

7.4 Discrete-Time Real Filters

Let us recall that in Chap. 4, we have got familiar with frequency response functions of analog real filters. We have already defined the 3 dB bandwidth for LP and BP analog filters. The definitions (4.16)–(4.17) are also valid for discrete-time filters, so we do not intend to repeat them here. However, we will illustrate this theory with some solved problems.

Problem 7.6 Given the "input-output" relation: $x(n) = \left(\frac{2}{3}\right)^n \cdot \mathbf{1}(n) \rightarrow y(n) = n\left(\frac{2}{3}\right)^n \cdot \mathbf{1}(n)$. Determine the frequency response function $H(\Omega)$ and sketch the magnitude response $A(F)$. Calculate the 3 dB-bandwidth. What is the input-output difference equation of this filter?

Solution Let us start with computing spectra $X(\Omega)$, $Y(\Omega)$ of $x(n)$ and $y(n)$. Using Table 6.1, we get

$$X(\Omega) = \frac{1}{1 - \frac{2}{3}e^{-j\Omega}}.$$

From Table 6.2 (derivation in frequency), we have

$$Y(\Omega) = j\frac{dX(\Omega)}{d\Omega} = j\frac{d}{d\Omega}\left(\frac{1}{1 - \frac{2}{3}e^{-j\Omega}}\right) = j\frac{\frac{2}{3}e^{-j\Omega}(-j)}{\left(1 - \frac{2}{3}e^{-j\Omega}\right)^2} = \frac{\frac{2}{3}e^{-j\Omega}}{\left(1 - \frac{2}{3}e^{-j\Omega}\right)^2}.$$

Then, applying (7.1), the frequency response function $H(\Omega)$ gets the form

$$H(\Omega) = \frac{Y(\Omega)}{X(\Omega)} = \frac{\frac{2}{3}e^{-j\Omega}}{1 - \frac{2}{3}e^{-j\Omega}}.$$

The magnitude response $A(\Omega)$ of the filter is

$$A(\Omega) = \left|\frac{\frac{2}{3}e^{-j\Omega}}{1 - \frac{2}{3}e^{-j\Omega}}\right| = \frac{\frac{2}{3}}{\left|1 - \frac{2}{3}e^{-j\Omega}\right|} = \frac{2}{|3 - 2\cos\Omega + 2j\sin\Omega|}$$
$$= \frac{2}{\sqrt{(3 - 2\cos\Omega)^2 + 4\sin^2\Omega}} = \frac{2}{\sqrt{13 - 12\cos\Omega}}$$

or

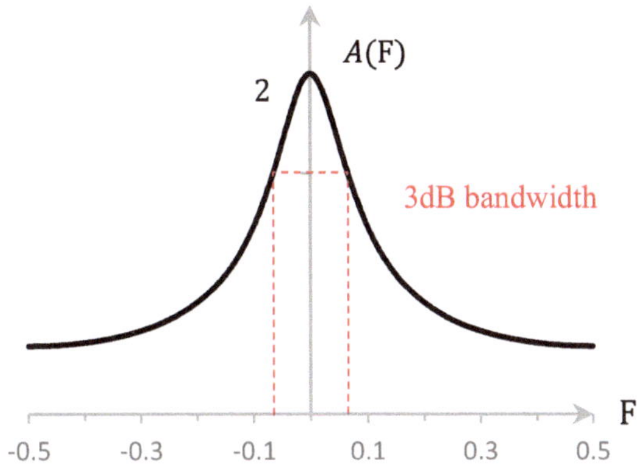

Fig. 7.10 Magnitude response of the LP filter from Problem 7.6. In red: the 3 dB-bandwidth

$$A(\mathrm{F}) = \frac{2}{\sqrt{13 - 12\cos(2\pi\mathrm{F})}}$$

The plot of $A(\mathrm{F})$ is shown in Fig. 7.10. We can see that we deal with an LP filter with cutoff frequency $\mathrm{F_c}$, defined by the 3 dB-bandwidth $[-\mathrm{F_c}, \mathrm{F_c}]$, calculated using the formula analogous to (4.16), that is, $A(\mathrm{F_c}) = \frac{A_{\max}}{\sqrt{2}}$, where $A_{\max} = A(0) = 2$. We have

$$A(\mathrm{F_c}) = \frac{2}{\sqrt{13 - 12\cos(2\pi\mathrm{F_c})}} = \frac{2}{\sqrt{2}},$$

$$\sqrt{13 - 12\cos(2\pi\mathrm{F_c})} = \sqrt{2} \Leftrightarrow \cos(2\pi\mathrm{F_c}) = \frac{11}{12} \Leftrightarrow \mathrm{F_c} \cong 0,0654.$$

The 3 dB-bandwidth [−0.0654,0.0654] is marked with red color in Fig. 7.10.

To obtain an input-output equation of the filter, we start with the frequency response function $H(\Omega)$ to first obtain a system equation in the domain Ω. We have

$$\frac{Y(\Omega)}{X(\Omega)} = \frac{\frac{2}{3}e^{-j\Omega}}{1 - \frac{2}{3}e^{-j\Omega}} \Leftrightarrow \left(1 - \frac{2}{3}e^{-j\Omega}\right)Y(\Omega) = \frac{2}{3}e^{-j\Omega} \cdot X(\Omega),$$

from where, using the inverse DTFT, we get the difference equation of the filter:

$$y(n) - \frac{2}{3}y(n-1) = \frac{2}{3}x(n-1).$$

7.4.1 Frequency Transformation

The method of *frequency transformation* in analog filter design has already been presented in Chap. 4. In digital filter design, it is also used and its principle is very similar. Starting from a prototype digital LP filter with the frequency response function $H(\Omega)$ and with the 3 dB cutoff normalized angular frequency Ω_{c_1}, we can design digital filters of different types (LP, HP, BP, BS) using formulas collected in Table 7.1. We only need to replace $e^{-j\Omega}$ in $H(\Omega)$ with the corresponding mathematical formula marked in red. We will illustrate this method solving Problem 7.6.

Problem 7.7 Given the digital LP filter from Problem 7.5 with the frequency response function $H(\Omega)=\frac{(2/3)e^{-j\Omega}}{1-(2/3)e^{-j\Omega}}$ and with the normalized cutoff angular frequency $\Omega_{c_1}=\mathrm{acos}(11/12)$. Using the frequency transformation method, design: (a) an LP filter with the cutoff angular frequency $\Omega_{c_2}=2\Omega_{c_1}$ and sketch its magnitude response $A_{LP}(F)$; (b) an HP filter with the cutoff angular frequency $\Omega_{c_2}=2\Omega_{c_1}$ and sketch its magnitude response $A_{HP}(F)$.

Solution (a) Let us first design an LP digital filter with the cutoff angular frequency $\Omega_{c_2}=2\Omega_{c_1}$ where $\Omega_{c_1}=\mathrm{acos}(11/12) \Leftrightarrow \cos\Omega_{c_1}=11/12$. Using some trigonometric equalities, we can calculate the value of the parameter α from Table 7.1.

Table 7.1 Frequency transformation formulas for digital filter design

Type of a designed filter	Frequency response function of a designed filter	
LP (Ω_{c_2}—3 dB normalized angular frequency)	$H_{LP}(\Omega)=H\left(\frac{e^{-j\Omega}-\alpha}{1-\alpha e^{-j\Omega}}\right)$	$\alpha=\frac{\sin\left(\frac{\Omega_{c_1}-\Omega_{c_2}}{2}\right)}{\sin\left(\frac{\Omega_{c_1}+\Omega_{c_2}}{2}\right)}$
HP (Ω_{c_2}—3 dB normalized angular frequency)	$H_{HP}(\Omega)=H\left(-\frac{e^{-j\Omega}+\alpha}{1+\alpha e^{-j\Omega}}\right)$	$\alpha=-\frac{\cos\left(\frac{\Omega_{c_1}+\Omega_{c_2}}{2}\right)}{\cos\left(\frac{\Omega_{c_1}-\Omega_{c_2}}{2}\right)}$
BP ($[\Omega_1,\Omega_2]$—3 dB passband)	$H_{BP}(\Omega)=H\left(\frac{e^{-j2\Omega}+\frac{k-1}{k+1}}{\frac{k-1}{k+1}e^{-j2\Omega}-\frac{2\alpha k}{k+1}e^{-j\Omega}+1}\right)$	$\alpha=\frac{\cos\left(\frac{\Omega_1+\Omega_2}{2}\right)}{\cos\left(\frac{\Omega_2-\Omega_1}{2}\right)}$ $k=\cot\left(\frac{\Omega_2-\Omega_1}{2}\right)\tan\left(\frac{\Omega_{c_1}}{2}\right)$
BS ($[\Omega_1,\Omega_2]$—3 dB stopband	$H_{BS}(\Omega)=H\left(\frac{e^{-j2\Omega}-\frac{2\alpha}{k+1}e^{-j\Omega}+\frac{1-k}{k+1}}{\frac{1-k}{k+1}e^{-j2\Omega}-\frac{2\alpha}{k+1}e^{-j\Omega}+1}\right)$	$\alpha=\frac{\cos\left(\frac{\Omega_1+\Omega_2}{2}\right)}{\cos\left(\frac{\Omega_2-\Omega_1}{2}\right)}$ $k=\tan\left(\frac{\Omega_2-\Omega_1}{2}\right)\tan\left(\frac{\Omega_{c_1}}{2}\right)$

$$\alpha = \frac{\sin\left(\dfrac{\Omega_{c_1} - 2\Omega_{c_1}}{2}\right)}{\sin\left(\dfrac{\Omega_{c_1} + 2\Omega_{c_1}}{2}\right)} = -\frac{\sin(\Omega_{c_1}/2)}{\sin(\Omega_{c_1} + \Omega_{c_1}/2)}$$

$$= -\frac{\sin(\Omega_{c_1}/2)}{\sin(\Omega_{c_1})\cos(\Omega_{c_1}/2) + \cos(\Omega_{c_1})\sin(\Omega_{c_1}/2)}$$

$$= -\frac{\sin(\Omega_{c_1}/2)}{2\sin(\Omega_{c_1}/2)\cos^2(\Omega_{c_1}/2) + \cos(\Omega_{c_1})\sin(\Omega_{c_1}/2)}$$

$$= -\frac{1}{2\cos^2(\Omega_{c_1}/2) + \cos(\Omega_{c_1})} = -\frac{1}{1 + 2\cos(\Omega_{c_1})},$$

then $\alpha = -6/17$. Since the frequency response function of the LP prototype filter is

$$H(\Omega) = \frac{(2/3)e^{-j\Omega}}{1 - (2/3)e^{-j\Omega}},$$

to determine the frequency response function $H_{\mathrm{LP}}(\Omega)$ of the designed filter, we use the formula from Table 7.1, replacing $e^{-j\Omega}$ in $H(\Omega)$ with

$$\frac{e^{-j\Omega} - \alpha}{1 - \alpha e^{-j\Omega}} = \frac{e^{-j\Omega} + 6/17}{1 + (6/17)e^{-j\Omega}},$$

That is,

$$H_{\mathrm{LP}}(\Omega) = \frac{\dfrac{2}{3}\left(\dfrac{e^{-j\Omega} + 6/17}{1 + (6/17)e^{-j\Omega}}\right)}{1 - \dfrac{2}{3}\left(\dfrac{e^{-j\Omega} + 6/17}{1 + (6/17)e^{-j\Omega}}\right)} = \frac{2(e^{-j\Omega} + 6/17)}{3(1 + (6/17)e^{-j\Omega}) - 2(e^{-j\Omega} + (6/17))}$$

$$= \frac{2(e^{-j\Omega} + 6/17)}{39/17 - (6/17)e^{-j\Omega}} = \frac{2(17e^{-j\Omega} + 6)}{39 - 16e^{-j\Omega}}.$$

Let us calculate the magnitude response $\mathrm{A_{LP}}(\Omega)$ of the designed filter as follows:

$$\mathrm{A_{LP}}(\Omega) = 2\left|\frac{17e^{-j\Omega} + 6}{39 - 16e^{-j\Omega}}\right| = 2\left|\frac{17(\cos\Omega - j\sin\Omega) + 6}{39 - 16(\cos\Omega - j\sin\Omega)}\right|$$

$$= 2\sqrt{\frac{(6 + 17\cos\Omega)^2 + 289\sin^2\Omega}{(39 - 16\cos\Omega)^2 + 256\sin^2\Omega}} = 2\sqrt{\frac{325 + 204\cos\Omega}{1777 - 1248\cos\Omega}}.$$

Then

$$\mathrm{A_{LP}}(\mathrm{F}) = 2\sqrt{\frac{325 + 204\cos(2\pi\mathrm{F})}{1777 - 1248\cos(2\pi\mathrm{F})}}.$$

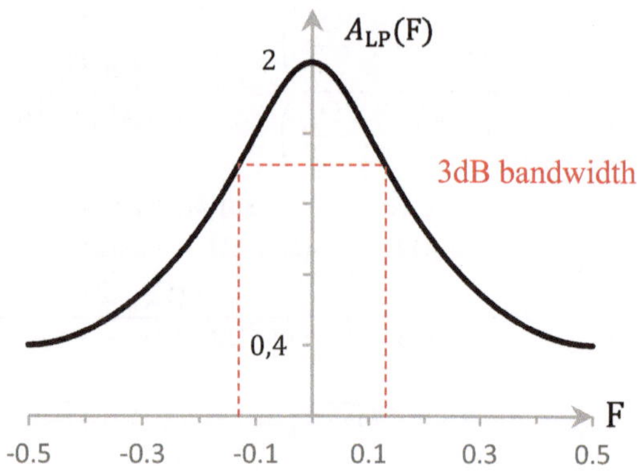

Fig. 7.11 Magnitude response of the designed LP filter from Problem 7.7. In red: the 3 dB-bandwidth

The plot of $A_{LP}(F)$ is shown in Fig. 7.11. The 3 dB-bandwidth $[-F_{c_2}, F_{c_2}] = [-0.1308, 0.1308]$ is marked in red.

(b) Now, let us design an HP filter with the cutoff angular frequency $\Omega_{c_2} = 2\Omega_{c_1}$ where $\cos\Omega_{c_1} = 11/12$. Using some trigonometric equalities, we calculate the value of the parameter α from Table 7.1.

$$\begin{aligned}\alpha &= -\frac{\cos\left(\dfrac{\Omega_{c_1} + 2\Omega_{c_1}}{2}\right)}{\cos\left(\dfrac{\Omega_{c_1} - 2\Omega_{c_1}}{2}\right)} = -\frac{\cos(\Omega_{c_1} + \Omega_{c_1}/2)}{\cos(\Omega_{c_1}/2)} \\ &= -\frac{\cos(\Omega_{c_1})\cos(\Omega_{c_1}/2) - \sin(\Omega_{c_1})\sin(\Omega_{c_1}/2)}{\cos(\Omega_{c_1}/2)} \\ &= -\frac{\cos(\Omega_{c_1})\cos(\Omega_{c_1}/2) - 2\sin^2(\Omega_{c_1}/2)\cos(\Omega_{c_1}/2)}{\cos\left(\dfrac{\Omega_{c_1}}{2}\right)} \\ &= -\cos(\Omega_{c_1}) + 1 - \cos(\Omega_{c_1}) = 1 - 2\cos(\Omega_{c_1}),\end{aligned}$$

then $\alpha = -5/6$. Let us now substitute the formula

$$-\frac{e^{-j\Omega} + \alpha}{1 + \alpha e^{-j\Omega}} = -\frac{e^{-j\Omega} - 5/6}{1 - (5/6)e^{-j\Omega}}$$

in place of $e^{-j\Omega}$ in $H(\Omega)$ to get the frequency response function $H_{HP}(\Omega)$ of the designed HP filter in the form

Fig. 7.12 Magnitude response of the designed HP filter from Problem 7.7. In red: the 3 dB-bandwidth

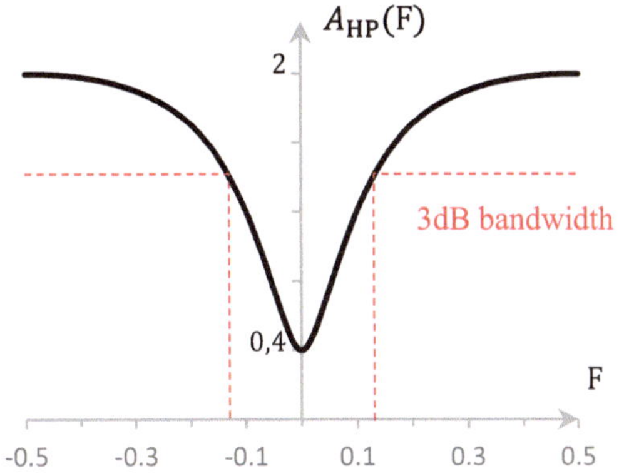

$$H_{HP}(\Omega) = \frac{\frac{2}{3}\left(-\frac{e^{-j\Omega}-5/6}{1-(5/6)e^{-j\Omega}}\right)}{1-\frac{2}{3}\left(-\frac{e^{-j\Omega}-5/6}{1-(5/6)e^{-j\Omega}}\right)} = \frac{-2(e^{-j\Omega}-5/6)}{3(1-(5/6)e^{-j\Omega})+2(e^{-j\Omega}-5/6)}$$

$$= \frac{-2(e^{-j\Omega}-5/6)}{4/3-(1/2)e^{-j\Omega}} = \frac{2(5-6e^{-j\Omega})}{8-3e^{-j\Omega}}.$$

The magnitude response $A_{HP}(\Omega)$ of the HP filter is

$$A_{HP}(\Omega) = 2\left|\frac{5-6e^{-j\Omega}}{8-3e^{-j\Omega}}\right| = 2\left|\frac{5-6(\cos\Omega - j\sin\Omega)}{8-3(\cos\Omega - j\sin\Omega)}\right|$$

$$= 2\sqrt{\frac{(5-6\cos\Omega)^2+36\sin^2\Omega}{(8-3\cos\Omega)^2+9\sin^2\Omega}} = 2\sqrt{\frac{61-60\cos\Omega}{73-48\cos\Omega}}.$$

Then

$$A_{HP}(F) = 2\sqrt{\frac{61-60\cos(2\pi F)}{73-48\cos(2\pi F)}}$$

The plot of $A_{HP}(F)$ is displayed in Fig. 7.12. The 3 dB-bandwidth $[-0.5, -0.1308] \cup [0.1308, 0.5]$ is marked in red.

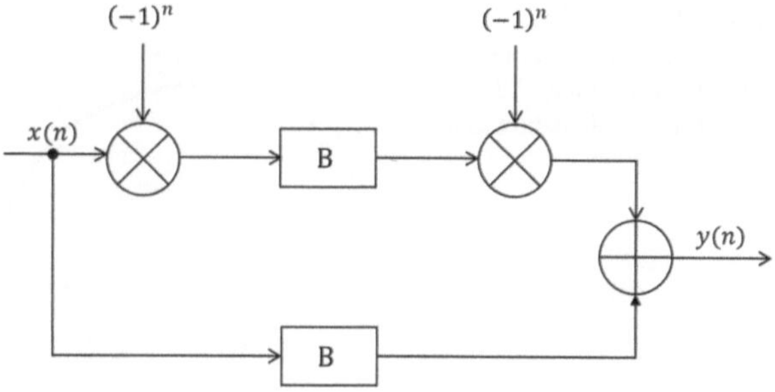

Fig. 7.13 Block diagram of the filter from Exercise 7.1

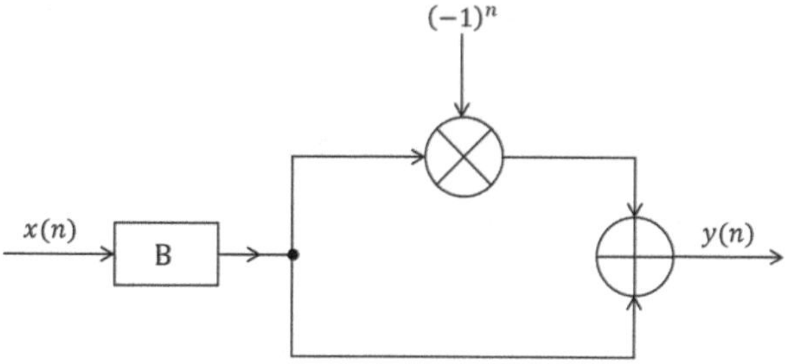

Fig. 7.14 Block diagram of the filter from Exercise 7.2

Exercises

7.1 Fig. 7.13 shows the block diagram of a discrete-time ideal filter. Block B represents an ideal LP filter with gain $K = 1$, cutoff normalized angular frequency $\Omega_c = \pi/6$ [rad] and zero phase response. Compute and sketch the frequency response function $H(\Omega)$ of this system. What type of filter is it?

7.2 Calculate the impulse response $h(n)$ of a system with the block diagram shown in Fig. 7.14. Block B represents an ideal HP filter with gain $K = 1$, cutoff normalized angular frequency $\Omega_c = \pi/2$ [rad] and zero phase response.

7.3 Given an ideal discrete-time BP filter with gain $K = 1$, $\Omega_1 = \pi/3$, $\Omega_2 = \pi/2$ (see formula (7.14)). Compute the filter response to the following input signals: (a) $x(n) = (-1)^n$; (b) $x(n) = 1 + \cos\left(\frac{3\pi}{8}n + \frac{\pi}{4}\right)$.

7.4 Given the signal $x(n) = \sin\left(\frac{\pi}{8}n\right) - 2\cos\left(\frac{\pi}{4}n\right)$ at the input of a discrete-time LTI system with the impulse response function $h(n)$. Calculate the system output if

$$\text{(a) } h(n) = \frac{\sin\left(\frac{\pi}{6}n\right)}{\pi n}; \qquad \text{(b) } h(n) = \frac{\sin\left(\frac{\pi}{6}n\right)}{\pi n} + \frac{\sin\left(\frac{\pi}{2}n\right)}{\pi n};$$

$$\text{(c) } \frac{\sin\left(\frac{\pi}{6}n\right)\sin\left(\frac{\pi}{3}n\right)}{\pi^2 n^2}.$$

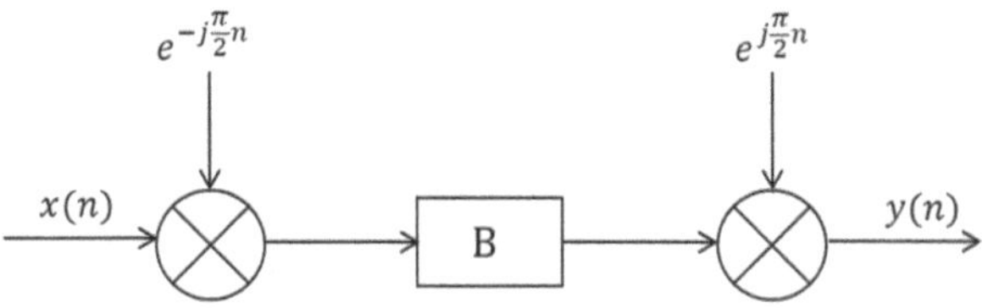

Fig. 7.15 Block diagram of the filter from Exercise 7.8

7.5 Given the impulse response of a discrete-time LTI system: $h(n) = (0,5)^n \cdot \cos(\pi/2)\mathbf{1}(n)$. Determine the system response to the following input signals: (a) $x(n) = (0,5)^n \cdot \mathbf{1}(n)$; (b) $x(n) = \cos\left(\frac{\pi}{2}n\right)$.

7.6 Given the input-output difference equation of a discrete-time LTI system

$$y(n) - \frac{1}{6}y(n-1) - \frac{1}{6}y(n-2) = x(n).$$

Determine the frequency response function $H(\Omega)$ of this system and compute its impulse response $h(n)$.

7.7 Given a discrete-time LTI system defined by the system of equations:

$$\begin{cases} y(n) + \frac{1}{4}y(n-1) + w(n) + \frac{1}{2}w(n-1) = \frac{2}{3}x(n), \\ y(n) - \frac{5}{4}y(n-1) + 2w(n) - 2w(n-1) = -\frac{5}{3}x(n), \end{cases}$$

where $w(n)$ is an internal system signal. Determine the input-output difference equation between $x(n)$ and $y(n)$. Calculate the frequency response function $H(\Omega)$ of this system and calculate its impulse response $h(n)$.

7.8 Given a discrete-time LTI system with the block diagram shown in Fig. 7.15. The impulse response of the block B is $h_1(n) = (1/4)^n \cdot \mathbf{1}(n)$. Determine the frequency response function $H(\Omega)$ and the impulse response $h(n)$ of the whole system. Sketch its magnitude response $A(F)$, $|F| \leq 0,5$.

7.9 Given a discrete-time LTI system with the frequency response function

$$H(\Omega) = \frac{2}{\left(1 - e^{-j2\Omega}\right)\left(1 - \frac{1}{2}e^{-j\Omega}\right)}.$$

Find the input-output difference equation of this system. Draw its block diagram. Then, assuming the input $x(n) = \delta(n) - 0,5\delta(n-1)$, compute the output signal $y(n)$ and sketch its amplitude spectrum $|Y(\Omega)|$, $|\Omega| \leq \pi$.

7.10 Given the input-output difference equation of a discrete-time LTI system: $y(n) = x(n) - 0,5y(n-2)$. Determine the frequency response function $H(\Omega)$. Sketch the magnitude response function $A(F)$, $|F| \leq 0,5$ and calculate the 3 dB-bandwidth of this filter.

7.11 Given the impulse response $h(n) = \left(\frac{1}{5}\right)^n \cdot \mathbf{1}(n)$ of a discrete-time LTI system. Compute and sketch the magnitude response *A*(F), $|F| \leq 0, 5$. Calculate its 3 dB-bandwidth. Find the system output *y*(*n*) to $x(n) = \left(\frac{2}{5}\right)^n \cdot \mathbf{1}(n)$.

7.12 Given a discrete-time LTI system defined by the input-output equation

$$y(n) - ay(n-1) = bx(n) + x(n-1),$$

where $|a| < 1$, $a, b \in \mathbb{R}$. Find a relation between parameters a and b such that $|H(\Omega)| = 1$, $|\Omega| \leq \pi$ (such a system is called an *all-pass filter*). Then assuming $a = -1/2$, compute the filter response *y*(*n*) to the input $x(n) = (-1/2)^n \cdot \mathbf{1}(n)$.

7.13 Given a discrete-time LTI system defined by the input-output equation

$$y(n) - \frac{1}{4}y(n-1) = -\frac{1}{4}x(n) + x(n-2).$$

Determine the frequency response function *H*(Ω) of this system. Compute and sketch its magnitude response *A*(F), $|F| \leq 0, 5$. Keeping a minimal possible number of elementary blocks, draw the block diagram.

7.14 The impulse response of a discrete-time LTI system satisfies: $h(0) = 1$, $h(1) = 0$, $h(2) = 2$, $h(n) = 0$ otherwise. Determine the frequency response function *H*(Ω) of this system. Write its input-output difference equation. Compute and sketch the magnitude response *A*(F), $|F| \leq 0, 5$.

7.15 Given a digital LP filter with the impulse response $h(n) = \left(\frac{1}{2}\right)^n \cdot \mathbf{1}(n)$. Calculate its normalized cutoff angular frequency Ω_{c_1}, and using the frequency transformation method, design an LP filter with a cutoff angular frequency $\Omega_{c_2} = 0, 5\Omega_{c_1}$. Sketch the magnitude response A_{LP}(F), $|F| \leq 0, 5$ of the designed filter.

7.16 Given a digital LP filter with the impulse response $h(n) = n\left(\frac{1}{2}\right)^n \cdot \mathbf{1}(n)$. Calculate its normalized cutoff angular frequency Ω_{c_1}, and, using the frequency transformation method, design a HP filter with cutoff angular frequency $\Omega_{c_2} = 2\Omega_{c_1}$ and sketch its magnitude response A_{HP}(F), $|F| \leq 0, 5$.

Chapter 8
$\mathcal{Z}$–Transformation in Discrete–Time LTI Systems

In this Chapter, we deal with the one-sided $\mathcal{Z}$–transformation, a very useful tool widely applied in digital filtering, for example, for solving difference input-output equations of discrete-time systems (see Chap. 3). We start with the definition and basic properties of the one-sided $\mathcal{Z}$–transformation. A table containing some useful $\mathcal{Z}$–transforms is also included. In the next part, using the convolution theorem, we introduce the *transfer function* of a digital filter and present formulas for transfer functions of FIR and IIR digital filters as well as different system interconnections presented in Chap. 3. We also show the relation between the transfer function and the frequency response function defined in Chap. 7. Then we present some basic criteria for the *stability* of digital filters.

8.1 The One-Sided $\mathcal{Z}$–Transformation

Definition 8.1 The *one-sided $\mathcal{Z}$–transform* $X(z)$, $z \in \mathbb{C}$ of a discrete-time signal$x(n)$, $n \in \mathbb{Z}$ is defined in the form of the following infinite power series:

$$X(z) = \sum_{n=0}^{\infty} x(n) z^{-n}. \tag{8.1}$$

The transformation (8.1) converts a discrete-time signal $x(n)$ into a function $X(z)$ of a complex variable z. The pair of $\mathcal{Z}$–transforms will be denoted as $x(n) \overset{\mathcal{Z}}{\leftrightarrow} X(z)$. In Table 8.1, we collected some useful pairs of $\mathcal{Z}$–transforms often applied in digital system theory.

The set of values for which the power series (8.1) converges absolutely is called the *region of convergence* (ROC). In the case of the one-sided $\mathcal{Z}$–transform, the ROC is outside the circle of radius ρ (*radius of convergence*), that is, $|z| > \rho$.

K. M. Snopek, *Hands-on Signals and Systems Theory*,
https://doi.org/10.1007/978-3-031-56081-1_8

Table 8.1 Pairs of $\mathcal{Z}$–transforms

$x(n),\ n \in \mathbb{Z}$	$X(z),\ z \in \mathbb{C}$
$\delta(n)$	1
$\mathbf{1}(n)$	$\frac{z}{z-1}$
$n \cdot \mathbf{1}(n)$	$\frac{z}{(z-1)^2}$
$a^n \cdot \mathbf{1}(n)$	$\frac{z}{z-a}$
$na^n \cdot \mathbf{1}(n)$	$\frac{za}{(z-a)^2}$
$(n+1)a^n \cdot \mathbf{1}(n)$	$\frac{z^2}{(z-a)^2}$
$\binom{n-1}{k-1} a^{n-k} \cdot \mathbf{1}(n-1)$	$\frac{1}{(z-a)^k}$
$\sin(\alpha n) \cdot \mathbf{1}(n)$	$\frac{z \sin \alpha}{z^2 - 2z \cos \alpha + 1}$
$\cos(\alpha n) \cdot \mathbf{1}(n)$	$\frac{z(z - \cos \alpha)}{z^2 - 2z \cos \alpha + 1}$
$a^n \cdot \sin(\alpha n) \cdot \mathbf{1}(n)$	$\frac{za \sin \alpha}{z^2 - 2za \cos \alpha + a^2}$
$a^n \cdot \cos(\alpha n) \cdot \mathbf{1}(n)$	$\frac{z(z - \cos \alpha)}{z^2 - 2za \cos \alpha + a^2}$
$n \cdot \sin(\alpha n) \cdot \mathbf{1}(n)$	$\frac{z\left(z^2-1\right) \sin \alpha}{\left(z^2 - 2z \cos \alpha + 1\right)^2}$
$n \cdot \cos(\alpha n) \cdot \mathbf{1}(n)$	$\frac{z\left[\left(z^2+1\right) \cos \alpha - 2z\right]}{\left(z^2 - 2z \cos \alpha + 1\right)^2}$

In digital systems, we often meet the task of calculating the inverse $\mathcal{Z}$–transform of $X(z)$—a rational function in the form of a quotient $X(z) = \frac{N(z)}{D(z)}$, where $N(z)$ and $D(z)$ are polynomials of variable z. There are some algebraic methods of calculation of the inverse $\mathcal{Z}$–transform. The most commonly used is the *partial fraction decomposition* or the *direct division method.* It will be explained in some solved problems in this chapter.

The properties of the one-sided $\mathcal{Z}$–transformation are collected in Table 8.2.

Let us show how to solve difference equations of digital filters using $\mathcal{Z}$–transformation. In Problem 8.1, we will show an alternative method of finding the response of the system from Problem 3.6.

Problem 8.1 The IIR filter is defined by the following equation:

$$y(n) + 3 \cdot y(n-1) + 5 \cdot y(n-2) = x(n-2) - 2 \cdot x(n-1).$$

Assuming this system initially at rest, calculate its response $y(n)$ to the input $x(n) = \left(\frac{1}{2}\right)^n \cdot 1(n)$.

Solution Let us apply the $\mathcal{Z}$–transformation to the system equation (see Table 8.2). Having in mind that $y(-1) = y(-2) = 0$ and $x(-1) = x(-2) = 0$ (causal input), we get

Table 8.2 Basic properties of the one-sided $\mathcal{Z}$-transformation

Property	Signal $x(n)$, $n \in \mathbb{Z}$	$X(z)$, $z \in \mathbb{C}$
Linearity	$\alpha x_1(n) + \beta x_2(n)$	$\alpha X_1(z) + \beta X_2(z)$
Time reversal	$x(-n)$	$X(z^{-1})$
Time delay	$x(n - n_0)$	$z^{-n_0}X(z) + \sum\limits_{m=1}^{n_0} x(-m)z^{-(n_0-m)}$
Time-shift (causal)	$x(n - n_0) \cdot \mathbf{1}(n - n_0)$	$z^{-n_0}X(z)$
Time advance	$x(n + n_0)$	$z^{n_0}\left(X(z) - \sum\limits_{m=0}^{n_0-1} x(m)z^{-m}\right)$
Multiplication by $z_0{}^n$, $z_0 \in \mathbb{C}$	$z_0{}^n \cdot x(n)$	$X(z_0{}^{-1}z)$
Multiplication by $e^{j\Omega_0 n}$	$e^{j\Omega_0 n}x(n)$	$X(e^{-j\Omega_0}z)$
Modulation	$x(n) \cos(\Omega_0 n)$	$\frac{1}{2}[X(e^{-j\Omega_0}z) + X(e^{j\Omega_0}z)]$
	$x(n) \sin(\Omega_0 n)$	$\frac{1}{2j}[X(e^{-j\Omega_0}z) - X(e^{j\Omega_0}z)]$
Multiplication by n	$n \cdot x(n)$	$-z\frac{dX(z)}{dz}$
Multiplication by n^2	$n^2 \cdot x(n)$	$z^2\frac{d^2X(z)}{dz^2} + z\frac{dX(z)}{dz}$
Accumulation	$\sum\limits_{l=-\infty}^{n} x(l)$	$\frac{X(z)}{1-z^{-1}}$
Time-convolution	$x_1(n) * x_2(n)$	$X_1(z) \cdot X_2(z)$
Initial value	$x(0) = \lim\limits_{z\to\infty} X(z)$ If $x(0) = 0$, then $x(1) = \lim\limits_{z\to\infty} zX(z)$.	
Final value	If $\lim\limits_{n\to\infty} x(n)$ exists, then $\lim\limits_{n\to\infty} x(n) = \lim\limits_{z\to 1}(z-1)X(z)$.	

$$
\begin{aligned}
Y(z) + 3z^{-1}Y(z) + 5z^{-2}Y(z) &= z^{-2}X(z) - 2z^{-1}X(z),\\
Y(z)(1 + 3z^{-1} + 5z^{-2}) &= X(z)(z^{-2} - 2z^{-1}),
\end{aligned}
$$

where $X(z) = \frac{z}{z-1/2}$ is the $\mathcal{Z}$–transform of the input $x(n) = \left(\frac{1}{2}\right)^n \cdot 1(n)$. Thus

$$
Y(z) = \frac{\frac{z}{z-1/2}(z^{-2} - 2z^{-1})}{1 + 3z^{-1} + 5z^{-2}}.
$$

Let us multiply the above transform by $\frac{z^2}{z^2}$ as follows:

$$
Y(z) = \frac{\frac{z}{z-1/2}(1 - 2z)}{z^2 + 3z + 5} = \frac{-2z}{z^2 + 3z + 5}.
$$

Our task is to calculate the inverse $\mathcal{Z}$–transform of $Y(z)$. To do this, we can, for example, perform the *direct division* of $(-2z)$ by $(z^2 + 3z + 5)$ as shown below:

$$
\begin{array}{ll}
(-2z) & :(z^2+3z+5) = -2z^{-1}+6z^{-2}-8z^{-3}-6z^{-4}+\cdots \\
\underline{2z+6+10z^{-1}} & \\
\quad 6+10z^{-1} & \\
\quad \underline{6-18z^{-1}-30z^{-2}} & \\
\qquad -8z^{-1}-30z^{-2} & \\
\qquad \underline{8z^{-1}+24z^{-2}+40z^{-3}} & \\
\qquad\quad -6z^{-2}+40z^{-3} & \\
\qquad\quad \underline{6z^{-2}+18z^{-3}+30z^{-4}} & \\
\qquad\qquad 58z^{-3}+30z^{-4}\ldots &
\end{array}
$$

We can see that $Y(z)$ has the form of the infinite sum $-2z^{-1} + 6z^{-2} - 8z^{-3} - 6z^{-4} + \ldots$. Applying linearity of the $\mathcal{Z}$–transformation, we immediately get the inverse transform $y(n)$, that is, the solution of our equation given by

$$y(n) = -2 \cdot \delta(n-1) + 6 \cdot \delta(n-2) - 8 \cdot \delta(n-3) - 6 \cdot \delta(n-4) + \ldots$$

We can notice that it is exactly the same signal as computed in Problem 3.6 (Chap. 3).

8.2 Zero-Input and Zero-State Responses

Let us recall the general form (3.7) of the input-output difference equation of a LTI discrete-time system:

$$
\begin{aligned}
&a_m y(n-m) + a_{m-1} y(n-m+1) + \ldots + a_1 y(n-1) + y(n) = b_l x(n-l) \\
&\quad + b_{l-1} x(n-l+1) + \ldots + b_1 x(n-1) + b_0 x(n), n \geq n_0
\end{aligned}
\tag{8.2}
$$

with initial conditions: $y(n_0), y(n_0-1), \ldots, y(n_0-m+1)$. The solution of (8.2) is the system response $y(n)$ expressed as the sum $y(n) = y_{0i}(n) + y_{0s}(n)$ of two components: the *zero-input (free) response* $y_{0i}(n)$ and the *zero-state (forced)* response $y_{0s}(n)$. The zero-input response $y_{0i}(n)$ is the solution of (8.2) when the input $x(n)$ is zero, that is,

$$a_m y(n-m) + a_{m-1} y(n-m+1) + \ldots + a_1 y(n-1) + y(n) = 0 \tag{8.3}$$

The zero-state response $y_{0s}(n)$ is the solution of (8.2) when initial conditions are zero, that is, $y(n_0) = 0$, $y(n_0-1) = 0, \ldots, y(n_0-m+1) = 0$.

In Problem 8.2, we will show how to compute zero-input and zero-state responses of a discrete-time LTI system. From a technical point of view, the procedure is very similar to that presented in Sect. 5.2 for analog systems. Here also, we have to transform the system equation (8.2) into the equation in the z-domain. Then we determine the $\mathcal{Z}$-transform $Y(z)$ of the output signal. At last, we compute the inverse $\mathcal{Z}$-transform of $Y(z)$ to get $y(n)$.

Problem 8.2 Given the equation of an IIR filter: $y(n) - 0,5y(n-1) = x(n)$. Find zero-input and zero-state responses if $y(-1) = 1$ and $x(n) = \mathbf{1}(n-1)$. Compute $y(n)$.

Solution Let us write the system equation in the z-domain as follows:

$$Y(z) - 0,5\left[z^{-1}Y(z) + y(-1)\right] = X(z).$$

Then

$$\begin{aligned} Y(z)(1 - 0,5z^{-1}) &= X(z) + 0,5y(-1) \\ Y(z) &= \frac{X(z)}{1 - 0,5z^{-1}} + \frac{0,5y(-1)}{1 - 0,5z^{-1}}. \end{aligned}$$

We can see that $Y(z)$ is the sum of two components: $\frac{X(z)}{1-0,5z^{-1}}$ representing the $\mathcal{Z}$-transform of the zero-state response denoted with $Y_{0s}(z)$ and $\frac{0,5y(-1)}{1-0,5z^{-1}}$ representing the $\mathcal{Z}$-transform of the zero-input response denoted with $Y_{0i}(z)$.

Let us start with calculation of the zero-state response $y_{0s}(n)$. We have the $\mathcal{Z}$-pair: $x(n) = \mathbf{1}(n-1) \overset{\mathcal{Z}}{\leftrightarrow} X(z) = z^{-1}\frac{z}{z-1} = \frac{1}{z-1}$, then

$$Y_{0s}(z) = \frac{X(z)}{1 - 0,5z^{-1}} = \frac{\frac{1}{z-1}}{1 - 0,5z^{-1}} = \frac{z}{(z-1)(z-0,5)}$$

Let us apply the *partial fraction decomposition* of $Y_{0s}(z)$ in the form

$$Y_{0s}(z) = \frac{A}{z-1} + \frac{B}{z-0,5}$$

From where $A(z - 0,5) + B(z - 1) = z$. For $z = 0,5$, we get $B(-0,5) = 0,5 \Leftrightarrow B = -1$. For $z = 1$, we have $A(0,5) = 1 \Leftrightarrow A = 2$.

Finally, we have

$$Y_{0s}(z) = \frac{2}{z-1} - \frac{1}{z-0,5} = z^{-1}\frac{2z}{z-1} - z^{-1}\frac{z}{z-0,5}.$$

Using linearity and time-shift property from Table 8.2, we get the zero-state response in the form

$$y_{0s}(n) = 2 \cdot \mathbf{1}(n-1) - 0,5^{n-1} \cdot \mathbf{1}(n-1) = \left(2 - 0,5^{n-1}\right) \cdot \mathbf{1}(n-1).$$

Now, let us compute the zero-input response $y_{0i}(n)$. Knowing that its $\mathcal{Z}$-transform is given by

$$Y_{0i}(z) = \frac{0,5y(-1)}{1-0,5z^{-1}} = \frac{0,5}{1-0,5z^{-1}} = 0,5\frac{z}{z-0,5}$$

and using the Table 8.1, we easily get

$$y_{0i}(n) = 0,5(0,5)^n \cdot \mathbf{1}(n) = (0,5)^{n+1} \cdot \mathbf{1}(n).$$

The system response $y(n)$ is the sum of $y_{0s}(n)$ and $y_{0i}(n)$, that is,

$$y(n) = (2 - (0,5)^{n-1}) \cdot \mathbf{1}(n-1) + (0,5)^{n+1} \cdot \mathbf{1}(n).$$

Let us calculate the first few values of $y(n)$ for $n = 0, 1, 2, 3$. We get $y(0) = 0,5$; $y(1) = 1,25$; $y(2) = 1,625$; $y(3) = 1,8125$, etc.

8.3 Transfer Function

In Chap. 3, we introduced the basics of the theory of discrete-time LTI system and stated that the response $y(n)$ to a given input signal $x(n)$ is a convolution of $x(n)$ with the impulse response $h(n)$ of the system (see (3.5)). Applying the convolution property of the $\mathcal{Z}$-transformation from Table 8.2, we obtain the following equality in the z-domain:

$$Y(z) = X(z)H(z) \tag{8.4}$$

where $y(n) \overset{\mathcal{Z}}{\leftrightarrow} Y(z)$, $x(n) \overset{\mathcal{Z}}{\leftrightarrow} X(z)$ and $h(n) \overset{\mathcal{Z}}{\leftrightarrow} H(z)$.

Definition 8.2 The *transfer function* $H(z)$ of a discrete-time LTI system is the $\mathcal{Z}$-transform of its impulse response $h(n)$: $h(n) \overset{\mathcal{Z}}{\leftrightarrow} H(z)$.

For the input-output relation $x(n) = \mathbf{1}(n) \rightarrow y(n) = k(n)$, we can write the input-output relation in the z-domain: $X(z) = z/(z-1) \rightarrow Y(z) = K(z)$. In this case, the transfer function $H(z) = Y(z)/X(z)$ is given by

$$H(z) = \frac{K(z)}{\frac{z}{z-1}} = (1 - z^{-1})K(z) \tag{8.5}$$

and we got the relation between the transfer function and the $\mathcal{Z}$-transform $K(z)$ of the step response.

The transfer function $H(z)$ is the ratio of the $\mathcal{Z}$-transform $Y(z)$ of the output signal to the $\mathcal{Z}$-transform $X(z)$ of the input signal under assumption that *all initial conditions are zero*. In other words, the transfer function is the ratio of the $\mathcal{Z}$-transform $Y_{0s}(z)$ of the *zero-state response* to $X(z)$. We have

$$H(z) = \left.\frac{Y(z)}{X(z)}\right|_{\text{initial conditions} = 0} = \frac{Y_{0s}(z)}{X(z)}. \tag{8.6}$$

In Problem 8.3, we will show how to calculate the step response of a discrete-time system using the formula (8.5). Then, in Problem 8.4, we will show the usefulness of the one-sided $\mathcal{Z}$-transformation in solving a set of difference equations describing a system.

Problem 8.3 Given the impulse response $h(n) = (0,5)^n \cdot \mathbf{1}(n)$ of a discrete-time LTI system. Determine its step response $k(n)$.

Solution In Problem 3.3, we presented the solution of the difference equation (3.3) in the time domain. Here, we can use (8.5) and show how to take advantage of the benefits of the $\mathcal{Z}$-transformation. The transfer function $H(z)$ of the filter is the $\mathcal{Z}$-transform of the impulse response $h(n) = (0,5)^n \cdot \mathbf{1}(n)$, that is, $H(z) = \frac{z}{z-0,5}$. Let us write (8.5) in the equivalent form

$$K(z) = \frac{z}{z-1} H(z).$$

Thus,

$$K(z) = \frac{z^2}{(z-1)(z-0,5)} = \frac{N(z)}{D(z)}.$$

We can see that $\deg N(z) = \deg D(z)$. In such a situation, we can apply *the fraction decomposition method of $K(z)/z$* as shown below:

$$\frac{K(z)}{z} = \frac{z}{(z-1)(z-0,5)} = \frac{A}{z-1} + \frac{B}{z-0,5}.$$

The solution is $A = 2$ and $B = -1$ (see Problem 8.2). Further,

$$\frac{K(z)}{z} = \frac{2}{z-1} - \frac{1}{z-0,5} \Leftrightarrow K(z) = 2\frac{z}{z-1} - \frac{z}{z-0,5}.$$

Using Table 8.1, we get the inverse $\mathcal{Z}$-transform of $K(z)$ in the form

$$k(n) = (2 - (0,5)^n) \cdot 1(n).$$

The step response $k(n)$ assumes exactly the same values as shown in the solution of Problem 3.3. The advantage is that here we obtained its closed mathematical formula and we can calculate $k(n)$ for every n directly and not in a recursive way.

Problem 8.4 Given the discrete-time LTI system defined by the system of equations:

$$\begin{cases} x(n) = 6w(n) - 5w(n-1) + w(n-2) \\ y(n) = 6w(n) + 4w(n-1) - 3w(n-2), \end{cases}$$

where $w(n)$ is an internal system signal. Compute the transfer function $H(z)$ of this system and determine its input-output equation, as well as its impulse response $h(n)$.

Solution Assuming the system is initially at rest, let us write the set of system equations in the z-domain as follows:

$$\begin{cases} X(z) = W(z)(6 - 5z^{-1} + z^{-2}) \\ Y(z) = W(z)(6 + 4z^{-1} - 3z^{-2}). \end{cases}$$

Dividing the second equation by the first one, we can easily get the transfer function

$$H(z) = \frac{Y(z)}{X(z)} = \frac{6 + 4z^{-1} - 3z^{-2}}{6 - 5z^{-1} + z^{-2}}.$$

Cross-multiplying produces the system equation in the z-domain in the form

$$Y(z)(6 - 5z^{-1} + z^{-2}) = X(z)(6 + 4z^{-1} - 3z^{-2}),$$

from which it follows the input-output equation:

$$6y(n) - 5y(n-1) + y(n-2) = 6x(n) + 4x(n-1) - 3x(n-2).$$

Now, let us calculate the impulse response $h(n)$ as the inverse $\mathcal{Z}$-transform of $H(z)$. First, let us write $H(z)$ in the equivalent form

$$H(z) = \frac{6z^2 + 4z - 3}{6z^2 - 5z + 1}.$$

Dividing the nominator by the denominator, we can get

$$H(z) = 1 + \frac{9z - 4}{6z^2 - 5z + 1} = 1 + \frac{9z - 4}{6(z - 1/2)(z - 1/3)}.$$

Now, we have to decompose the second component into fractions as follows:

$$\frac{9z - 4}{(z - 1/2)(z - 1/3)} = \frac{A}{z - 1/2} + \frac{B}{z - 1/3}.$$

We have $A\left(z - \frac{1}{3}\right) + B\left(z - \frac{1}{2}\right) = 9z - 4$ and for $z = \frac{1}{3}$: $-\frac{1}{6}B = -1 \Leftrightarrow B = 6$ and for $z = \frac{1}{2} : \frac{1}{6}A = \frac{1}{2} \Leftrightarrow A = 3$. We have

$$H(z) = 1 + \frac{3}{6\left(z - \frac{1}{2}\right)} + \frac{6}{6\left(z - \frac{1}{3}\right)} = 1 + \frac{1}{2\left(z - \frac{1}{2}\right)} + \frac{1}{z - \frac{1}{3}}.$$

The impulse response *h*(*n*) is given by

$$h(n) = \delta(n) + \frac{1}{2}\left(\frac{1}{2}\right)^{n-1} \cdot \mathbf{1}(n-1) + \left(\frac{1}{3}\right)^{n-1} \cdot \mathbf{1}(n-1).$$

8.3.1 Relation Between the Transfer Function and the Frequency Response of a Discrete-Time LTI System

Let us notice that the DTFT $X(\Omega)$ of a discrete-time causal signal $x(n)$, $n \in \mathbb{Z}$ is

$$X(\Omega) = \sum_{n=0}^{\infty} x(n)e^{-jn\Omega} \tag{8.7}$$

(compare with (6.15)). We can see that (8.7) is a special case of the one-sided $\mathcal{Z}$-transform (8.1) for $z = e^{j\Omega}$. Having this in mind, we can write the following relation between the transfer function $H(z)$ and the frequency response $H(\Omega)$ of a discrete-time LTI system:

$$H(\Omega) = H(z)|_{z=e^{j\Omega}}. \tag{8.8}$$

8.3.2 Transfer Functions of System Interconnections

In Sect. 3.5, we presented formulas (3.9) and (3.10) for impulse responses of cascade and parallel system interconnections. Using properties of the one-sided $\mathcal{Z}$-transformation from Table 8.2, we can write the following formulas for their transfer functions:

$$\text{cascade}: H(z) = H_1(z) \cdot H_2(z), \tag{8.9}$$

$$\text{parallel}: H(z) = H_1(z) + H_2(z). \tag{8.10}$$

The transfer function of the *negative feedback* system presented in Fig. 3.8 is given by

$$H(z) = \frac{H_1(z)}{1 + H_1(z) \cdot H_2(z)}. \tag{8.11}$$

Let us notice that the formulas (8.9) and (8.11) are analog to (5.8), (5.9) and (5.10) presented for analog systems.

Problem 8.5 Given the cascade interconnection S of two blocks S_1 and S_2 with impulse responses $h_1(n)$ and $h_2(n)$. The impulse response $h(n)$ of the system S is

$$h(n) = \delta(n) + 2 \cdot \delta(n-1) + 3 \cdot \delta(n-2) + \delta(n-4) + 3 \cdot \delta(n-5).$$

Assuming $h_1(n) = \mathbf{1}(n) - \mathbf{1}(n-2)$, compute $h_2(n)$.

Solution First, let us determine the transfer function $H(z)$ of the system S as the $\mathcal{Z}$ -transform of $h(n)$, that is,

$$H(z) = 1 + 2z^{-1} + 3z^{-2} + z^{-4} + 3z^{-5}.$$

Knowing that $h_1(n) = \mathbf{1}(n) - \mathbf{1}(n-2)$, we can express it as $h_1(n) = \delta(n) + \delta(n-1)$; thus, the transfer function $H_1(z)$ of the system S_1 is

$$H_1(z) = 1 + z^{-1}.$$

Using (8.9), we can calculate the transfer function $H_2(z)$ of the system S_2 in the form

$$H_2(z) = \frac{H(z)}{H_1(z)} = \frac{1 + 2z^{-1} + 3z^{-2} + z^{-4} + 3z^{-5}}{1 + z^{-1}}.$$

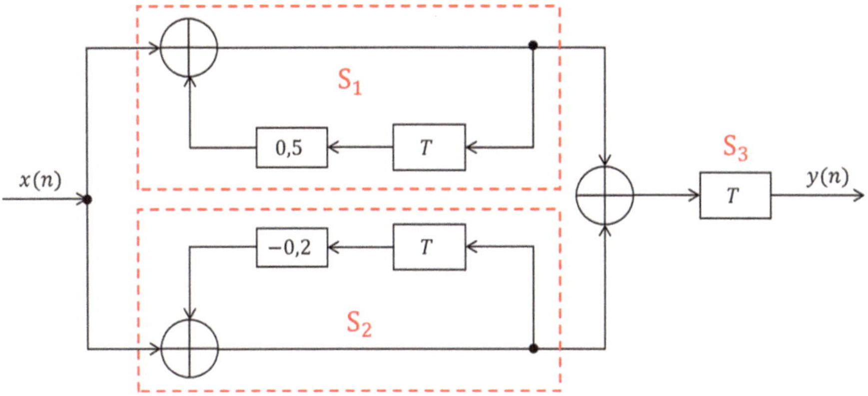

Fig. 8.1 Block diagram of the system from Problem 8.6

Further, performing the following direct division:

$$
\begin{array}{l}
(1 + 2z^{-1} + 3z^{-2} + z^{-4} + 3z^{-5}) : (1 + z^{-1}) = 1 + z^{-1} + 2z^{-2} - 2z^{-3} + 3z^{-4} \\
\underline{-1 - z^{-1}} \\
\quad z^{-1} + 3z^{-2} + z^{4} + 3z^{-5} \\
\underline{-z^{-1} - z^{-2}} \\
\qquad 2z^{-2} + z^{-4} + 3z^{-5} \\
\qquad \underline{-2z^{-2} - 2z^{-3}} \\
\qquad\qquad -2z^{-3} + z^{-4} + 3z^{-5} \\
\qquad\qquad \underline{2z^{-3} + 2z^{-4}} \\
\qquad\qquad\qquad 3z^{-4} + 3z^{-5} \\
\qquad\qquad\qquad \underline{-3z^{-4} - 3z^{-5}} \\
\qquad\qquad\qquad\quad 0
\end{array}
$$

we obtain $H_2(z) = 1 + z^{-1} + 2z^{-2} - 2z^{-3} + 3z^{-4}$. In consequence, the impulse response $h_2(n)$ is

$$h_2(n) = \delta(n) + \delta(n-1) + 2 \cdot \delta(n-2) - 2 \cdot \delta(n-3) + 3 \cdot \delta(n-4).$$

Problem 8.6 Given the system interconnection with the block diagram presented in Fig. 8.1. Determine the transfer function $H(z)$ of the whole system.

Solution We can notice that the system is a cascade interconnection of system S_3 with the parallel interconnection of two feedback systems S_1 and S_2. Let us denote with $H_i(z)$, $i = 1, 2, 3$ the transfer functions of systems S_1, S_2, and S_3.

From (8.11), we have

$$
\begin{aligned}
H_1(z) &= \frac{1}{1-0,5z^{-1}} \quad \text{(positive feedback)},\\
H_2(z) &= \frac{1}{1+0,2z^{-1}} \quad \text{(negative feedback)},\\
H_3(z) &= z^{-1}
\end{aligned}
$$

where $0,5z^{-1}$ is the transfer function of the feedback loop of S_1 and $-0,2z^{-1}$ is the transfer function of the feedback loop of S_2 (see Table 3.1). The transfer function of the parallel interconnection S_1–S_2 is $H_1(z) + H_2(z)$ (see formula (8.10)) and the transfer function $H(z)$ of the whole system (cascade interconnection) is

$$H(z) = (H_1(z) + H_2(z)) \cdot H_3(z),$$

Then

$$
\begin{aligned}
H(z) &= \left(\frac{1}{1-0,5z^{-1}} + \frac{1}{1+0,2z^{-1}}\right) z^{-1},\\
H(z) &= \left(\frac{z}{z-0,5} + \frac{z}{z+0,2}\right) z^{-1} = \frac{2z-0,3}{(z-0,5)(z+0,2)}.
\end{aligned}
$$

8.3.3 *Zeros and Poles of the Transfer Function*

Zeros and poles of a rational transfer function $H(z) = N(z)/D(z)$ are defined in an analogous way as in Sect. 5.3 for continuous-time systems. The *zeros* of $H(z)$ are roots of $N(z)$, while the *poles* are roots of $D(z)$. The *pole-zero plot* is a graphical representation of poles and zeros of the transfer function $H(z)$ in the domain of the complex variable z. It is very useful in the context of the stability of a digital system.

8.4 Stability of Discrete-Time LTI Systems

Definition 8.3 We say that a discrete-time LTI system defined by the input-output relation $x(n) \rightarrow y(n)$ is *BIBO–stable* (BIBO—Bounded Input Bounded Output) if the following condition is fulfilled:

$$\forall n \in \mathbb{Z}, \exists C_1, C_2 \in \mathbb{R} : |x(n)| \le C_1 < \infty \Rightarrow |y(n)| \le C_2 < \infty. \tag{8.12}$$

The condition (8.12) is equivalent to the absolute summability of the impulse response $h(n)$ expressed as

$$\sum_{n \in \mathbb{Z}} |h(n)| < \infty \tag{8.13}$$

For causal systems, the condition (8.13) gets the form $\sum_{n=0}^{\infty} |h(n)| < \infty$. The system BIBO–stable is simply called "*stable.*"

Theorem 8.1 A discrete-time LTI system is BIBO stable if and only if all poles of its transfer function $H(z)$ are *inside the unit circle in the z-plane* and $H(z)$ is proper (compare with Definition 5.3).

To study stability of digital systems with rational transfer functions $H(z) = N(z)/D(z)$, we have to calculate all roots of $D(z)$. It is an easy task for polynomials of degree 1 or 2. For higher–order polynomials, we can apply some algebraic criteria like the *Jury*– or *Bistritz criterion.*

Definition 8.4 The polynomial $D(z) = a_0 z^m + a_1 z^{m-1} + \ldots + a_{m-1} z + a_m$, $a_m > 0$, $a_i \in \mathbb{R}$, $i = 0, \ldots, m$, is called *the Schur polynomial* (*stable polynomial*) if *all its roots are inside the unit circle in the z-plane.*

8.4.1 *Jury Criterion*

The Jury criterion (Theorem 8.2) is based on a table with the structure shown in Fig. 8.2.

Theorem 8.2 The polynomial $D(z) = a_0 z^m + a_1 z^{m-1} + \ldots + a_{m-2} z^2 + a_{m-1} z + a_m$ with the coefficient $a_0 > 0$ and all $a_i \in \mathbb{R}$ is the Schur polynomial if and only if all elements of the first column of the *Jury array* in odd-indexed rows are positive, that is, $a_0 > 0$, $b_0 > 0$, $c_0 > 0$, $d_0 > 0$, $e_0 > 0, \ldots$

The elements b_i, c_i are calculated using the following formulas:

$$b_i = \frac{a_0 a_i - a_m a_{m-i}}{a_0} = a_i - \frac{a_m}{a_0} a_{m-i}, i = 0, 1, \ldots, m-1 \tag{8.14}$$

$$c_i = \frac{b_0 b_i - b_{m-1} b_{m-1-i}}{b_0} = b_i - \frac{b_{m-1}}{b} b_{m-1-i}, i = 0, 1, \ldots, m-2 \tag{8.15}$$

and so on. Let us demonstrate how to use the Jury criterion in practice solving Problem 8.7.

No.	z^m	z^{m-1}	z^{m-2}	...	z^2	z	z^0
1	a_0	a_1	a_2	...	a_{m-2}	a_{m-1}	a_m
2	a_m	a_{m-1}	a_{m-2}	...	a_2	a_1	a_0
3	b_0	b_1	b_2	...	b_{m-2}	b_{m-1}	0
4	b_{m-1}	b_{m-2}	b_{m-3}	...	b_1	b_0	
5	c_0	c_1	c_2	...	c_{m-2}	0	
6	c_{m-2}	c_{m-3}	c_{m-4}	...	c_0		
...	...	...	...				
$2m-5$	d_0	d_1	d_2	d_3	0		
$2m-4$	d_3	d_2	d_1	d_0			
$2m-3$	e_0	e_1	e_2	0			
$2m-2$	e_2	e_1	e_0				
$2m-1$	f_0	f_1	0				
$2m$	f_1	f_0					
$2m+1$	g_0						

Fig. 8.2 The Jury array

Problem 8.7 Using the Jury criterion, check stability of a discrete-time system with the transfer function $H(z) = \frac{z^2}{z^4 - 1,2z^3 + 0,07z^2 + 0,3z - 0,08}$.

Solution We have to check if the polynomial $D(z) = z^4 - 1,2z^3 + 0,07z^2 + 0,3z - 0,08$ is the Schur polynomial. Let us construct the first two rows of the Jury array as follows:

1	$a_0 = 1$	$a_1 = -1,2$	$a_2 = 0,07$	$a_3 = 0,3$	$a_4 = -0,08$
2	$-0,08$	$0,3$	$0,07$	$-1,2$	1

Then using (8.14), let us calculate b_0, b_1, b_2, and b_3 in the following way:

$$b_0 = \frac{a_0a_0 - a_4a_4}{a_0} = 0,9936; \quad b_1 = \frac{a_0a_1 - a_4a_3}{a_0} = -1,176;$$
$$b_2 = \frac{a_0a_2 - a_4a_2}{a_0} = 0,0756; \quad b_3 = \frac{a_0a_3 - a_4a_1}{a_0} = 0,204.$$

The next two rows of the Jury array are shown below:

3	$b_0 = 0,9936$	$b_1 = -1,176$	$b_2 = 0,0756$	$b_3 = 0,204$
4	0, 204	0, 0756	−1, 176	0, 9936

Then using (8.15), we can calculate the elements of the next row, that is, c_0, c_1, and c_2 as follows:

$$c_0 = \frac{b_0 b_0 - b_3 b_3}{b_0} = 0,9517; \ c_1 = \frac{b_0 b_1 - b_3 b_2}{b_0} = -1,1915;$$
$$c_2 = \frac{b_0 b_2 - b_3 b_1}{b_0} = 0,317.$$

The rows 5 and 6 are:

5	$c_0 = 0,9517$	$c_1 = -1,1915$	$c_2 = 0,317$
6	0, 317	−1, 1915	0, 9517

Let us now calculate d_0, d_1 and e_0 in an analogous way as above, that is,

$$d_0 = \frac{c_0 c_0 - c_2 c_2}{c_0} = 0,8461; \ d_1 = \frac{c_0 c_1 - c_2 c_1}{c_0} = -0,7946;$$
$$e_0 = \frac{d_0 d_0 - d_1 d_1}{d_0} = 0,0999.$$

The last three rows of the Jury table are as follows:

7	$d_0 = 0,8461$	$d_1 = -0,7946$
8	−0, 7946	0, 8461
9	$e_0 = 0,0999$	

Applying the Theorem 8.2, we can see that $a_0 > 0$, $b_0 > 0$, $c_0 > 0$, $d_0 > 0$, $e_0 > 0$, which means that the system is stable.

8.4.2 Bistritz Criterion

For the purpose of the Bistritz criterion, let us denote the denominator polynomial $D(z)$ of the transfer function $H(z)$ with $D_m(z)$. We have

$$D_m(z) = a_0 z^m + a_1 z^{m-1} + \ldots + a_{m-2} z^2 + a_{m-1} z + a_m. \tag{8.16}$$

Let us also introduce another polynomial $D_m^{-1}(z) = z^m D_m(z^{-1})$.

The recursive algorithm is presented in Fig. 8.3. The variable v is the number of sign (+/ −) changes of values of polynomials $T_i(1)$, $i = m, m-1, \ldots, 0$.

$T_m(z) = D_m(z) + D_m^{-1}(z);$

$T_{m-1}(z) = \dfrac{D_m(z) - D_m^{-1}(z)}{z-1}$

for $i = m-1, \ldots, 1$

{

$\delta_{i+1} = \dfrac{T_{i+1}(0)}{T_i(0)}$

$T_{i-1}(z) = \dfrac{\delta_{i+1}(1+z)T_i(z) - T_{i+1}(z)}{z}$

}

$v = \mathrm{Var}\{T_m(1), T_{m-1}(1), T_{m-2}(1), \ldots T_1(1), T_0(1)\}$

Fig. 8.3 The Bistritz stability algorithm

According to the Bistritz criterion, if $v = 0$, then the polynomial $D_m(z)$ is *stable*. Let us show how to use this criterion in practice solving Problem 8.8.

Problem 8.8 Given the polynomial $D_3(z) = 2z^3 + \frac{1}{2}z^2 + \frac{1}{2}z + 1$. Check its stability using the Bistritz criterion.

Solution The first step is to determine the polynomial $D_3(z^{-1})$, that is,

$$D_3\left(z^{-1}\right) = 2z^{-3} + \frac{1}{2}z^{-2} + \frac{1}{2}z^{-1} + 1.$$

Then we define

$$D_3^{-1}(z) = z^3 D_3\left(z^{-1}\right) = z^3\left(2z^{-3} + \frac{1}{2}z^{-2} + \frac{1}{2}z^{-1} + 1\right) = 2 + \frac{1}{2}z + \frac{1}{2}z^2 + z^3.$$

We need to calculate $T_3(z)$ and $T_2(z)$ as follows:

$$T_3(z) = D_3(z) + D_3^{-1}(z) = 2z^3 + \frac{1}{2}z^2 + \frac{1}{2}z + 1 + 2 + \frac{1}{2}z + \frac{1}{2}z^2 + z^3 = 3z^3 + z^2 + z + 3;$$

$$T_2(z) = \frac{D_3(z) - D_3^{-1}(z)}{z-1} = \frac{z^3 - 1}{z-1} = z^2 + z + 1.$$

Then we start to perform calculations in the loop for $i = 2$ as shown below:

$$\delta_3 = \frac{T_3(0)}{T_2(0)} = 3;$$

$$T_1(z) = \frac{3(1+z)T_2(z) - T_3(z)}{z} = \frac{3(1+z)(z^2+z+1) - (3z^3+z^2+z+3)}{z}$$

$$= \frac{3z^3+6z^2+6z+3-(3z^3+z^2+z+3)}{z} = \frac{5z^2+5z}{z} = 5z+5.$$

For $i = 1$, calculations in the second loop are performed in the following way:

$$\delta_2 = \frac{T_2(0)}{T_1(0)} = \frac{1}{5};$$

$$T_0(z) = \frac{\delta_2(1+z)T_1(z) - T_2(z)}{z} = \frac{\frac{1}{5}(1+z)(5z+5) - (z^2+z+1)}{z}$$

$$= \frac{z^2+2z+1-(z^2+z+1)}{z} = 1.$$

The last step is calculating the value of v as

$$v = \mathrm{Var}\{T_3(1), T_2(1), T_1(1), T_0(1)\} = \mathrm{Var}\{8, 3, 10, 1\}$$

Since all values of $T_i(1)$, $i = 3, 2, 1, 0$ are positive, then $v = 0$ and the polynomial $D_3(z)$ is stable.

Exercises

8.1. Given the input-output equation of a discrete-time LTI system: $y(n) = 0,25y(n-1) - 0,125x(n-1) + 1,5x(n)$. Compute the system response $y(n)$ if $x(n) = \delta(n)$ and $y(-1) = 1$.

8.2. Given the input-output equation of a discrete-time LTI system: $y(n) + 3y(n-1) + 5y(n-2) = x(n-2) - 2x(n-1)$. Assuming the system is initially at rest, compute $y(n)$ if $x(n) = 2^n \cdot \mathbf{1}(n)$.

8.3. Given the input-output equation of a discrete-time LTI system: $y(n) - 0,25y(n-1) = x(n) - 0,5\,x(n-1)$. Find zero-input and zero-state responses, if $x(n) = \mathbf{1}(n)$ and $y(-1) = 1$.

8.4. Given the input-output equation of a discrete-time LTI system: $2y(n) + y(n-1) = x(n)$. Find zero-input and zero-state responses if $x(n) = \left(\frac{1}{4}\right)^n \cdot \mathbf{1}(n)$ and $y(-1) = 2$.

8.5. Given the input-output equation of a discrete-time LTI system: $y(n) - 2y(n-1) + y(n-2) = x(n)$. Determine the transfer function $H(z)$, the impulse response $h(n)$, and the step response $k(n)$. Compute the system response $y(n)$ to the input signal $x(n) = 0,5^n \cdot \mathbf{1}(n)$.

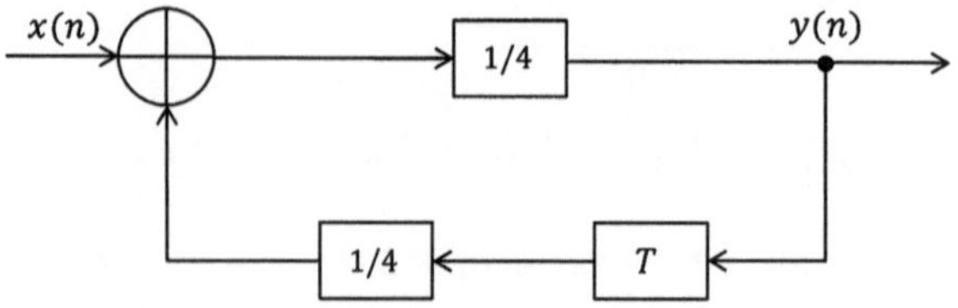

Fig. 8.4 Block diagram of the system from Exercise 8.6

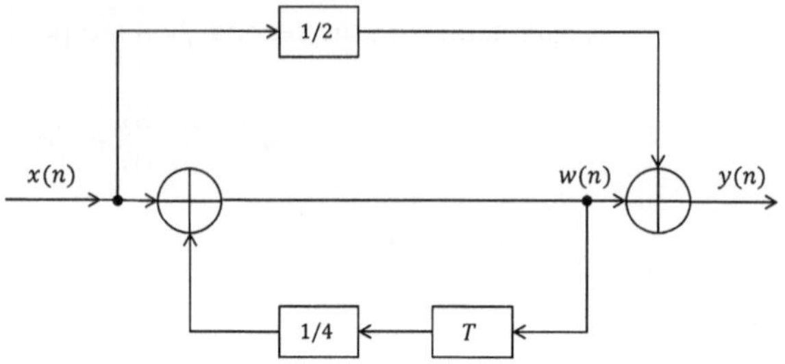

Fig. 8.5 Block diagram of the system from Exercise 8.7

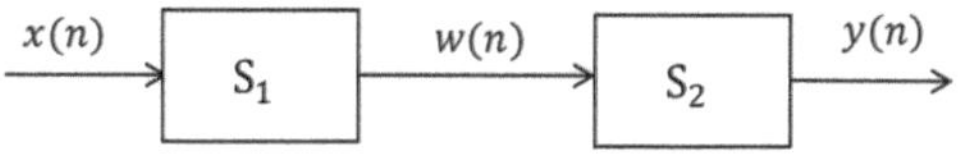

Fig. 8.6 Cascade realization from Exercise 8.9

8.6. Figure 8.4 shows the block diagram of a discrete-time LTI system. Determine the transfer function $H(z)$ and the impulse response $h(n)$. Compute and sketch the magnitude response $A(\mathrm{F})$, $|\mathrm{F}| \leq 0, 5$.

8.7. Given a discrete-time LTI system with the block diagram shown in Fig. 8.5. Determine the system response if $x(n) = \delta(n)$ and $w(-1) = 1$. Compute the transfer function $H(z)$. Compute and sketch the magnitude response $A(\mathrm{F})$, $|\mathrm{F}| \leq 0, 5$.

8.8. Given the impulse response $h(n) = \frac{1}{2}\left[\left(\frac{1}{2}\right)^n + \left(-\frac{1}{2}\right)^n\right] \cdot \mathbf{1}(n)$ of a discrete-time LTI system. Determine its transfer function $H(z)$ and the input-output system equation. Compute and sketch the magnitude response $A(\mathrm{F})$, $|\mathrm{F}| \leq 0, 5$.

8.9. The system S (Fig. 8.6) is a cascade realization of two blocks S_1 and S_2 defined by the following equations:

$$\begin{aligned} \mathrm{S1}: &\quad w(n) = \frac{1}{2}w(n-1) + x(n), \\ \mathrm{S2}: &\quad y(n) = \alpha y(n-1) + \beta w(n), \end{aligned}$$

where $\alpha, \beta \in \mathbb{R}$. The input-output equation of the system S is

$$y(n) = -\frac{1}{8}y(n-2) + \frac{3}{4}y(n-1) + x(n).$$

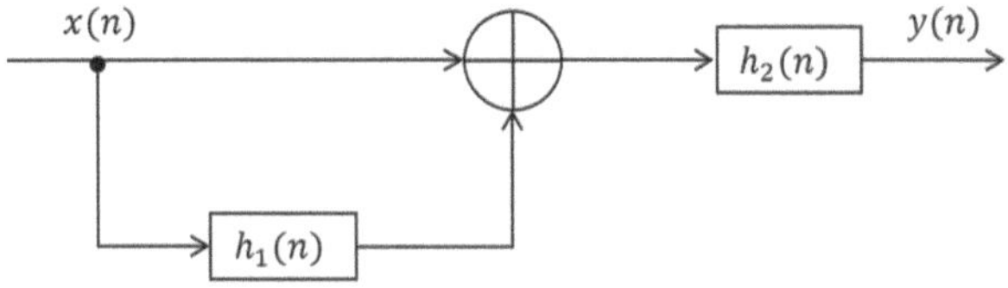

Fig. 8.7 Block diagram of the system from Exercise 8.11

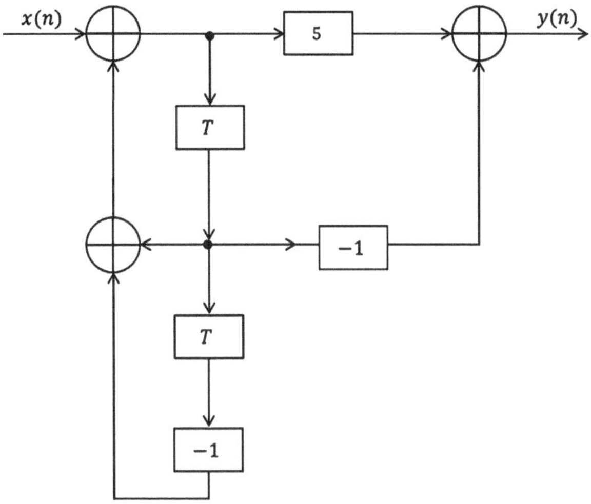

Fig. 8.8 Block diagram of the system from Exercise 8.12

Calculate values of parameters α, β. Find the impulse response $h(n)$ of the whole system.

8.10. Given the transfer function of a discrete-time LTI system $H(z) = \frac{1-1,5z^{-1}}{1-0,5z^{-1}}$. This system can be realized as: (a) a cascade interconnection; (b) a parallel interconnection of two blocks S_1 and S_2. For cases (a) and (b), determine the transfer functions $H_1(z)$ and $H_2(z)$ of blocks S_1 and S_2. Compute the impulse response $h(n)$ of the whole system.

8.11. Figure 8.7 shows the block diagram of a discrete-time LTI system. The impulse responses $h_1(n) = \frac{1}{2}\delta(n-1)$ and $h_2(n) = \left(\frac{1}{3}\right)^n \cdot \mathbf{1}(n)$. Determine the transfer function $H(z)$ and the impulse response $h(n)$ of the system. Find its input-output equation.

8.12. Figure 8.8 shows the block diagram of a discrete-time LTI system. Determine the input-output equation of the system. Compute the transfer function $H(z)$ and its zeros and poles. Plot the zero-pole diagram. Is this system stable? What is the system response $y(n)$ in case $x(n) = (-0,5)^n \cdot \mathbf{1}(n)$?

8.13. Figure 8.9 shows the block diagram of a discrete-time LTI system. Determine the input-output equation of the system. Calculate the transfer function $H(z)$ and its zeros and poles. Plot the zero-pole diagram in the z-plane. Is this system stable? Compute and sketch the magnitude response $A(\mathrm{F})$, $|\mathrm{F}| \leq 0,5$ of the system.

Fig. 8.9 Block diagram of the system from Exercise 8.13

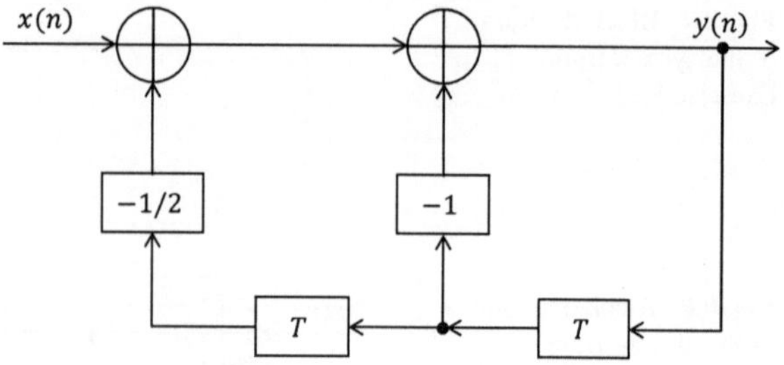

8.14. Using the Jury criterion, check the stability of two discrete-time systems with the following transfer functions:

(a) $H(z) = \frac{z+1}{2z^3+0,5z+0,1}$; (b) $H(z) = \frac{z+1}{6z^4+4z^3-1,5z^2+z+0,1}$.

8.15. Using the Bistritz criterion, check the stability of the systems from Exercise 8.14.

Chapter 9
Digital Simulation of LTI Analog Systems

This Chapter is devoted to relations between the s-domain and the z-domain. First, we develop the formula relating the one-sided Laplace transform and the one-sided $\mathcal{Z}$-transform and define the transformation called *bilinear*. Then we introduce the notion of the digital simulator of an analog system and present some methods of discretization of ordinary differential equations. Two methods of design of IIR filters will be presented: the *impulse invariance method* and the *bilinear transformation method*. This chapter recapitalizes the theory of LTI systems presented in Chaps. 3, 4, 5, 6 and 7.

9.1 Relation Between One-Sided Laplace- and $\mathcal{Z}$–Transforms

Let us recall the formula (6.1) of an ideally sampled continuous-time signal:

$$x_s(t) = \sum_{n\in\mathbb{Z}} x(nT_s)\delta(t - nT_s). \tag{9.1}$$

The one-sided Laplace transform (5.1) of $x_s(t)$, given by (9.1), is

$$\begin{gathered}\mathcal{L}\{x_s(t)\} = \mathcal{L}\left\{\sum_{n\in\mathbb{Z}} x(nT_s)\delta(t - nT_s)\right\} = \sum_{n\in\mathbb{Z}} x(nT_s)\mathcal{L}\{\delta(t - nT_s)\},\\ \mathcal{L}\{x_s(t)\} = \sum_{n\in\mathbb{Z}} x(nT_s)e^{-snT_s}.\end{gathered} \tag{9.2}$$

If we substitute in (9.2) $x(n) = x(nT_s)$, we can notice that the one-sided Laplace transform of an ideally sampled signal equals the one-sided $\mathcal{Z}$-transform of a discrete-time signal $x(n)$ for $z = e^{sT_s}$, that is,

K. M. Snopek, *Hands-on Signals and Systems Theory*,
https://doi.org/10.1007/978-3-031-56081-1_9

$$\mathcal{L}\{x_s(t)\} = \mathcal{Z}\{x(n)\}|_{z=e^{sT_s}}. \tag{9.3}$$

The above-defined transformation of the s-domain into the z-domain in the form

$$z = e^{sT_s} \Leftrightarrow s = \frac{1}{T_s} \ln z \tag{9.4}$$

is called the *bilinear transformation*. It plays a key role in the design of IIR digital filters (see next Sect. 9.3). Figure 9.1 shows the effects of this transformation.

The top graph shows that the imaginary axis in the s-plane is transformed into the boundary of the unit circle in the z-plane. It is easy to check because every point s of the imaginary axis satisfies $s = j\omega$. Its "image" in the z-domain is $z = e^{j\omega T_s}$ and $|z| = |e^{j\omega T_s}| = 1$. If $s = \sigma_0 + j\omega$, then the "image" is $z = e^{(\sigma_0 + j\omega)T_s} = e^{\sigma_0 T_s} e^{j\omega T_s}$. For $\mathrm{Re}\, s = \sigma_0 < 0$ (the middle graph of Fig. 9.1), $e^{\sigma_0 T_s} < 1$, so the left half-plane $\mathrm{Re}\, s < 0$ is transformed into the inside of the unit circle. Reciprocally, the right half-plane is transformed into the outside of the unit circle (bottom graph of Fig. 9.1).

9.2 Discretization of Linear Differential Equations

Given the continuous-time LTI system S_a with the impulse response $h_a(t)$ defined by the input-output relation $x_a(t) \to y_a(t)$. The discrete-time LTI system S_d is defined by its impulse response $h_d(n)$ and the input-output relation $x_d(n) \to y_d(n)$. Let us assume that the input signal $x_d(n)$ is formed of samples $x_a(nT_s)$, where T_s is the sampling period, that is, $x_d(n) = x_a(nT_s)$, $n \in \mathbb{Z}$.

Definition 9.1 The system S_d is called the *digital simulator* of the system S_a if its output $y_d(n)$ satisfies: $y_d(n) = y_a(nT_s)$, $n \in \mathbb{Z}$.

The digital simulator of a given analog system can easily be determined using discretization of the input-output differential equation defining the analog system. Its result is a linear difference equation of a digital simulator. Here below, we recall the discretization technique known as the *finite difference method*. It is based on approximating derivatives of different order with formulas involving finite differences.

9.2.1 Finite Difference Method

Definition 9.2 The *finite difference* $(\Delta y)_n$ of the 1st order of a signal $y(n)$ is defined as

$$(\Delta y)_n = y(n) - y(n-1). \tag{9.5}$$

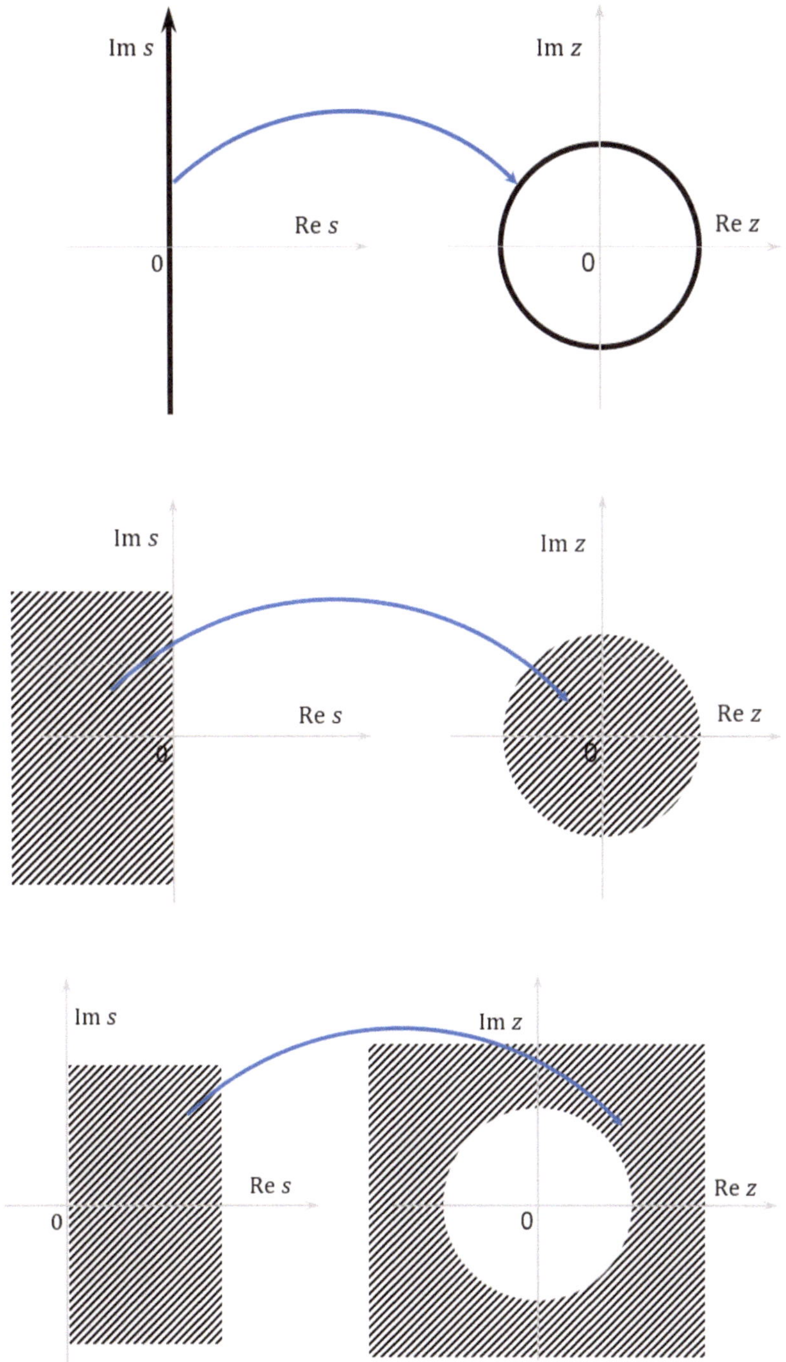

Fig. 9.1 The idea of the bilinear transformation $z = e^{sT_s}$

The *finite differences of higher orders* are defined in a recursive way as follows:

$$(\Delta^2 y)_n = (\Delta(\Delta y))_n = (\Delta y)_n - (\Delta y)_{n-1}, \tag{9.6}$$

$$(\Delta^3 y)_n = (\Delta(\Delta(\Delta y)))_n = (\Delta(\Delta y))_n - (\Delta(\Delta y))_{n-1}, \tag{9.7}$$

etc. So we have

$$(\Delta^2 y)_n = y(n) - 2y(n-1) + y(n-2), \tag{9.8}$$

$$(\Delta^3 y)_n = y(n) - 3y(n-1) + 3y(n-2) - y(n-3). \tag{9.9}$$

The *discretization method* of differential equations based on formulas (9.5), (9.6), (9.7), (9.8) and (9.9) is as follows:

- The first derivative $\frac{dy(t)}{dt}$ can be approximated by the quotient

$$\frac{dy(t)}{dt} \approx \frac{y(nT_s) - y((n-1)T_s)}{T_s} = \frac{y(n) - y(n-1)}{T_s} \tag{9.10}$$

where, in the nominator, we can recognize the finite difference of the 1st order defined by (9.5).

- The second derivative $\frac{d^2y(t)}{dt^2}$ can be approximated by the quotient

$$\frac{d^2y(t)}{dt^2} \approx \frac{y(n) - 2y(n-1) + y(n-2)}{T_s^2}, \tag{9.11}$$

where, in the nominator, we have the finite difference of the 2nd order given by (9.8).

- The third derivative $\frac{d^3y(t)}{dt^3}$ can be approximated as

$$\frac{d^3y(t)}{dt^3} \approx \frac{y(n) - 3y(n-1) + 3y(n-2) - y(n-3)}{T_s^3}, \tag{9.12}$$

where the nominator is defined by the finite difference of the 3rd order (9.9). We can see that using Definition 9.1, we are able to approximate derivatives of different orders. In Problem 9.1, we will find a digital simulator of a given analog system using the above-presented method.

Problem 9.1 Given the input-output differential equation of an analog system S_a in the form $\frac{d^2y(t)}{dt^2}+3\frac{dy(t)}{dt}+2y(t)=x(t)$, where $x(t) = \mathbf{1}(t)$ and $y(0) = 0$ and $y'(0) = 0$. Using the *finite difference method*, find the equation of its digital simulator S_d. Compare the analog system response $y_a(t)$ with the output of the digital simulator $y_d(n)$. Assume $T_s = 0{,}1$ s.

Solution Using (9.10) and (9.11), we can transform the differential equation of the analog system into the following difference equation:

$$\frac{y(n)-2y(n-1)+y(n-2)}{T_s^2}+3\frac{y(n)-y(n-1)}{T_s}+2y(n)=x(n),$$
$$y(n)-2y(n-1)+y(n-2)+3T_s[y(n)-y(n-1)]+2T_s^2y(n)=T_s^2x(n),$$
$$(1+3T_s+2T_s^2)y(n)+(-2-3T_s)y(n-1)+y(n-2)=T_s^2x(n).$$

Since $T_s = 0{,}1$ s, the above equation gets the final form:

$$1,32\cdot y(n)-2,3\cdot y(n-1)+y(n-2)=0,01\cdot x(n),$$
$$132\cdot y(n)-230\cdot y(n-1)+100\cdot y(n-2)=x(n).$$

Now, using the Laplace transformation, let us find the solution $y_a(t)$ of the differential equation $\frac{d^2y(t)}{dt^2}+3\frac{dy(t)}{dt}+2y(t)=x(t)$ with zero initial conditions. From Table 5.2, we have $X(s) = 1/s$. Then

$$s^2Y(s)+3sY(s)+2Y(s)=\frac{1}{s},$$
$$Y(s)\left(s^2+3s+2\right)=\frac{1}{s},$$
$$Y(s)=\frac{1}{s(s^2+3s+2)}=\frac{1}{s(s+1)(s+2)}.$$

We can decompose $Y(s)$ into the following sum of partial fractions:

$$Y(s)=\frac{1}{s(s+1)(s+2)}=\frac{A}{s}+\frac{B}{s+1}+\frac{C}{s+2}.$$

Thus, $A(s + 1)(s + 2) + Bs(s + 2) + Cs(s + 1) = 1$. If $s = -1$, then $-B = 1 \Leftrightarrow B = -1$; if $s = -2$, then $2C = 1 \Leftrightarrow C = 1/2$; if $s = 0$, then $2A = 1 \Leftrightarrow A = 1/2$. Finally,

$$Y(s)=\frac{1}{2s}-\frac{1}{s+1}+\frac{1}{2}\frac{1}{s+2}.$$

The inverse Laplace transform of $Y(s)$ (see Table 5.1) is given by

$$y_a(t) = \left(\frac{1}{2} - e^{-t} + \frac{1}{2}e^{-2t}\right) \cdot \mathbf{1}(t).$$

Now, using the one-sided $\mathcal{Z}$-transform, we can find $y_d(n)$—the solution of the difference eq. $132y(n) - 230y(n-1) + 100y(n-2) = x(n)$ with zero initial conditions and $x(n) = \mathbf{1}(n) \overset{\mathcal{Z}}{\leftrightarrow} X(z) = z/(z-1)$ (see Table 8.1). We can write the system equation in the z-domain as

$$132 \cdot Y(z) - 230 \cdot z^{-1}Y(z) + 100 \cdot z^{-2}Y(z) = \frac{z}{z-1},$$
$$Y(z)(132 - 230 \cdot z^{-1} + 100 \cdot z^{-2}) = \frac{z}{z-1},$$

from where

$$Y(z) = \frac{z}{(z-1)(132 - 230 \cdot z^{-1} + 100 \cdot z^{-2})} = \frac{z^3}{(z-1)(132 \cdot z^2 - 230 \cdot z + 100)}.$$

Let us notice that $132z^2 - 230z + 100 = 132\left(z - \frac{5}{6}\right)\left(z - \frac{10}{11}\right)$, so we can decompose $Y(z)/z$ into the following partial fractions:

$$\frac{Y(z)}{z} = \frac{z^2}{132(z-1)\left(z - \frac{10}{11}\right)\left(z - \frac{5}{6}\right)} = \frac{1}{132}\left(\frac{A}{z-1} + \frac{B}{z - \frac{10}{11}} + \frac{C}{z - \frac{5}{6}}\right).$$

We have $A\left(z - \frac{5}{6}\right)\left(z - \frac{10}{11}\right) + B(z-1)\left(z - \frac{10}{11}\right) + C(z-1)\left(z - \frac{5}{6}\right) = z^2$.

In order to find A, B, C, we can substitute $z = 1$ in the above equality and get $A = 66$. If $z = \frac{10}{11}$, then $B = -120$ and if $z = \frac{5}{6}$, then $C = 55$. Then we can write $Y(z)$ in the form

$$Y(z) = \frac{1}{132}\left(66\frac{z}{z-1} - 120\frac{z}{z - \frac{10}{11}} + 55\frac{z}{z - \frac{5}{6}}\right) = \frac{1}{2}\frac{z}{z-1} - \frac{10}{11}\frac{z}{z - \frac{10}{11}} + \frac{5}{12}\frac{z}{z - \frac{5}{6}}.$$

The inverse one-sided $\mathcal{Z}$-transform of $Y(z)$ gives the solution $y_d(n)$ of the equation of the digital simulator in the form

$$y_d(n) = \left[\frac{1}{2} - \left(\frac{10}{11}\right)^{n+1} + \frac{1}{2}\left(\frac{5}{6}\right)^{n+1}\right] \cdot \mathbf{1}(n).$$

Figure 9.2 shows the plots of $y_a(t)$ and of $y_d(n)$. Let us notice that both signals represent step responses of the analog filter S_a and of its digital simulator S_d. They slightly differ but for much smaller values of the sampling period this discrepancy would be less visible (here $T_s = 0, 1$ s).

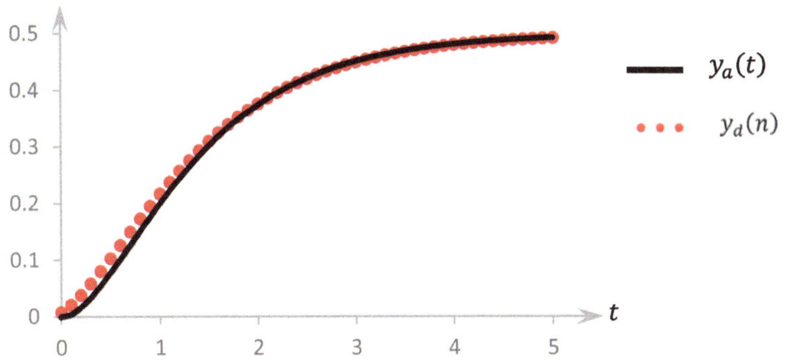

Fig. 9.2 Outputs $y_a(t)$ and $y_d(n)$ of systems S_a and S_d from Problem 9.1

9.3 Design Methods of IIR Filters

Here we present two methods applied in the design of IIR filters. These are the *impulse invariance method* and the method based on *bilinear transformation* introduced in Sect. 9.1.

9.3.1 Impulse Invariance Method

Let us denote with $h_a(t)$ the impulse response of the analog system S_a. The impulse response $h_d(n)$ of its digital simulator S_d is defined by the formula

$$h_d(n) = T_s \cdot h_a(nT_s), \tag{9.13}$$

that is, it equals the sampled analog system impulse response multiplied by T_s (sampling period) as a normalizing factor. Let us illustrate this method in Problem 9.2.

Problem 9.2 Given the impulse response $h_a(t) = (e^{-t} - e^{-2t}) \cdot \mathbf{1}(t)$ of the analog system S_a. Using the *impulse invariance method* determine the impulse response $h_d(n)$ of its digital simulator S_d. Find the input-output equation of S_d. Plot and compare the magnitude responses of both systems. Assume $T_s = 0, 1$ *s*.

Solution Using (9.13), we get

$$h_d(n) = T_s \cdot \left(e^{-nT_s} - e^{-2nT_s}\right) \cdot \mathbf{1}(nT_s).$$

Since $T_s = 0, 1$ s and $\mathbf{1}(nT_s) = \mathbf{1}(n)$, the impulse response of the digital simulator is

$$h_d(n) = 0, 1 \cdot \left(e^{-0,1n} - e^{-0,2n}\right) \cdot \mathbf{1}(n).$$

Let us now calculate the transfer function $H(z)$ of S_d as the one-sided $\mathcal{Z}$-transform of $h_d(n)$ (see Definition 8.2) as follows:

$$H(z) = 0, 1\left(\frac{z}{z - e^{-0,1}} - \frac{z}{z - e^{-0,2}}\right) = 0, 1\left(\frac{z(z - e^{-0,2}) - z(z - e^{-0,1})}{(z - e^{-0,1})(z - e^{-0,2})}\right)$$
$$= 0, 1\frac{z(e^{-0,1} - e^{-0,2})}{(z - e^{-0,1})(z - e^{-0,2})}.$$

Since $H(z) = Y(z)/X(z)$, then

$$\frac{Y(z)}{X(z)} = \frac{0, 1z(e^{-0,1} - e^{-0,2})}{(z - e^{-0,1})(z - e^{-0,2})} = \frac{0, 1z(e^{-0,1} - e^{-0,2})}{z^2 - (e^{-0,1} + e^{-0,2})z + e^{-0,3}},$$
$$\frac{Y(z)}{X(z)} = \frac{0, 1z^{-1}(e^{-0,1} - e^{-0,2})}{1 - (e^{-0,1} + e^{-0,2})z^{-1} + e^{-0,3}z^{-2}}.$$

The system equation in the z-domain is

$$Y(z) \cdot \left[1 - \left(e^{-0,1} + e^{-0,2}\right)z^{-1} + e^{-0,3}z^{-2}\right] = 0, 1\left(e^{-0,1} - e^{-0,2}\right) \cdot X(z)z^{-1}.$$

The inverse $\mathcal{Z}$-transform of the above equation gets the difference equation of the system S_d in the form

$$y(n) - \left(e^{-0,1} + e^{-0,2}\right)y(n-1) + e^{-0,3}y(n-2) = 0, 1\left(e^{-0,1} - e^{-0,2}\right)x(n-1).$$

Since $e^{-0,\,1} + e^{-0,\,2} \cong 1, 724$, $e^{-0,\,3} \cong 0, 741$ and $0, 1(e^{-0,\,1} - e^{-0,\,2}) \cong 0, 009$, the above equation can be written as

$$y(n) - 1,724 \cdot y(n-1) + 0,741 \cdot y(n-2) = 0,009 \cdot x(n-1).$$

Let us now compute the frequency response functions of both systems S_a and S_d. Let us recall that the frequency response function $H(\omega)$ of the system S_a is the Fourier transform of $h_a(t)$ (see Sect. 4.1 and Table 2.1). We have ($\omega = 2\pi f$)

$$H(\omega) = \frac{1}{1 + j\omega} - \frac{1}{2 + j\omega} = \frac{1}{(1 + j\omega)(2 + j\omega)},$$

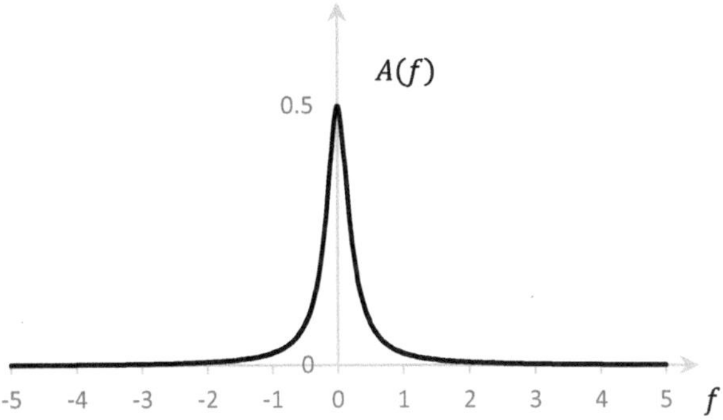

Fig. 9.3 Magnitude response $A(f)$ of the analog system S_a from Problem 9.2

then the magnitude response $A(\omega) = |H(\omega)|$ gets the form

$$A(\omega) = \left|\frac{1}{(1+j\omega)(2+j\omega)}\right| = \frac{1}{\sqrt{(1+\omega^2)(4+\omega^2)}}$$

or in the f-domain

$$A(f) = \frac{1}{\sqrt{(1+4\pi^2 f^2)(4+4\pi^2 f^2)}}$$

The plot of $A(f)$ is displayed in Fig. 9.3.

Now, let us calculate the frequency response function $H(\Omega)$ of the system S_d as follows:

$$H(\Omega) = 0,1\frac{e^{j\Omega}(e^{-0,1} - e^{-0,2})}{(e^{j\Omega} - e^{-0,1})(e^{j\Omega} - e^{-0,2})}.$$

The magnitude response of S_d, that is, $A(\Omega) = |H(\Omega)|$, is

$$\begin{aligned} A(\Omega) &= \frac{0,1(e^{-0,1} - e^{-0,2})}{|e^{j\Omega} - e^{-0,1}||e^{j\Omega} - e^{-0,2}|} = \frac{0,1(e^{-0,1} - e^{-0,2})}{|\cos\Omega - e^{-0,1} + j\sin\Omega||\cos\Omega - e^{-0,2} + j\sin\Omega|} \\ &= \frac{0,1(e^{-0,1} - e^{-0,2})}{\sqrt{\left[(\cos\Omega - e^{-0,1})^2 + \sin^2\Omega\right]\left[(\cos\Omega - e^{-0,2})^2 + \sin^2\Omega\right]}} \\ &= \frac{0,1(e^{-0,1} - e^{-0,2})}{\sqrt{(1+e^{-0,2} - 2e^{-0,1}\cos\Omega)(1+e^{-0,4} - 2e^{-0,2}\cos\Omega)}}. \end{aligned}$$

Since $0,1(e^{-0,1} - e^{-0,2}) \cong 0,009$, $1 + e^{-0,2} \cong 1,819$, $2e^{-0,1} \cong 1,81$, $1 + e^{-0,4} \cong 1,67$ and $2e^{-0,2} \cong 1,637$, we can write the magnitude response $A(F)$ of the digital simulator as

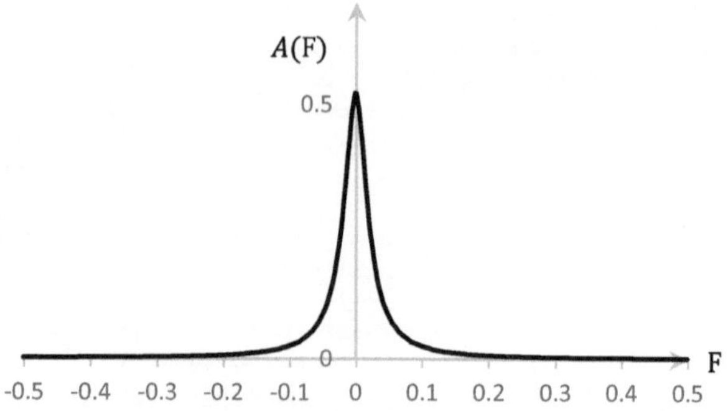

Fig. 9.4 Magnitude response $A(\mathrm{F})$ of the digital simulator S_d from Problem 9.2

$$A(\mathrm{F}) = \frac{0,009}{\sqrt{(1,819 - 1,81 \cdot \cos(2\pi \mathrm{F}))(1,67 - 1,637 \cdot \cos(2\pi \mathrm{F}))}}.$$

The plot of $A(\mathrm{F})$ for $|\mathrm{F}| \leq 0, 5$ is shown in Fig. 9.4.

Both plots from Figs. 9.3 and 9.4 apparently seem to be identical. However, they slightly differ, especially for limit values $\mathrm{F} = \pm 0{,}5$. For example, the gain of the analog system for $f = 5$ [Hz] equals approx. 0,001, while the gain of the digital simulator for $\mathrm{F} = 0{,}5$ (corresponding to the frequency $\frac{\mathrm{F}}{T_s} = 5$) is approx. 0,0026. This effect is caused by the *aliasing* of repeating segments of the frequency response function of a discrete-time system.

9.3.2 Bilinear Transformation Method

This method of IIR filter design is based on bilinear transformation (9.4) between the domain of the complex variable s and the z-domain. Let us express it in the equivalent form as

$$z = e^{sT_s} = \frac{e^{s\frac{T_s}{2}}}{e^{-s\frac{T_s}{2}}}.$$

Applying approximation of this expression by the first two terms of the Taylor series for $e^x = 1 + x + \frac{x^2}{2!} + \frac{x^3}{3!} + \ldots$ for $e^{\pm s\frac{T_s}{2}}$, we can get the formula $z \approx \frac{1+s\frac{T_s}{2}}{1-s\frac{T_s}{2}}$. After some elementary calculations, we obtain

$$s = \frac{2}{T_s}\frac{z-1}{z+1} \Leftrightarrow z = \frac{2 + sT_s}{2 - sT_s} \tag{9.14}$$

The idea of this method of IIR filter design is very simple. In order to determine the transfer function $H(z)$ of the designed IIR filter, we have to substitute the variable

s in the transfer function $H(s)$ of the analog prototype with (9.14). It is described in Problems 9.3 and 9.4.

Problem 9.3 Using the *bilinear transformation method*, design the digital simulator of the ideal differentiator from Table 3.1. Write its input-output difference equation and draw the block diagram. Assume $T_s = 0{,}1\ s$.

Solution The input-output equation of the ideal differentiator is $y(t) = \frac{dx(t)}{dt}$. Its transfer function $H(s) = s$. Using the formula (9.14), we immediately get the transfer function $H(z)$ of the digital simulator expressed as

$$H(z) = \frac{2}{T_s}\frac{z-1}{z+1} = \frac{2}{T_s}\frac{1-z^{-1}}{1+z^{-1}}.$$

Since $H(z) = Y(z)/X(z)$ and

$$\frac{Y(z)}{X(z)} = \frac{2}{T_s}\frac{1-z^{-1}}{1+z^{-1}},$$

we easily get the equation of the digital simulator in the z-domain in the form

$$y(n) = \frac{2}{T_s}[x(n) - x(n-1)] - y(n-1).$$

As $T_s = 0,\ 1\ s$, the above equation gets the final form:

$$y(n) = 20[x(n) - x(n-1)] - y(n-1).$$

The block diagram of the digital simulator in the form of a direct realization of type II (see Sect. 3.5) is shown in Fig. 9.5.

Let us now determine the relation between the frequency response functions of systems S_a and S_d. Substituting $s = j\omega$ and $z = e^{j\Omega}$ in (9.14), we can get the following relation:

$$j\omega = \frac{2}{T_s}\frac{e^{j\Omega} - 1}{e^{j\Omega} + 1}$$

which can be expressed as

$$j\omega = \frac{2}{T_s}\frac{e^{j\frac{\Omega}{2}}e^{j\frac{\Omega}{2}} - e^{j\frac{\Omega}{2}}e^{-j\frac{\Omega}{2}}}{e^{j\frac{\Omega}{2}}e^{j\frac{\Omega}{2}} + e^{j\frac{\Omega}{2}}e^{-j\frac{\Omega}{2}}} = \frac{2}{T_s}\frac{e^{j\frac{\Omega}{2}}\left(e^{j\frac{\Omega}{2}} - e^{-j\frac{\Omega}{2}}\right)}{e^{j\frac{\Omega}{2}}\left(e^{j\frac{\Omega}{2}} + e^{-j\frac{\Omega}{2}}\right)} = \frac{2}{T_s}\frac{e^{j\frac{\Omega}{2}}\left(2j\sin\frac{\Omega}{2}\right)}{e^{j\frac{\Omega}{2}}\left(2\cos\frac{\Omega}{2}\right)}.$$

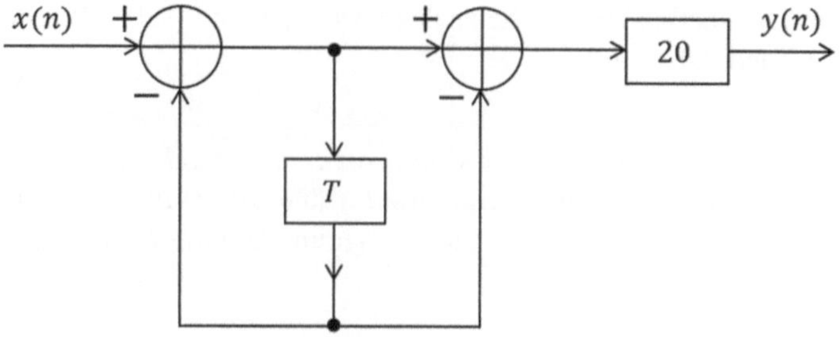

Fig. 9.5 Block diagram (direct realization type II) of the digital simulator of the ideal differentiator block from Problem 9.3

Thus,

$$\omega = \frac{2}{T_s} \tan\left(\frac{\Omega}{2}\right) \Leftrightarrow \Omega = 2 \cdot \mathrm{atan}\left(\frac{\omega T_s}{2}\right). \tag{9.15}$$

The transformation defined by (9.15) is nearly linear only for small values of ω. For bigger values of ω, we can observe a nonlinearity. In consequence, the frequency response functions of S_a and S_d are very similar only for lower frequencies. As a consequence, the method based on bilinear transformation is perfect for design of low-pass digital IIR filters. Let us analyze this case solving the next Problem 9.4.

Problem 9.4 Using *bilinear transformation method*, find the digital simulator of the second order low-pass Butterworth filter with cutoff angular frequency $\omega_c = 1$ rad/s. Plot the magnitude response $A(\Omega)$, $|\Omega| \leq \pi$, of the digital simulator. Assume $T_s = 0, 1$ s.

Solution In Problem 5.6, we computed the frequency response functions of Butterworth LP filters (of order 1, 2, and 3) and plotted them in Fig. 5.8. Now, let us consider the 2nd order low-pass Butterworth filter with the transfer function $H(s) = \frac{1}{s^2 + \sqrt{2}s + 1}$ and apply the bilinear transformation method to find its digital simulator. We have

$$\begin{aligned} H(z) &= \frac{1}{\frac{4}{T_s^2}\left(\frac{z-1}{z+1}\right)^2 + \sqrt{2}\frac{2}{T_s}\left(\frac{z-1}{z+1}\right) + 1} \\ &= \frac{T_s^2(z+1)^2}{4(z^2 - 2z + 1) + 2\sqrt{2} \cdot T_s(z^2 - 1) + T_s^2 \cdot (z+1)^2} \\ &\frac{T_s^2(z+1)^2}{(4 + 2\sqrt{2} \cdot T_s + T_s^2)z^2 + (-8 + 2 \cdot T_s^2)z + (4 - 2\sqrt{2} \cdot T_s + T_s^2)}. \end{aligned}$$

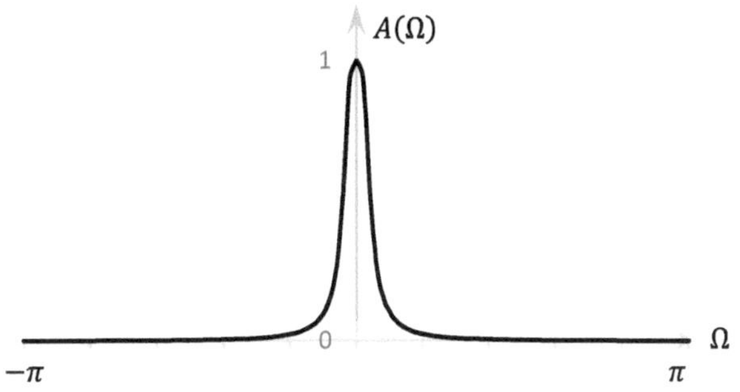

Fig. 9.6 The magnitude response $A(\Omega)$ of the digital simulator from Problem 9.4

For $T_s = 0,\ 1$ s, we have $4 + 2\sqrt{2}T_s + T_s^2 \cong 4,29$, $-8 + 2T_s^2 = -7,98$, $4 - 2\sqrt{2}T_s + T_s^2 \cong 3,73$. Finally, we obtain

$$H(z) = \frac{0,01 \cdot (z+1)^2}{4,29 \cdot z^2 - 7,98 \cdot z + 3,73}.$$

Then the magnitude response is

$$A(\Omega) = |H(e^{j\Omega})| = \left| \frac{0,01(e^{j\Omega}+1)^2}{4,29 \cdot e^{j2\Omega} - 7,98 \cdot e^{j\Omega} + 3,73} \right|$$
$$= 0,01 \frac{|e^{j\Omega}+1|^2}{|4,29 \cdot e^{j2\Omega} - 7,98 \cdot e^{j\Omega} + 3,73|}$$

Next,

$$A(\Omega) = \frac{0,01 \cdot (2 + 2\cos\Omega)}{\sqrt{(4,29 \cdot \cos 2\Omega - 7,98 \cdot \cos\Omega + 3,73)^2 + (4,29 \cdot \sin 2\Omega - 7,98 \cdot \sin\Omega)^2}}.$$

The plot of $A(\Omega)$ is shown in Fig. 9.6.

Exercises

9.1. Using the *finite difference method*, find the digital simulator S_d of the analog system S_a defined by the eq. $2\frac{dy(t)}{dt} + y(t) = x(t)$, where $x(t) = \mathbf{1}(t)$ and $y(0) = 0$. Assuming $T_s = 0,1$ [s], determine the difference equation of S_d. Compute and sketch the responses $y_a(t)$ and $y_d(n)$ to the input signals $x(t)$ and $x(nT_s)$.

9.2. Using the *finite difference method*, find the digital simulator S_d of the analog system S_a defined by the equation $\frac{d^2y(t)}{dt^2} + 4\frac{dy(t)}{dt} + 4y(t) = x(t)$. Assuming $T_s = 0, 1$ [s], determine the difference equation of S_d. Draw the block diagrams of systems S_a and S_d. Sketch the magnitude responses of systems S_a and S_d.

9.3. Using the *impulse invariance method*, find the digital simulator S_d of the analog system S_a defined by the eq. $2\frac{dy(t)}{dt} + y(t) = x(t)$, where $x(t) = \mathbf{1}(t)$ and $y(0) = 0$. Assuming $T_s = 0, 1$ [s], determine the difference equation of S_d. Compute and sketch plot the response $y_d(n)$ to the input $x(nT_s)$. Plot the magnitude response of the system S_d.

9.4. Using the *impulse invariance method*, find the digital simulator S_d of the analog system S_a defined by the equation $\frac{d^2y(t)}{dt^2} + 4\frac{dy(t)}{dt} + 4y(t) = x(t)$, where $x(t) = e^{-2t} \cdot \mathbf{1}(t)$ and $y(0) = y'(0) = 0$. Assuming $T_s = 0, 1$ [s], determine the difference equation of S_d. Compute and sketch the response $y_d(n)$ to the input signal $x(nT_s)$. Plot the magnitude response of the system S_d.

9.5. Using the *bilinear transformation method*, find the digital simulator of the ideal integrating block from Table 3.1. Write its input-output difference equation and draw the block diagram. Assume $T_s = 0, 1$ s.

9.6. Using the *bilinear transformation method*, find the digital simulator S_d of the analog system S_a defined by the equation $\frac{dy(t)}{dt} + 2y(t) = x(t)$, where $x(t) = e^{-t} \cdot \mathbf{1}(t)$ and $y(0) = 0$. Assuming $T_s = 0, 1$ [s], determine the difference equation of S_d. Compute and sketch the responses $y_a(t)$ and $y_d(n)$ to the input signals $x(t)$ and $x(nT_s)$. Compare the magnitude responses of systems S_a and S_d.

9.7. Using the *bilinear transformation method*, find the digital simulator S_d of the analog system S_a defined by the equation $\frac{d^2y(t)}{dt^2} + 4\frac{dy(t)}{dt} + 4y(t) = x(t)$, where $x(t) = e^{-2t} \cdot \mathbf{1}(t)$ and $y(0) = y'(0) = 0$. Assuming $T_s = 0, 1$ [s], determine the difference equation of S_d. Compute and plot the response $y_d(n)$ to the input signal $x(nT_s)$. Sketch the magnitude response of the system S_d.

Answers to Exercises

Chapter 1

1.1. **(a)** Periodic signal, $T_0 = 20$ [s], $P_x = 5/2$; **(b)** nonperiodic signal, $P_x = 1/8$; **(c)** periodic signal, $N_0 = 8$, $P_x = 1/4$.

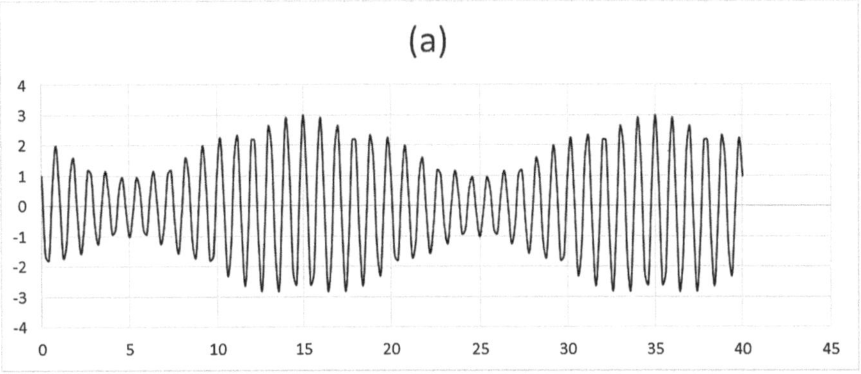

K. M. Snopek, *Hands-on Signals and Systems Theory*,
https://doi.org/10.1007/978-3-031-56081-1

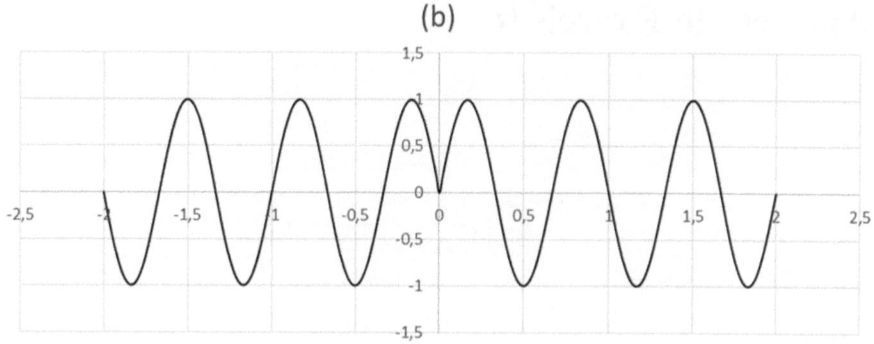

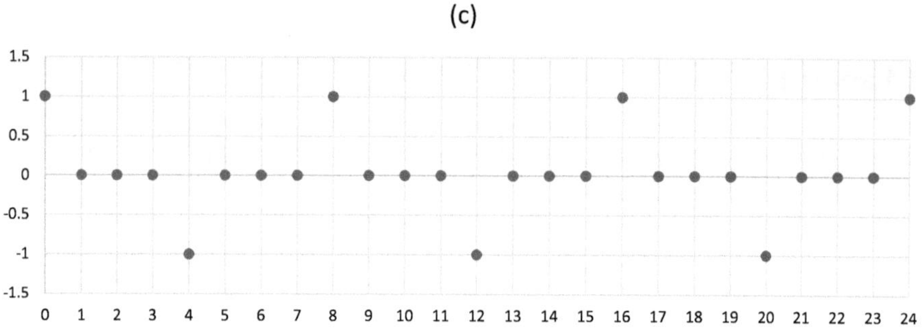

1.2. $\overline{x} = 3/2$; $P_x = 11/4$.

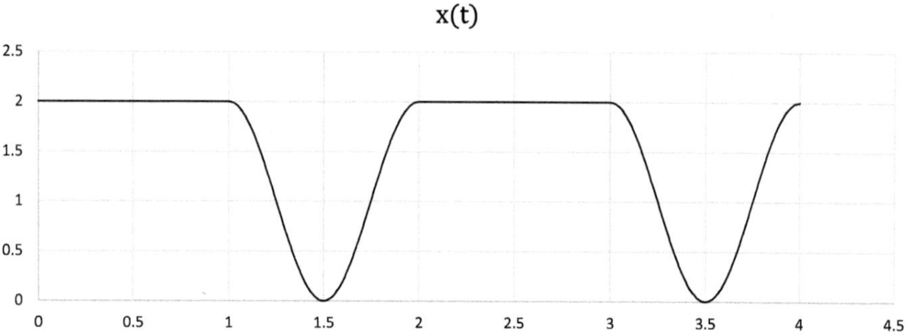

1.3. **(a)**

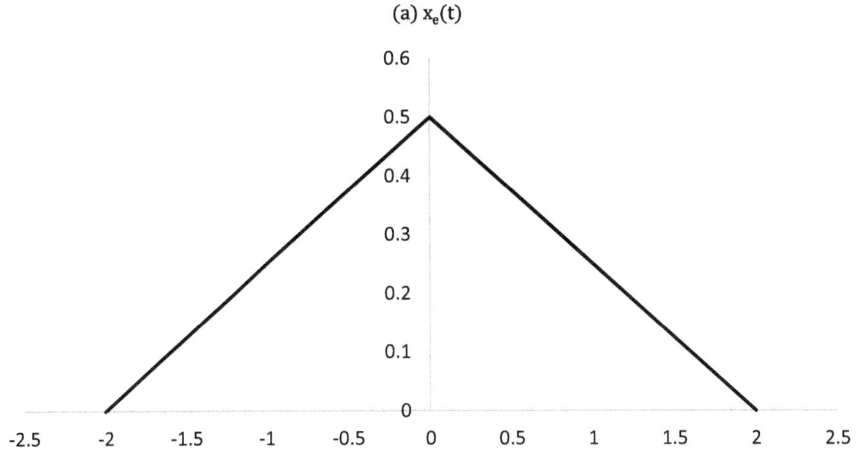

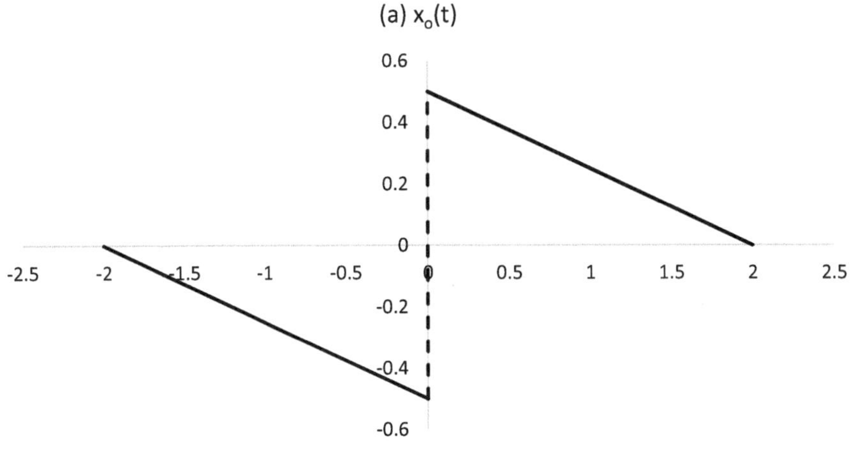

$x_e(t) = \frac{1}{2}\Lambda\left(\frac{t}{2}\right)$, $E_e = 1/3$; $x_o(t) = \frac{1}{2}\Lambda\left(\frac{t}{2}\right)\operatorname{sgn}(t)$, $E_o = 1/3$; $E_x = 2/3$.

(b)

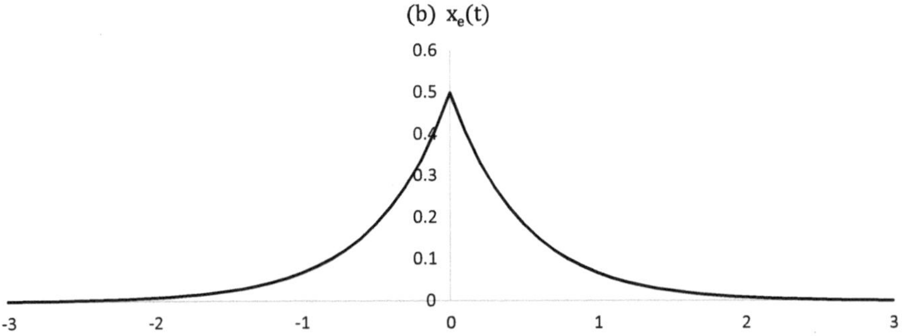

$x_e(t) = \frac{1}{2}e^{-2t}\cdot\mathbf{1}(t) + \frac{1}{2}e^{2t}\cdot\mathbf{1}(-t)$, $E_e = 1/4$.

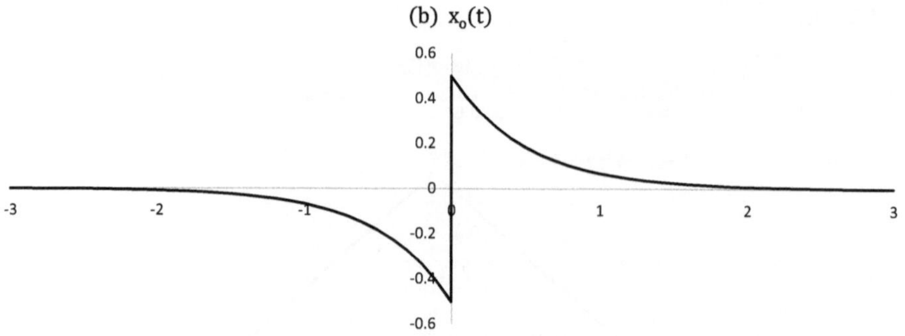

$x_o(t) = \frac{1}{2}e^{-2t} \cdot \mathbf{1}(t) - \frac{1}{2}e^{2t} \cdot \mathbf{1}(-t)$, $E_o = 1/4$; $E_x = 1/2$.

(c)

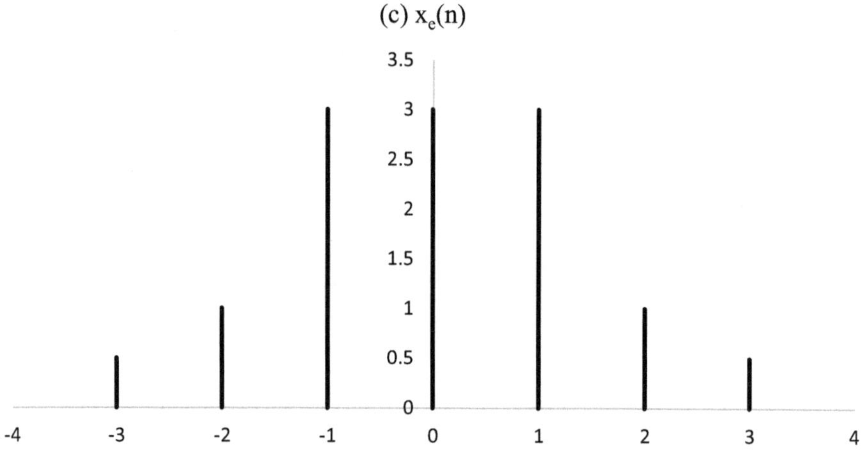

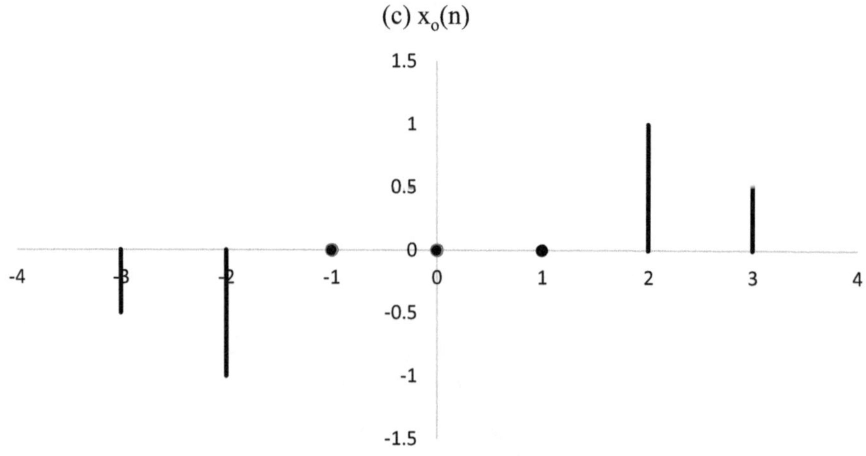

$x_e(n) = \frac{1}{2}\delta(n+3) + \delta(n+2) + 3\delta(n+1) + 3\delta$
$(n) + 3\delta(n-1) + \delta(n-2) + \frac{1}{2}\delta(n-3)$, $E_e = 29, 5$;

$x_o(n) = -\frac{1}{2}\delta(n+3) - \delta(n+2) + \delta(n-2) + \frac{1}{2}\delta(n-3)$; $E_o = 2, 5$; $E_x = 32$.

1.4. **(a)**

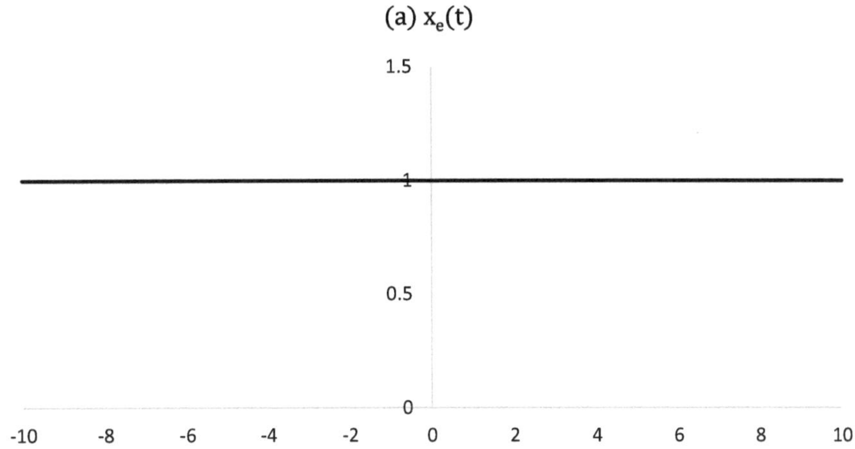

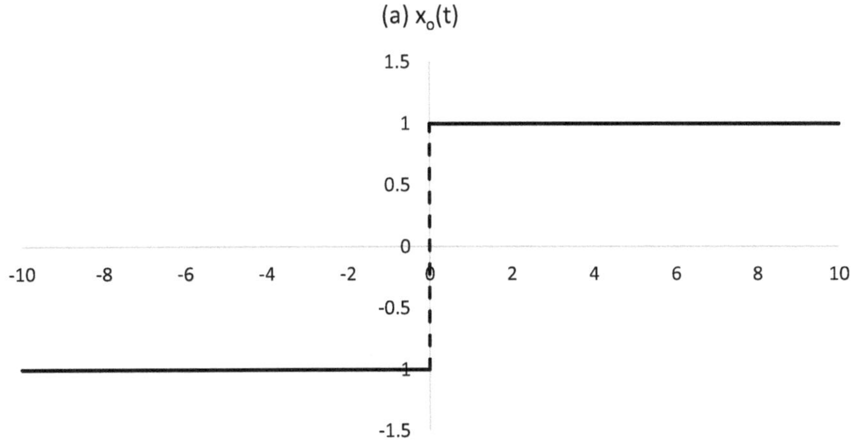

$x_e(t) = 1 = const$, $P_e = 1$; $x_o(t) =$ sgn (t), $P_o = 1$; $P_x = 2$.

(b)

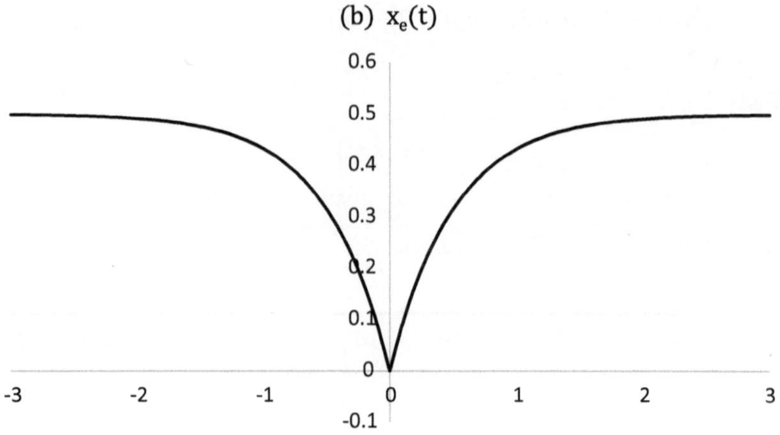

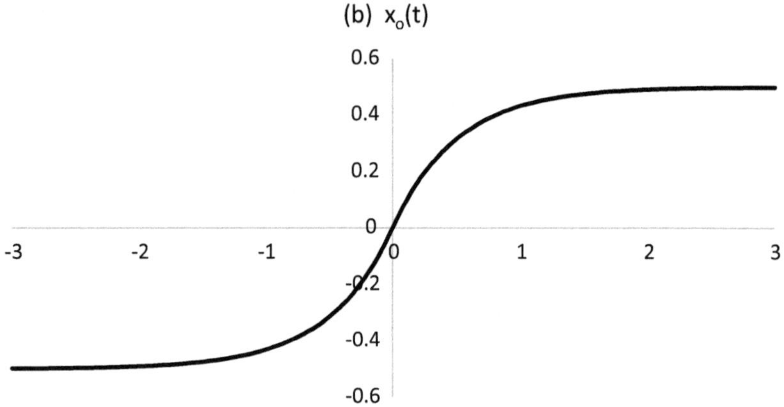

$x_e(t) = \frac{1}{2}(1 - e^{-2t}) \cdot \mathbf{1}(t) + \frac{1}{2}(1 - e^{2t}) \cdot \mathbf{1}(-t)$, $P_e = 1/4$;
$x_o(t) = \frac{1}{2}(1 - e^{-2t}) \cdot \mathbf{1}(t) - \frac{1}{2}(1 - e^{2t}) \cdot \mathbf{1}(-t)$, $P_o = 1/4$; $P_x = 1/2$.

(c)

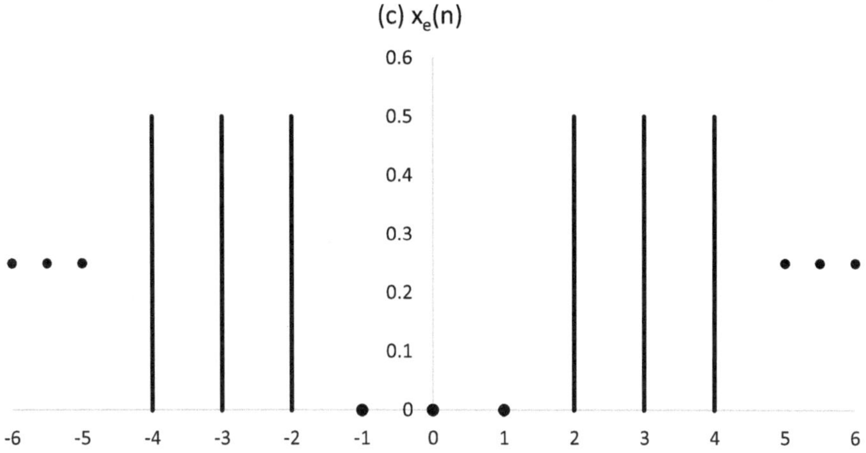

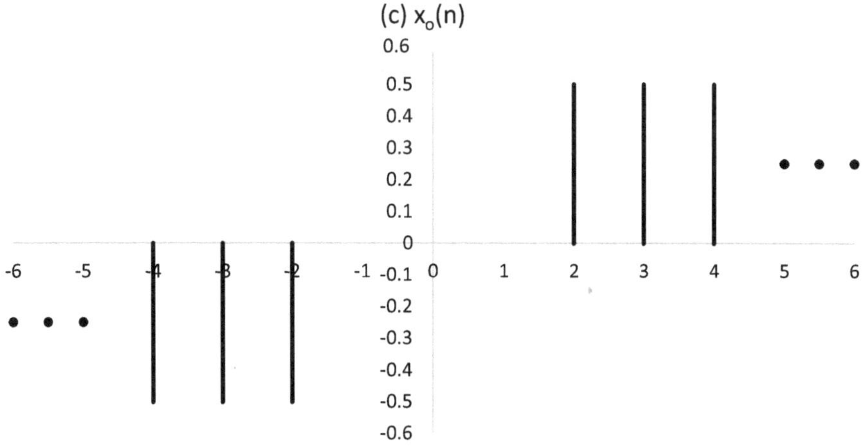

$x_e(n) = \frac{1}{2}\mathbf{1}(n-2) + \frac{1}{2}\mathbf{1}(-n-2)$, $P_e = 1/4$;
$x_o(n) = \frac{1}{2}\mathbf{1}(n-2) - \frac{1}{2}\mathbf{1}(-n-2)$, $P_o = 1/4$; $P_x = 1/2$.

1.5. $x(t) = (-2t-1)\cdot\mathbf{1}(t+1) + \frac{5}{2}t\cdot\mathbf{1}(t) + \left(-\frac{1}{2}t+1\right)\cdot\mathbf{1}(t-2)$;
$\frac{dx(t)}{dt} = \delta(t+1) - 2\cdot\mathbf{1}(t+1) + \frac{5}{2}\cdot\mathbf{1}(t) - \frac{1}{2}t\cdot\mathbf{1}(t-2)$.

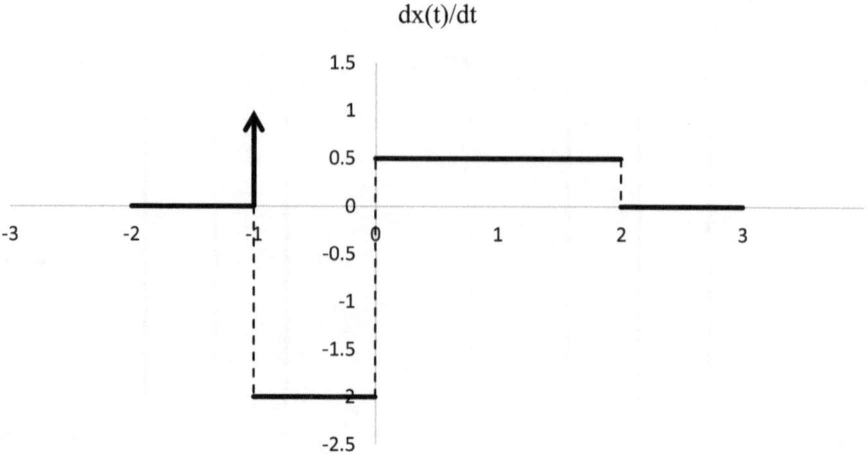

1.6. $x_1(t) * x_2(t) = \begin{cases} 2t, & t \in (0,1] \\ -4t+6, & t \in (1,2] \\ 2t-6, & t \in (2,3] \\ 0, & \text{elsewhere} \end{cases}$

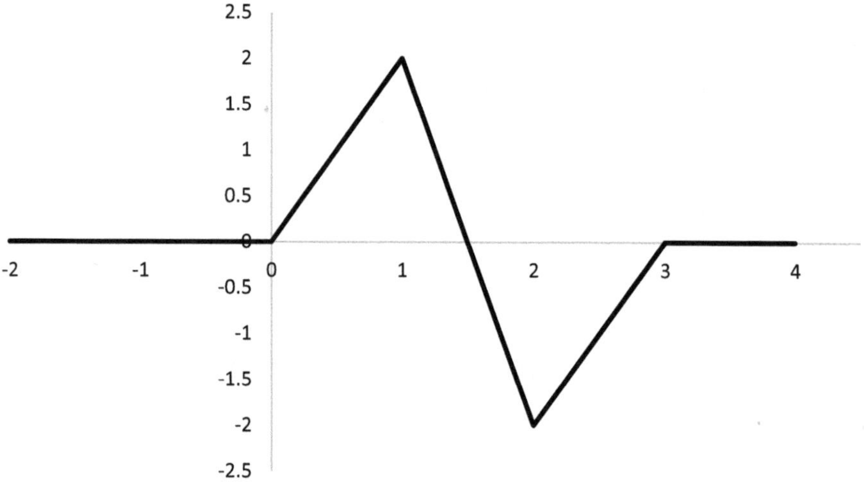

1.7. $x_1(n) \; * \; x_2(n) \; = \; [1; 1,5; 1,75; 1{,}875; 1{,}875; 0{,}875; 0{,}375; 0.125]$ for $n \in \{0, 1, \ldots, 7\}$.

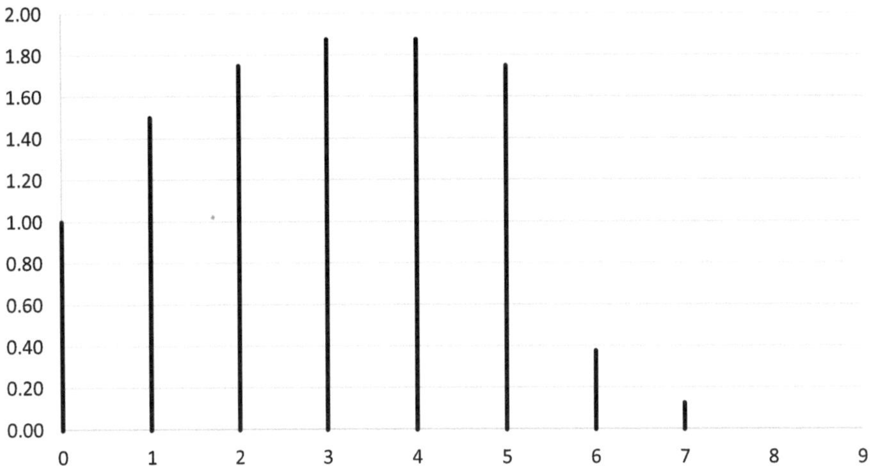

1.8. $$x(t) * x(t) = \begin{cases} t, & t \in (0,1] \\ 2-t, & t \in (1,2] \\ 0, & t \in (2,3] \\ 4t-12, & t \in (3,4] \\ -4t+20, & t \in (4,5] \\ 0, & t \in (5,6] \\ 4t-24, & t \in (6,7] \\ -4t+32, & t \in (7,8] \\ 0, & \text{elsewhere} \end{cases}.$$

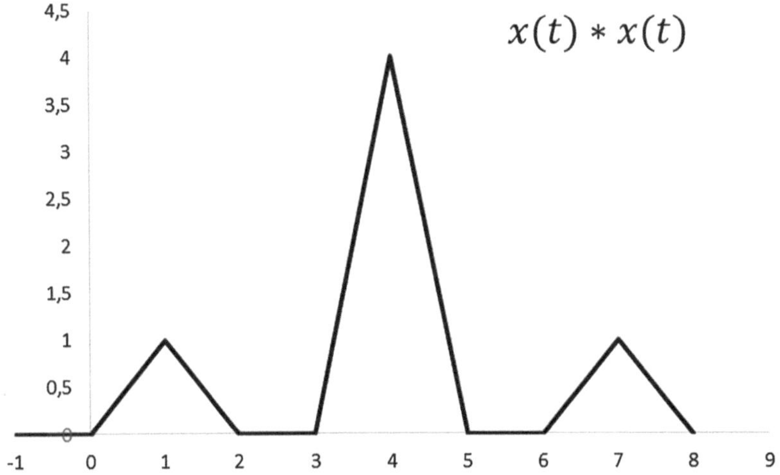

1.9. $z(t) = 2\Lambda(t)$, $\overline{z} = 1/2$, $E_z = \frac{2}{3}$, $P_z = 0$;
$x(t) = \sum_{n\in\mathbb{Z}} \Lambda(t-4n)$, $\overline{x} = 1/4$, $E_x = \infty$ $P_x = 1/12$.

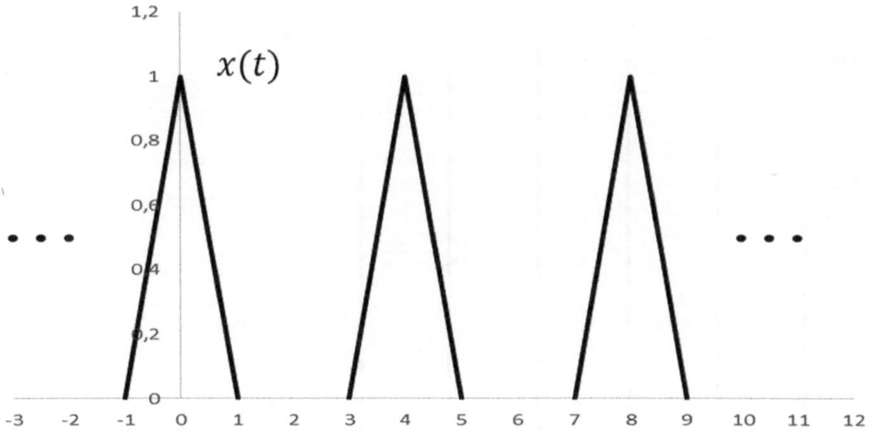

1.10. $R_x(\tau)=\begin{cases} -3\tau+2, & \tau\in(0,1] \\ \tau-2, & \tau\in(1,2] \\ 0, & \text{elsewhere} \end{cases}$; $R_x(0)=2=E_x$.

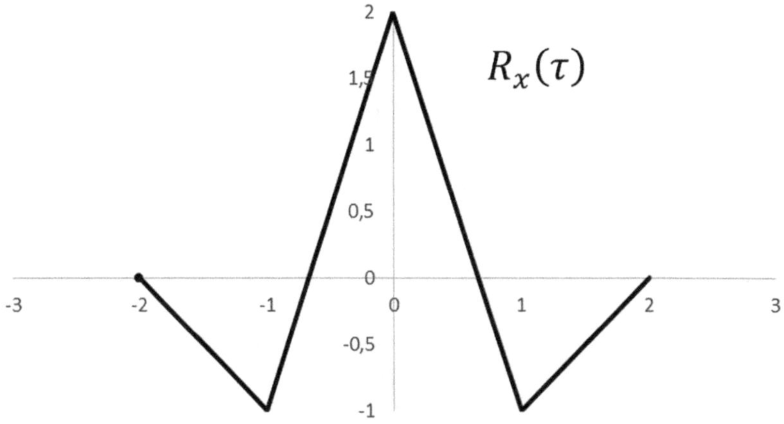

1.11. $R_{12}(\tau)=\begin{cases} 2\tau+2, & \tau\in(-1,0] \\ -4\tau+2, & \tau\in(0,1] \\ 2\tau-4, & \tau\in(1,2] \\ 0, & \text{elsewhere} \end{cases}$; $R_{21}(\tau)=R_{12}(-\tau)$.

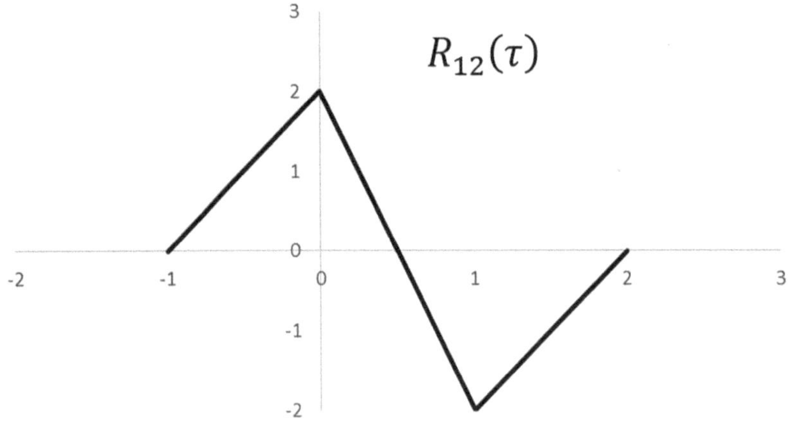

1.12. $R_{12}(k) = \left[\frac{1}{8}; \frac{3}{8}; \frac{7}{8}; \frac{15}{8}; \frac{15}{8}; \frac{7}{4}; \frac{3}{2}; 1\right]$ for $k \in \{-3, -2, \ldots, 4\}$; $R_{21}(k) = R_{12}(-k)$.

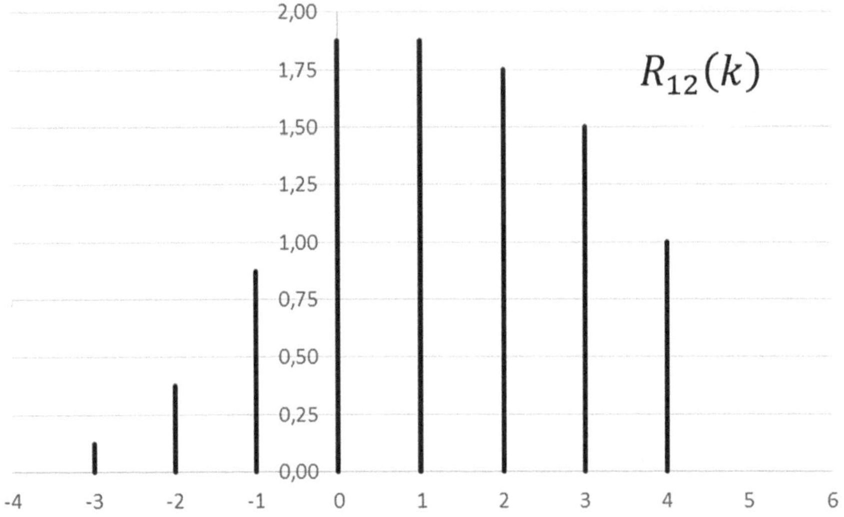

1.13. $c_0 = 1/2$; $c_1 = 0$; $c_2 = \; - \; 1/4$; $c_3 = 0$; $c_4 = 0$.

$$x(t) = \tfrac{1}{2}\mathrm{wal}(0,t) - \tfrac{1}{4}\mathrm{wal}(2,t) = \begin{cases} \dfrac{1}{4}, t \in \left(0, \dfrac{1}{4}\right) \\ \dfrac{3}{4}, t \in \left(\dfrac{1}{4}, \dfrac{3}{4}\right) \\ \dfrac{1}{4}, t \in \left(\dfrac{1}{4}, 1\right) \end{cases}.$$

1.14. $\phi_1(t) = \frac{1}{3}s_1(t)$; $\phi_2(t) = \frac{\sqrt{2}}{2}s_2(t) - \frac{\sqrt{2}}{6}s_1(t)$;
$\phi_3(t) = \sqrt{2}s_3(t) + \frac{\sqrt{2}}{2}s_2(t) + \frac{\sqrt{2}}{6}s_1(t)$.

Chapter 2

2.1. **(a)** $f_0 = \frac{1}{2}$ [Hz];
trigonometric Fourier series: $x(t) = 2 + \cos\left(\pi t - \frac{\pi}{2}\right) + \cos\left(2\pi t + \frac{\pi}{3}\right)$;
$A_0 = 2, A_1 = A_2 = 1; \varphi_1 = -\frac{\pi}{2}, \varphi_2 = \frac{\pi}{3}$;
exponential Fourier series: $x(t) = 2 - \frac{1}{2}je^{j\pi t} + \frac{1}{2}je^{-j\pi t} + \frac{1}{2}e^{j\frac{\pi}{3}}e^{j2\pi t} + \frac{1}{2}e^{-j\frac{\pi}{3}}e^{-j2\pi t}$;
$|c_0| = 2, |c_{-2}| = |c_2| = \frac{1}{2}, |c_{-1}| = |c_1| = \frac{1}{2}$; $\arg c_{-2} = -\frac{\pi}{3}, \arg c_{-1} = \frac{\pi}{2}, \arg c_0 = 0, \arg c_1 = -\frac{\pi}{2}, \arg c_2 = \frac{\pi}{3}$;
$P_x = 5$.
(b) $f_0 = \frac{3}{2}$ [Hz];
trigonometric Fourier series: $x(t) = 3\sqrt{2}\cos\left(3\pi t - \frac{\pi}{4}\right) + \cos(9\pi t)$;
$A_1 = 3\sqrt{2}, A_3 = 1; \varphi_1 = -\frac{\pi}{4}, \varphi_3 = 0$;
exponential Fourier series: $x(t) = \left(\frac{3}{2} - \frac{3}{2}j\right)e^{j3\pi t} + \left(\frac{3}{2} + \frac{3}{2}j\right)e^{-j3\pi t} + \frac{1}{2}e^{j9\pi t} + \frac{1}{2}e^{-j9\pi t}$;
$|c_0| = 0, |c_{-1}| = |c_1| = \frac{1}{2}, |c_{-3}| = |c_3| = \frac{3}{2}\sqrt{2}$; $\arg c_{-3} = 0, \arg c_{-1} = \frac{\pi}{4}$, $\arg c_0 = 0$, $\arg c_1 = -\frac{\pi}{4}, \arg c_3 = 0$;
$P_x = 19/2$.
(c) $f_0 = \frac{1}{2}$ [Hz];
trigonometric Fourier series: $x(t) = \frac{1}{2}\sin(\pi t) + \frac{1}{2}\sin(5\pi t)$;
$A_1 = A_5 = \frac{1}{2}; \varphi_1 = \varphi_5 = -\frac{\pi}{2}$;
exponential Fourier series: $x(t) = \frac{j}{4}e^{-j5\pi t} + \frac{j}{4}e^{-j\pi t} - \frac{j}{4}e^{j\pi t} - \frac{j}{4}e^{j5\pi t}$;
$|c_0| = 0, |c_{-1}| = |c_1| = |c_{-5}| = |c_5| = \frac{1}{4}$; $\arg c_{-5} = \arg c_{-1} = \frac{\pi}{2}, \arg c_0 = 0$, $\arg c_1 = \arg c_3 = -\frac{\pi}{2}$;
$P_x = 1/4$.
(d) $f_0 = \frac{1}{2}$ [Hz];
trigonometric Fourier series: $x(t) = \frac{1}{2} + \cos(\pi t) + \cos(2\pi t)$;
$A_0 = \frac{1}{2}, A_1 = A_2 = 1; \varphi_1 = \varphi_2 = 0$;
exponential Fourier series: $x(t) = \frac{1}{2}e^{-j2\pi t} + \frac{1}{2}e^{-j\pi t} + \frac{1}{2} + \frac{1}{2}e^{j\pi t} + \frac{1}{2}e^{j2\pi t}$;
$|c_0| = |c_{-1}| = |c_1| = |c_{-2}| = |c_2| = \frac{1}{2}$; $\arg c_{-2} = \arg c_{-1} = \arg c_0 = \arg c_1 = \arg c_2 = 0$;
$P_x - 5/4$.

2.2. $f_0 = \frac{1}{2}$ [Hz];
$A_0 = 2, A_1 = 1, A_2 = \frac{1}{2}; \varphi_1 = \frac{3\pi}{4}, \varphi_2 = -\frac{\pi}{3}$; THD $= 1/\sqrt{5}$.

2.3. exponential Fourier series: $x(t) = \frac{1}{4}e^{j\frac{\pi}{3}}e^{-j6\pi t} + \frac{1}{4}e^{j\frac{\pi}{3}}e^{-j2\pi t} + \frac{1}{4}e^{-j\frac{\pi}{3}}e^{j2\pi t} + \frac{1}{4}e^{-j\frac{\pi}{3}}e^{j6\pi t}$;
$|c_{-1}| = |c_1| = |c_{-3}| = |c_3| = \frac{1}{4}$;
$\arg c_{-3} = \arg c_{-1} = \frac{\pi}{3}, \arg c_1 = \arg c_3 = -\frac{\pi}{3}$;
$P_x = 1/4$.

2.4. $c_0 = \frac{1}{2}, c_n = \frac{j}{2n\pi}$.
2.5. $c_0 = 0, c_n = \frac{j}{2n\pi}(-1)^n, A_0 = 0, b_1 = \frac{1}{\pi}, b_2 = -\frac{1}{2\pi}, b_3 = \frac{1}{3\pi}, b_4 = -\frac{1}{4\pi}, b_5 = \frac{1}{5\pi}$ (see formula (2.4a)).
2.6. $x_5(t) = \frac{1}{\pi}\sin(\pi t) - \frac{1}{2\pi}\sin(2\pi t) + \frac{1}{3\pi}\sin(3\pi t) - \frac{1}{4\pi}\sin(4\pi t) + \frac{1}{5\pi}\sin(5\pi t)$.

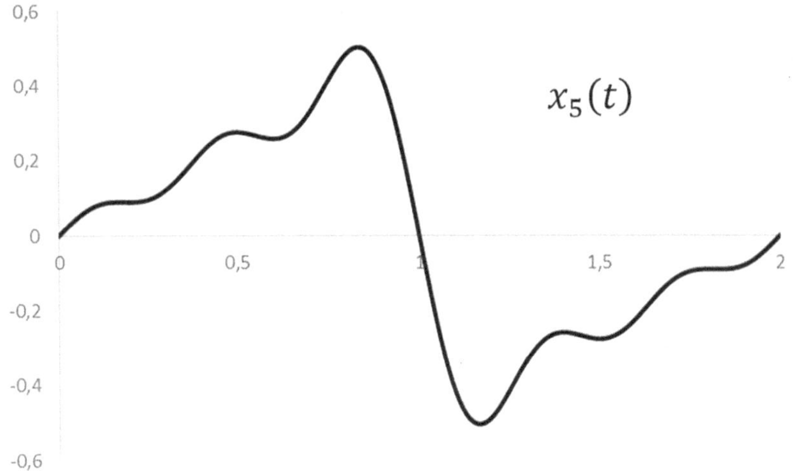

$P_x = \frac{1}{12}$, $\varepsilon_1 \cong 0{,}033$, $\varepsilon_2 \cong 0, 02$, $\varepsilon_3 \cong 0{,}014$, $\varepsilon_4 \cong 0{,}011$, $\varepsilon_5 \cong 0{,}009$.
$x(t) = -\frac{1}{2} + 4\sin(40\pi t) - \sin(80\pi t)$.

2.7. **(a)** *Hint*: Determine the Fourier transform of $y(\alpha, t)$ and calculate its limit when $\alpha \to 0$ for $\omega = 0$ and $\omega \neq 0$. Additionally, you should calculate the integral $\int_{-\infty}^{\infty} \frac{2\alpha}{\alpha^2+\omega^2} d\omega$.
(b) See Problem 2.9.

2.8. *Hint*: Write *x*(*t*) as a product of a cosine function and a rectangular pulse and then apply the modulation property of the FT from Table 2.2; in order to determine the spectrum of *y*(*t*), you can write *y*(*t*) as a difference of two cosine pulses shifted in time and then apply the linearity of FT and time-shift theorem.
(a) $X(\omega) = \frac{4\pi}{\pi^2 - \omega^2}\cos\left(\frac{\omega}{2}\right)$, **(b)**$X(\omega) = \frac{4\pi j}{\omega^2 - \pi^2}\sin(\omega)$.

2.9. $x(t) = \frac{1}{2}\Pi\left(\frac{t}{4}\right)\sin\left(\frac{\pi}{2}t\right)$.

2.10. $X(\omega) = \mathrm{Sa}^2\left(\frac{\omega}{2}\right)\frac{1 - e^{-j200\omega}}{1 - e^{-j4\omega}}$.

2.11. $X(\omega) = \frac{1}{2}\Pi\left(\frac{\omega - 4\pi}{\pi}\right) + \frac{1}{2}\Pi\left(\frac{\omega + 4\pi}{\pi}\right) - \frac{1}{2}\Pi\left(\frac{\omega - 12\pi}{\pi}\right) - \frac{1}{2}\Pi\left(\frac{\omega + 12\pi}{\pi}\right)$.
$E_x = 1/2$.

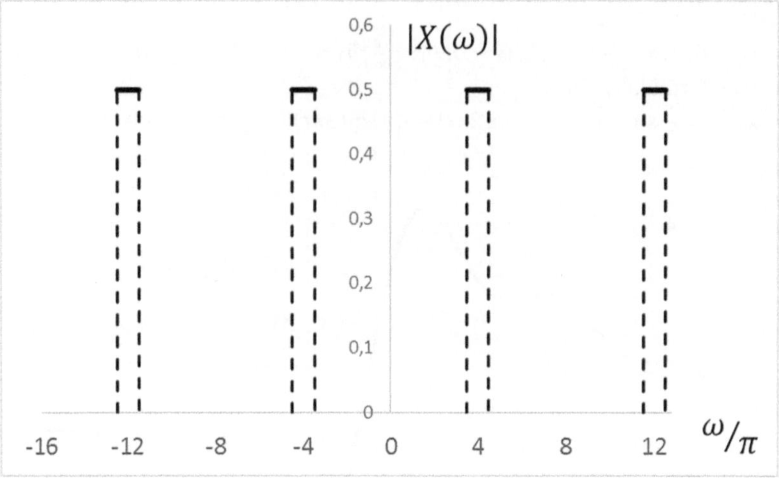

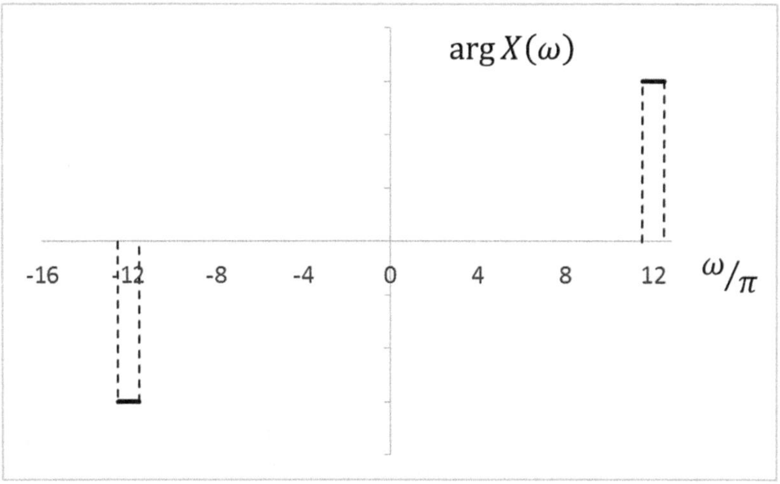

2.12. $X^2(\omega) = \mathrm{Sa}^2\left(\frac{\omega}{2}\right)\left(e^{-j\omega} + 4e^{-j4\omega} + 4e^{-j7\omega}\right)$;
$x(t) * x(t) = \Lambda(t-1) + 4\Lambda(t-4) + 4\Lambda(t-7)$.

2.13. $|X(\omega)|^2 = 256t_0^2\mathrm{Sa}^2(2t_0\omega)$; $R_x(\tau) = 64t_0\Lambda\left(\frac{\tau}{4t_0}\right)$.

2.14. $c_n = \frac{1}{4}\mathrm{Sa}^2\left(n\frac{\pi}{4}\right)$; $c_0 = \frac{1}{4}, c_1 = c_{-1} = \frac{2}{\pi^2}, c_2 = c_{-2} = \frac{1}{\pi^2}$, $c_3 = c_{-3} = \frac{2}{9\pi^2}$,
$c_4 = c_{-4} = 0$, $c_5 = c_{-5} = \frac{2}{25\pi^2}$.

2.15. $x(t) = \sum_{n\in\mathbb{Z}} \Pi(t - 4n - 1)$.

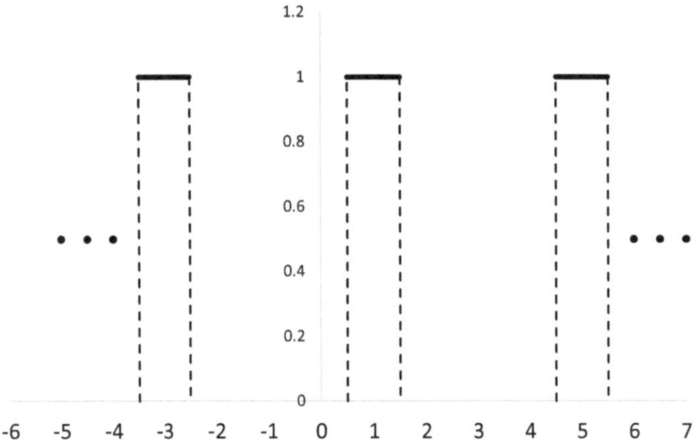

Chapter 3

3.1. System nonlinear, time-invariant, memoryless, causal.

(a) System nonlinear, time-invariant, memoryless, causal.
(b) System linear, non-time-invariant, with memory, causal.
(c) System linear, time-invariant, with memory, noncausal.
(d) System linear, non-time-invariant, with memory, causal.
(e) System nonlinear, time-invariant, with memory, causal.

3.2. $\forall T_1, T_2 \in \mathbb{R}, T_1 + T_2 > 0$ – system linear, time-invariant, with memory. For $T_2 \leq 0$ system is causal.
3.3. $\forall M, N \in \mathbb{R}, M + N > 0$ – system linear, time-invariant, with memory. For $M \leq 0$ system is causal.
3.4. $y_1(t) = 2[k(t) - k(t-2)]$;

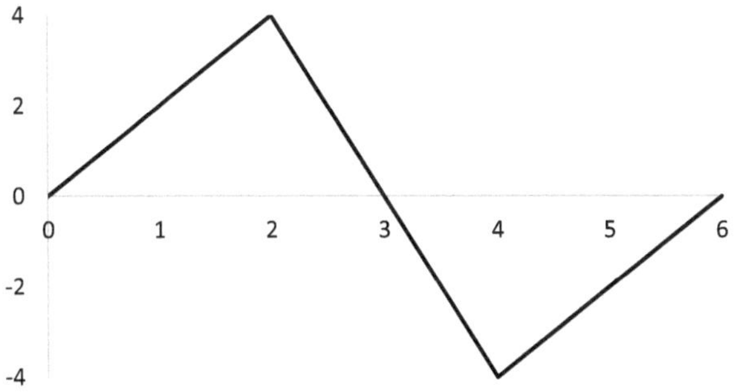

$y_2(t) = k(t) - 2 \cdot k(t-2) + k(t-4)$;

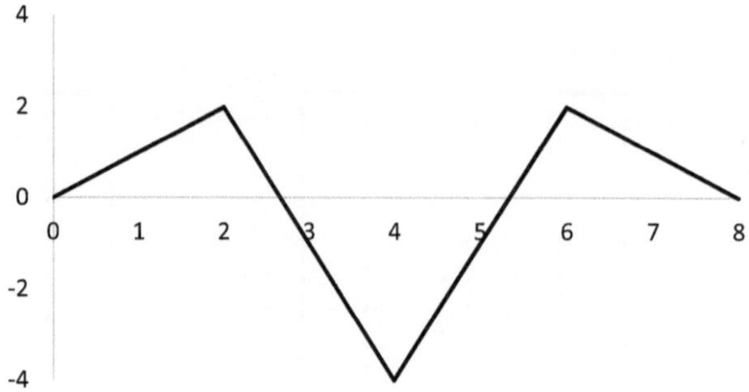

$y_3(t) = k(t) + k(t-1) - k(t-2) - k(t-3);$

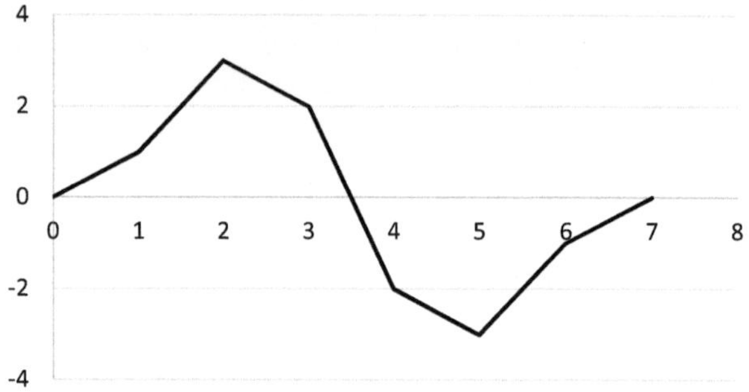

3.5. $y(t) = e^{-(t-2)} \cdot \mathbf{1}(t-2).$

3.6. *Hint*: Solve the difference equation: $k(n) = h(n) + k(n-1)$ in a recursive way; $k(0) = 0$; $k(1) = 0, 3$; $k(2) = 0, 51$; $k(3) = 0{,}627$, etc.

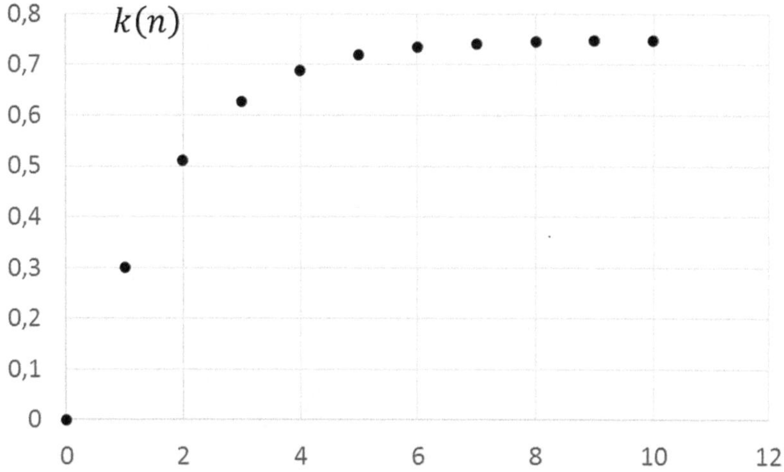

3.7. System is linear, non-time-invariant.

(a) $y(n) = \mathbf{1}(n-4) - \mathbf{1}(n-7) = \delta(n-4) + \delta(n-5) + \delta(n-6)$;

(b) $y(n) = \sum_{k=0}^{\infty} g(n-2k)$

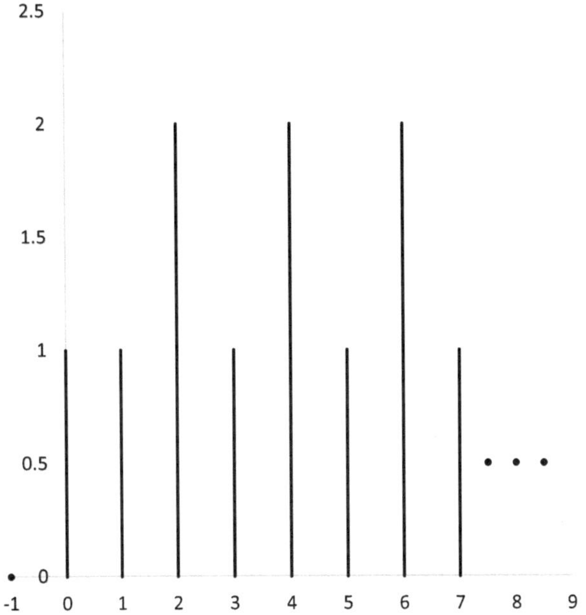

3.8. $A = -\frac{1}{4}, B = -\frac{1}{4}, C = \frac{1}{2}$.

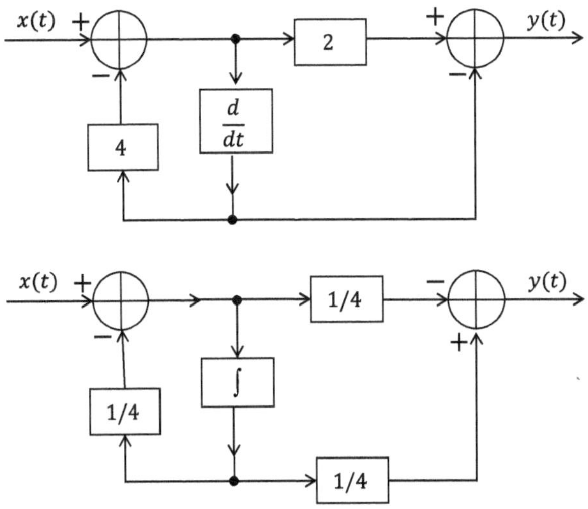

3.9. Direct realization type I:

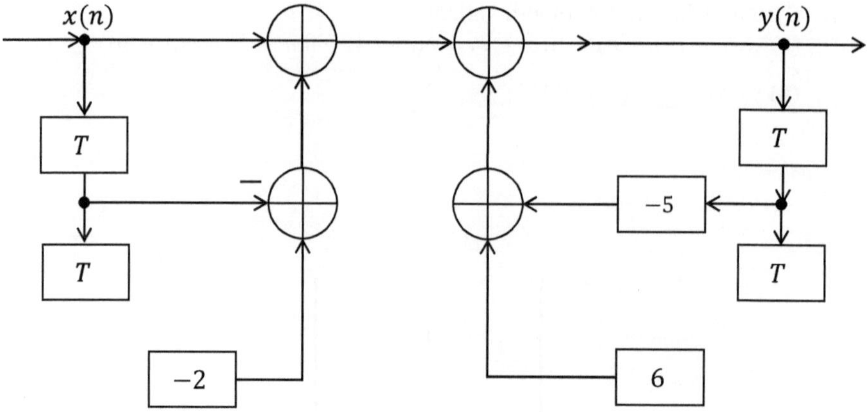

Direct realization type II:

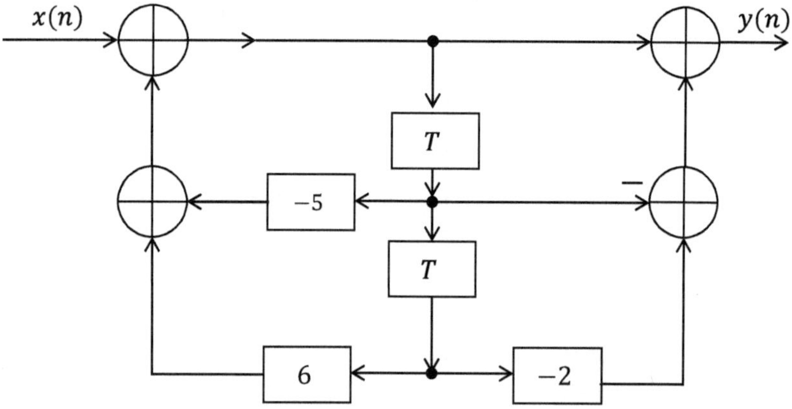

3.10. $y_1(t) = t \cdot \mathbf{1}(t) - 2(t-2) \cdot \mathbf{1}(t-2) + (t-4) \cdot \mathbf{1}(t-4)$;

$$y(t) = y_1(t) * \mathbf{1}(t) = \begin{cases} \dfrac{t^2}{2}, & t \in (0,2] \\ -\dfrac{t^2}{2} + 4t - 4, & t \in (2,4] \\ 4, & t \in (4,+\infty) \\ 0, & \text{elsewhere} \end{cases};$$

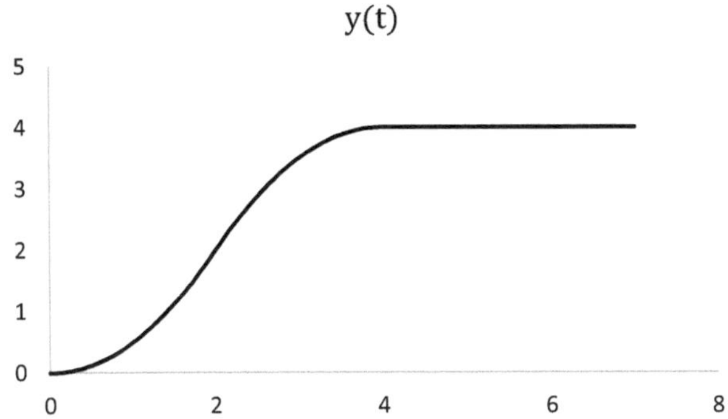

3.11. **(a)** $y(n) = x(n-2) + \frac{5}{2}x(n-3) + x(n-4)$;
$h(n) = \delta(n-2) + \frac{5}{2}\delta(n-3) + \delta(n-4)$.
(b) $y(n) = 2x(n) + 2x(n-1) + x(n-2) + \frac{1}{2}x(n-3)$;
$h(n) = 2\delta(n) + 2\delta(n-1) + \delta(n-2) + \frac{1}{2}\delta(n-3)$.

3.12. *Hint*: Apply the cascade system interconnection S–S^{-1}. The output of the system S^{-1}, that is, $y(t) = y_1(n) - y_1(n-1)$ where $y_1(n) = \sum_{k=-\infty}^{n} x(k)$.

Chapter 4

4.1. **(a)** $y_1(t) = 5f_s \cdot \mathrm{Sa}(5\pi f_s t)$;
(b) $y_2(t) = f_s[1 + 2\cos(2\pi f_s t) + 2\cos(4\pi f_s t)]$.

4.2. $A(\omega) = \begin{cases} 1, |\omega| \le 5 \\ 0, \text{elsewhere} \end{cases}, \varphi(\omega) = \begin{cases} \omega, |\omega| \le 5 \\ 0, \text{elsewhere} \end{cases}$;
$y(t) = \sin[4(t+1)] + \frac{3}{2}\mathrm{Sa}\left[\frac{3}{2}(t+1)\right]\sin\left[\frac{7}{2}(t+1)\right]$.

4.3. $A(\omega) = \begin{cases} 4, & 2 \le |\omega| \le 4 \\ 0, & \text{elsewhere} \end{cases}, \varphi(\omega) = \begin{cases} -3\omega, & 2 \le |\omega| \le 4 \\ 0, & \text{elsewhere} \end{cases}$;
$y_1(t) = \frac{4}{3}\mathrm{Sa}\left[\frac{1}{2}(t-3)\right]\cos\left[\frac{5}{2}(t-1)\right]$; $y_2(t) = \mathrm{Sa}(t-3)\cos[3(t-3)]$.

4.4. $f_0 = 450$ Hz, $f_c \in (400{,}500)$ [Hz], $K = 5$.

4.5. $y(t) = \frac{3}{4\pi}\mathrm{Sa}\left[\frac{1}{2}(t-1)\right]\cos\left[\frac{7}{2}(t-1)\right]$.

4.6. $X(\omega) = 2\sum_{n\in\mathbb{Z}} \frac{1-(-1)^n}{jn}\delta\left(\omega - n\frac{\pi}{5}\omega_c\right)$; $y(t) = \frac{4}{\pi}\sin\left(\frac{\pi}{5}\omega_c t\right)$.

4.7. $X(f) = \sum_{n\in\mathbb{Z}} \frac{\sin\left(n\frac{\pi}{2}\right)}{n\pi}\delta(f - n\cdot 1)$;
(a) $y(t) = \frac{1}{2} + \frac{2}{\pi}\cos(2\pi t)$; **(b)** $y(t) = \frac{1}{2} + \frac{2}{\pi}\cos(2\pi t) - \frac{2}{3\pi}\cos(6\pi t)$.

4.8. $Y(f) = \frac{1}{2}[\delta(f-10) + \delta(f+10)] + \frac{1}{8}[\delta(f-9) + \delta(f+9) + \delta(f-11) + \delta(f+11)]$ $y(t) = \left[1 + \frac{1}{2}\cos(2\pi t)\right]\cos(20\pi t)$.

4.9. **(a)** $H_1(\omega) = \pi\delta(\omega) + \frac{1}{j\omega}$; $y_1(t) = \sin t$; **(b)** $H_2(\omega) = -2 + \frac{5}{2+j\omega}$;
$y_2(t) = \sin t$; **(c)** $H_3(\omega) = \frac{2}{(1+j\omega)^2}$; $y_3(t) = \sin t$.

4.10. $h(t) = 5e^{-5t} \cdot \mathbf{1}(t)$; $k(t) = (1 - e^{-5t}) \cdot \mathbf{1}(t)$;
$A(\omega) = \frac{5}{\sqrt{\omega^2+25}}$; $\varphi(\omega) = -\operatorname{atan}(\omega/5)$; 3dB bandwidth: $[-5, 5]$ [rad/s];
$y(t) = 2 + \frac{3\sqrt{2}}{5}\cos(10t) + \frac{\sqrt{2}}{5}\sin(10t)$. (*Hint*: Use (4.8)–(4.10) and linearity of the system to calculate the responses to: the constant signal 2, $\sqrt{2}\cos(10t)$, $(-\sqrt{2}\sin(10t))$.)

4.11. $h(t) = \delta(t) - 5e^{-5t} \cdot \mathbf{1}(t)$; $k(t) = e^{-5t} \cdot \mathbf{1}(t)$;
$A(\omega) = \frac{|\omega|}{\sqrt{\omega^2+25}}$; 3dB bandwidth: $(-\infty, -5] \cup [5, +\infty)$ [rad/s];

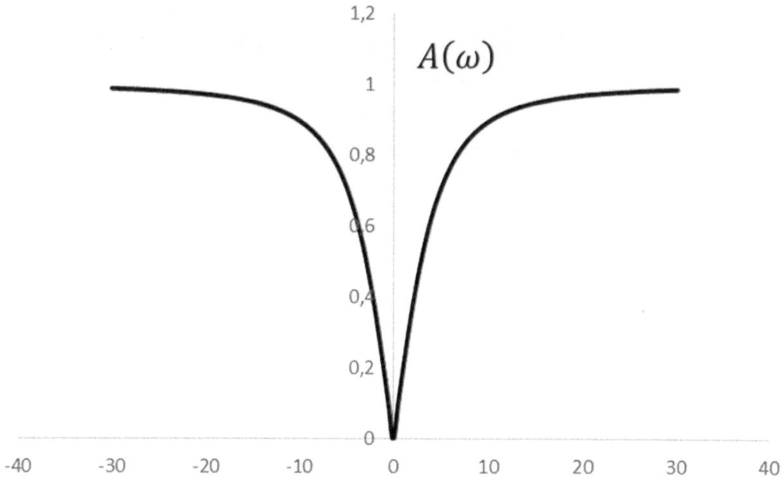

$$\varphi(\omega) = \begin{cases} \frac{\pi}{2} - \operatorname{atan}(\omega/5), & \omega \geq 0 \\ -\frac{\pi}{2} - \operatorname{atan}(\omega/5), & \omega < 0 \end{cases};$$

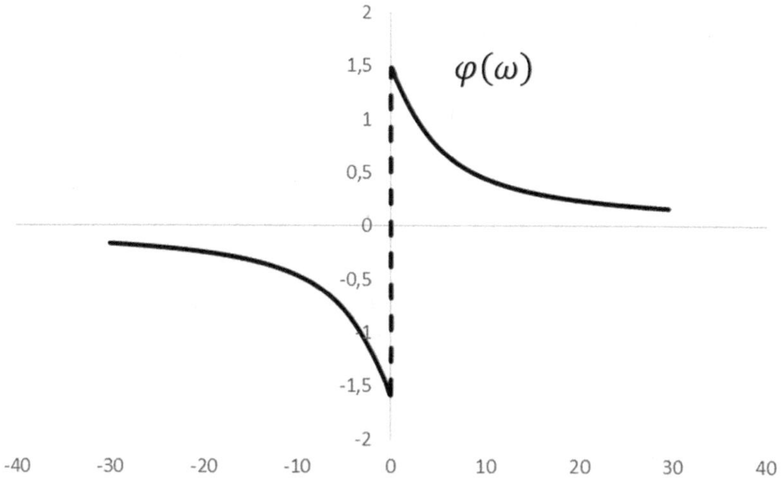

$y(t) = e^{-5\left(t-\frac{1}{2}\right)} \cdot \mathbf{1}\left(t - \frac{1}{2}\right) - e^{-5\left(t-\frac{3}{2}\right)} \cdot \mathbf{1}\left(t - \frac{3}{2}\right)$.

4.12. $H(\omega) = \frac{1}{1+4j\omega}$; $h(t) = \frac{1}{4}e^{-\frac{1}{4}t} \cdot \mathbf{1}(t)$; $A(\omega) = \frac{1}{\sqrt{1+16\omega^2}}$;

3dB bandwidth: $\left[-\frac{1}{4}, \frac{1}{4}\right]$ [rad/s]; $y(t) = -\frac{1}{7}e^{-2t} \cdot \mathbf{1}(t) + \frac{1}{7}e^{-\frac{1}{4}t} \cdot \mathbf{1}(t)$.

4.13. $H(\omega) = \frac{1}{(2+j\omega)^2}$; $h(t) = te^{-2t} \cdot \mathbf{1}(t)$; $A(\omega) = \frac{1}{4+\omega^2}$;

3dB bandwidth: $\left[-\sqrt{\sqrt{2}-1}, \sqrt{\sqrt{2}+1}\right]$ [rad/s]; $y(t) = \frac{1}{2}t^2e^{-2t} \cdot \mathbf{1}(t)$.

4.14. $p_{in} = 0, 1$; $\tau = 1$;

$y(t) = \frac{1}{2} + \frac{1}{2}\cos(2t) + \frac{1}{2}\sin(2t) + \frac{1}{10}\cos(4t) + \frac{1}{5}\sin(4t)$.

4.15. **(a)** $A(\omega) = \frac{2}{\sqrt{1+16\omega^2}}$; $\omega_1 = \frac{1}{4}$ rad/s; 3dB bandwidth: $\left[-\frac{1}{4}, \frac{1}{4}\right]$ [rad/s].

(b) $A_{\mathrm{HP}}(\omega) = \frac{2|\omega|}{\sqrt{1+\omega^2}}$.

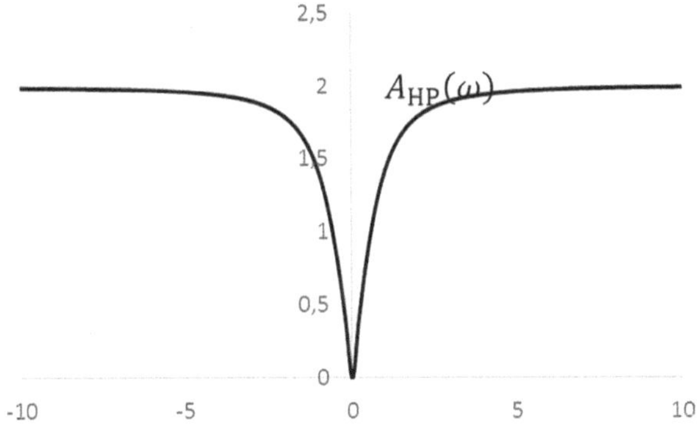

(c) $A_{\mathrm{LP}}(\omega) = \frac{2}{\sqrt{1+\omega^2}}$.

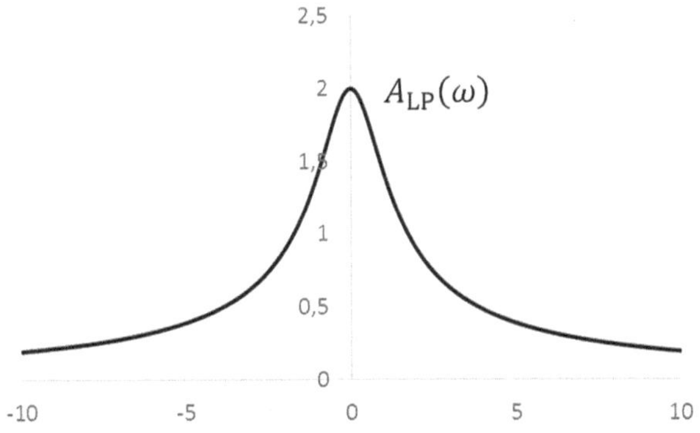

4.16. **(a)** $A(\omega) = \frac{2}{\sqrt{9+\omega^2}}$; $\omega_1 = 3$ rad/s; 3dB bandwidth: $[-3, 3]$ [rad/s].

(b) $A_{\mathrm{BP}}(\omega) = \frac{2}{3}\frac{|\omega|}{\sqrt{\omega^4 - 3\omega^2 + 4}}$.

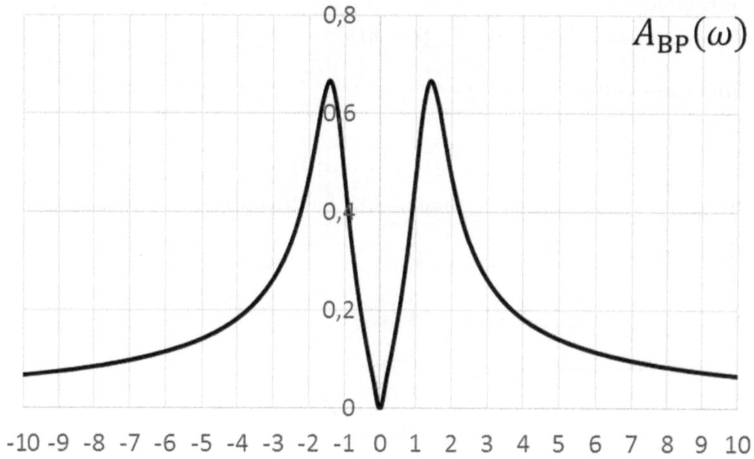

(c) $A_{\mathrm{BS}}(\omega) = \frac{2}{3}\,\frac{\left|2-\omega^2\right|}{\sqrt{\omega^4-3\omega^2+4}}$.

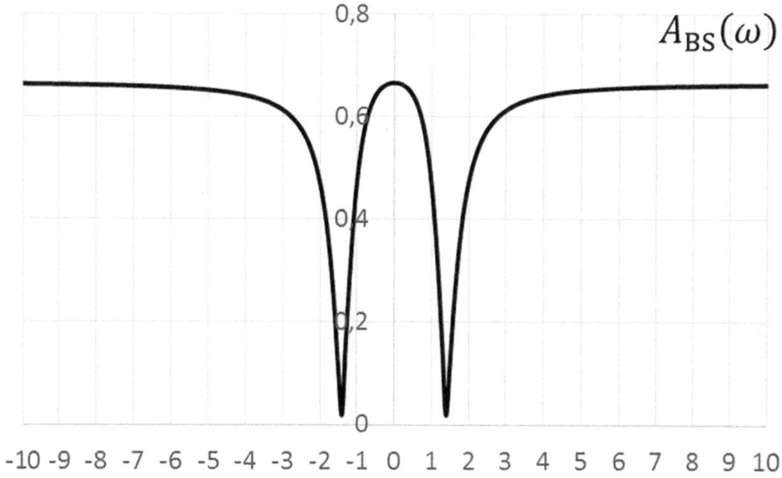

Chapter 5

5.1. $y_{0\mathrm{s}}(t) = \left(\frac{1}{2} - e^{-(t-1)} + \frac{1}{2}e^{-2(t-1)}\right)\cdot \mathbf{1}(t-1)$;
$y_{0\mathrm{i}}(t) = (e^{-t} - e^{-2t})\cdot \mathbf{1}(t)$; $y(t) = y_{0\mathrm{s}}(t) + y_{0\mathrm{i}}(t)$.

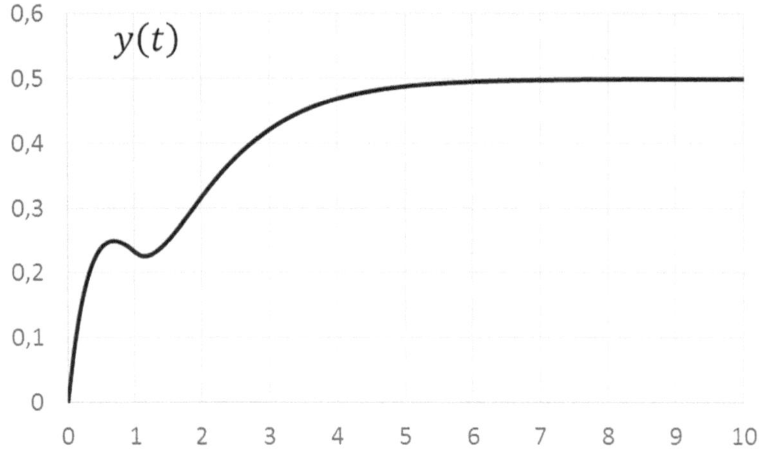

5.2. $y(t)=\left[\frac{21}{20}e^{t}-\frac{29}{20}e^{-t}-\frac{1}{2}\sin(t)-\frac{3}{5}\cos(2t)\right]\cdot\mathbf{1}(t).$

5.3. $H(s)=3\frac{s+3}{(s+2)(s+4)}$; $h(t)=\frac{3}{2}(e^{-2t}+e^{-4t})\cdot\mathbf{1}(t)$;

$y(t)=\frac{1}{8}\left[3\frac{dx(t)}{dt}+9x(t)-6\frac{dy(t)}{dt}-\frac{d^2y(t)}{dt^2}\right].$

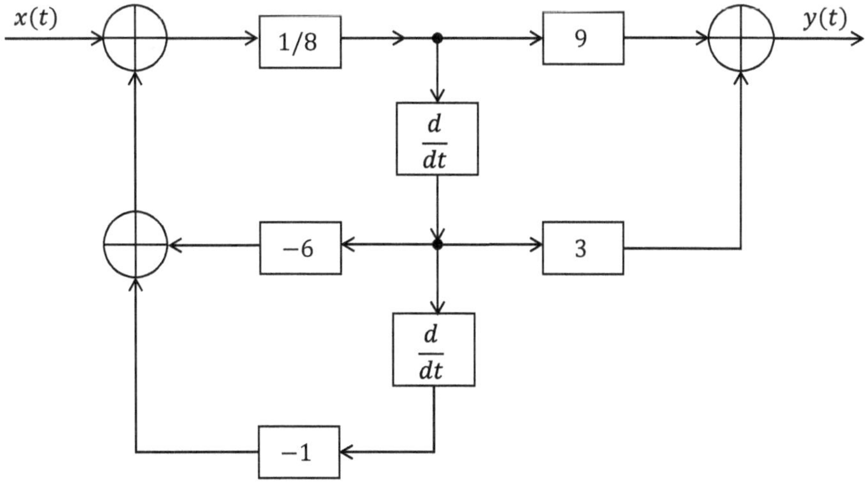

5.4. $H(s)=2\frac{e^{-\frac{1}{2}s}}{s+2}$; $h(t)=2e^{-2\left(t-\frac{1}{2}\right)}\cdot\mathbf{1}\left(t-\frac{1}{2}\right)$; $y(t)=\frac{1}{2}\cos(2t-1)+\frac{1}{2}\sin(2t-1)$, $t\in\mathbb{R}$.

5.5. $\frac{d^2y(t)}{dt^2}+2\frac{dy(t)}{dt}+y(t)=\frac{dx(t)}{dt}$; $H(s)=\frac{s}{(s+1)^2}$; $h(t)=(1-t)e^{-t}\cdot\mathbf{1}(t)$;

(a) $y(t)=te^{-t}\left(1-\frac{1}{2}t\right)\cdot\mathbf{1}(t)$; **(b)**$y(t)=\frac{1}{2}\cos t, t\in\mathbb{R}$.

5.6. $\frac{d^2y(t)}{dt^2}+4\frac{dy(t)}{dt}+4y(t)=\frac{d^2x(t)}{dt^2}+4x(t)$;

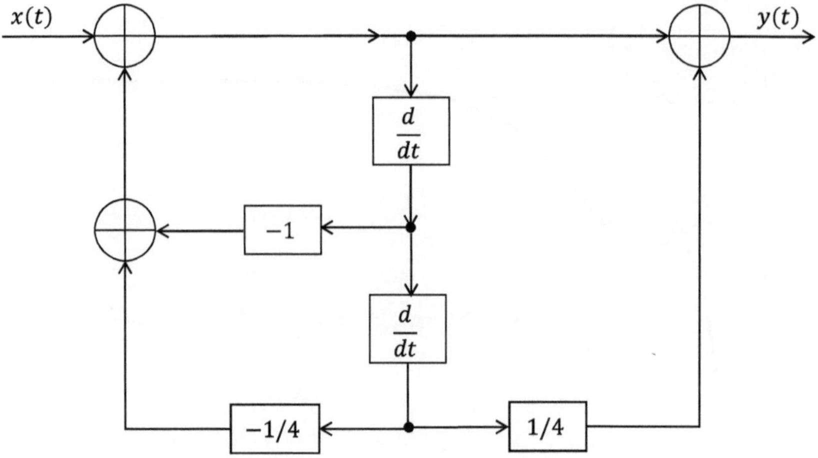

$A(\omega) = \frac{|4-\omega^2|}{4+\omega^2}$; $y(t) = (1 - 2t)e^{-2t} \cdot \mathbf{1}(t)$.

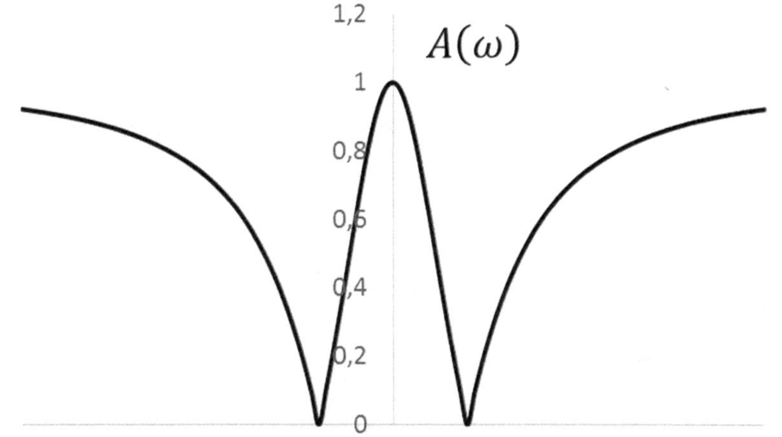

5.7. **(a)** $H_1(s) = \frac{1}{s+1}$, $H_2(s) = \frac{1}{s+2}$; $H(s) = H_1(s)H_2(s)$.

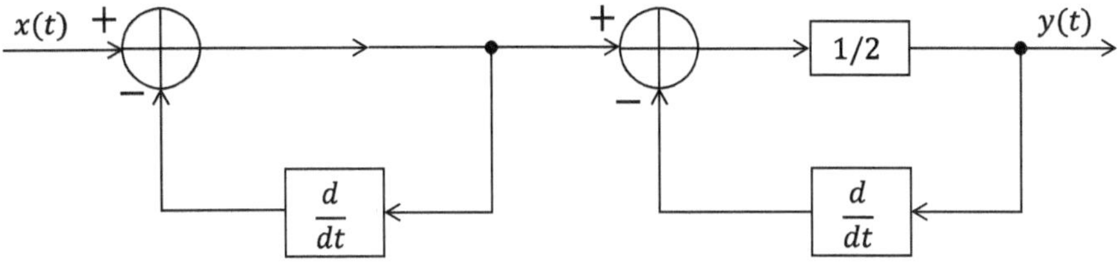

(b) $H_1(s) = \frac{1}{s+1}$, $H_2(s) = -\frac{1}{s+2}$; $H(s) = H_1(s) + H_2(s)$.

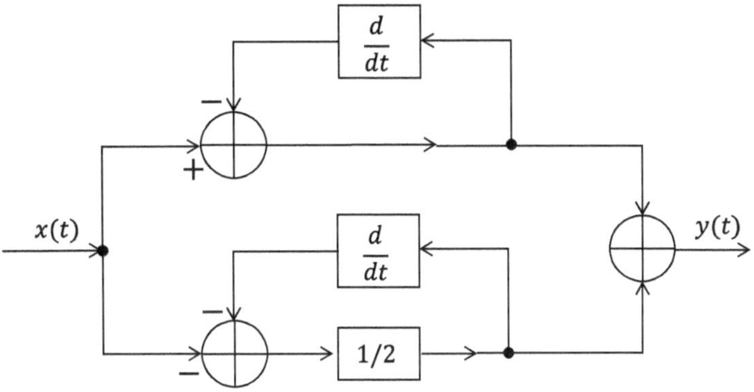

(c) $H_1(s) = 1$, $H_2(s) = s^2 + 3s + 1$; $H(s) = \frac{H_1(s)}{1+H_1(s)H_2(s)}$.

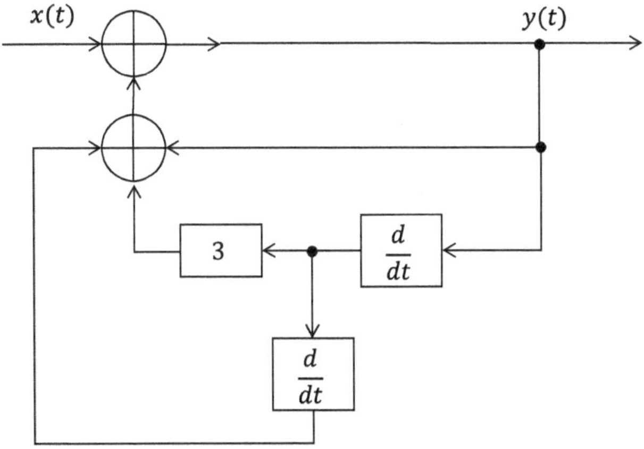

5.8. S_1: $H(s) = \frac{1}{s-a}$; a pole $\lambda = a \Rightarrow$ system S_1 is stable for $a < 0$; $H(s) = \frac{1}{(s-a)} \frac{1}{(s-b)} \frac{1}{s}$; three poles: $\lambda_1 = a$, $\lambda_2 = b$, $\lambda_3 = 0 \Rightarrow$ system is <u>not stable</u> since the λ_3 is not in the half plane Re $s < 0$.

5.9. $H_1(s) = \frac{1}{s}$, $H_2(s) = \frac{b}{s+2}$; $H(s) = \frac{s+2}{s^2+2s+b}$; $b = 1$. (*Hint*: Notice that the input-output relation of S_2 is the convolution integral.)

5.10. $\frac{dy(t)}{dt} + 2y(t) = \frac{dx(t)}{dt} + x(t)$; **(a)**$y(t) = 1$; **(b)**$y(t) = \frac{1}{2}(1 + e^{-2t}) \cdot \mathbf{1}(t)$.

5.11. $H(s)=\frac{s^2}{2s+1}$; $G(s)=-\frac{s^2}{s^2-2s-1}$; $2\frac{dy(t)}{dt}+y(t)=\frac{d^2x(t)}{dt^2}$;
$y(t)=\frac{1}{2}\delta(t)-\frac{1}{4}e^{-\frac{t}{2}}\cdot\mathbf{1}(t)-\frac{1}{2}\delta(t-1)+\frac{1}{4}e^{-\frac{(t-1)}{2}}\cdot\mathbf{1}(t-1)$.

5.12. $H(s)=\frac{s}{s^2+(a+1)s+(a+b)}$; stability region: $\begin{cases} a>-1 \\ b>-a \end{cases}$.

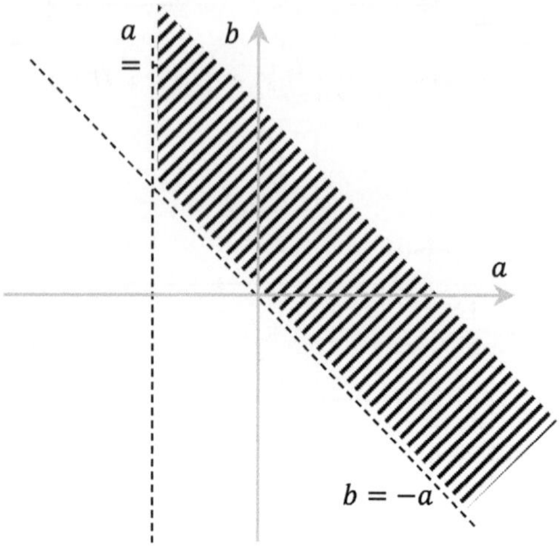

5.13. $K\in(35.2,76)$.

5.14. Stability region: $\begin{cases} K_1>0 \\ K_2<0 \end{cases}$; $k(t)=\left(\frac{1}{2}+\frac{1}{2}e^{-2000t}-e^{-1000t}\right)\cdot\mathbf{1}(t)$.

5.15. **(a)** Stable system, **(b)** nonstable system.

Chapter 6

6.1. $f_N=20$ Hz; **(b)**$f_N=40$ Hz; **(b)**$f_N=40$ hz.

6.2. $f_N=4$ Hz; $X(f)=\delta(f)+\frac{1}{2}\delta(f-\frac{1}{2})+\frac{1}{2}\delta(f+\frac{1}{2})-\frac{1}{4}\delta(f-2)-\frac{1}{4}\delta(f+2)$.

(a) $\left(X_s(f)=5\sum_{n\in\mathbb{Z}}X(f-5n)=\right.$

$5\sum_{n\in\mathbb{Z}}\left[\delta(f-5n)+\frac{1}{2}\delta\left(f-5n-\frac{1}{2}\right)+\frac{1}{2}\delta\left(f-5n+\frac{1}{2}\right)-\frac{1}{4}\delta(f-5n-2)-\right.$

$\left.\frac{1}{4}\delta(f-5n+2)\right]$; no aliasing;

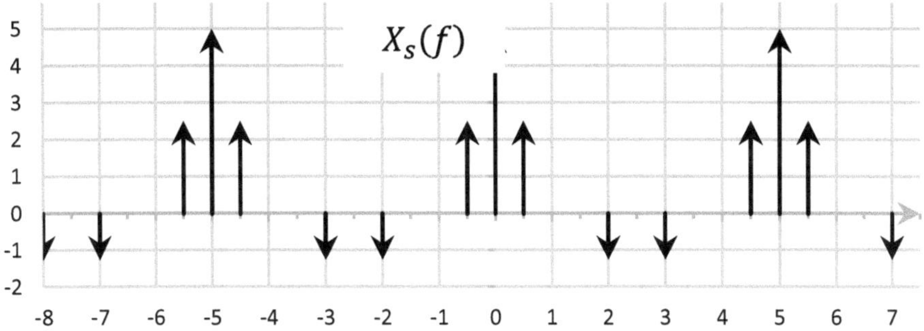

$y(t) = 1 + \cos(\pi t)$.

(b) $X_s(f) = 3\sum_{n\in\mathbb{Z}} X(f-3n) =$

$3\sum_{n\in\mathbb{Z}} \left[\delta(f-3n) + \frac{1}{2}\delta\left(f-3n-\frac{1}{2}\right) + \frac{1}{2}\delta\left(f-3n+\frac{1}{2}\right) - \frac{1}{4}\delta(f-3n-2) - \frac{1}{4}\delta(f-3n+2)\right]$; aliasing;

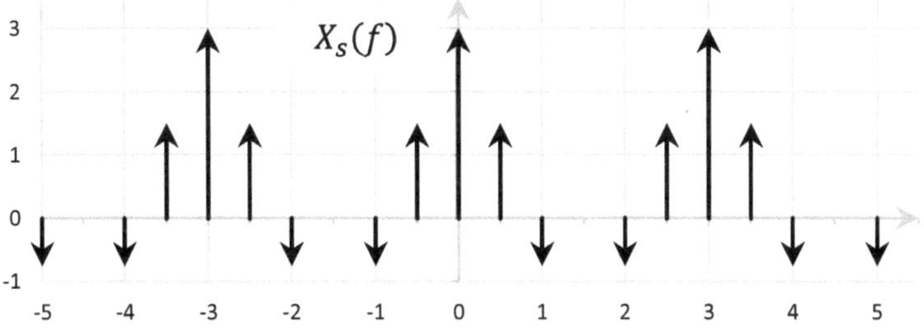

$y(t) = 1 + \cos(\pi t) - \frac{1}{2}\cos(2\pi t)$.

6.3. **(a)** Aliasing;

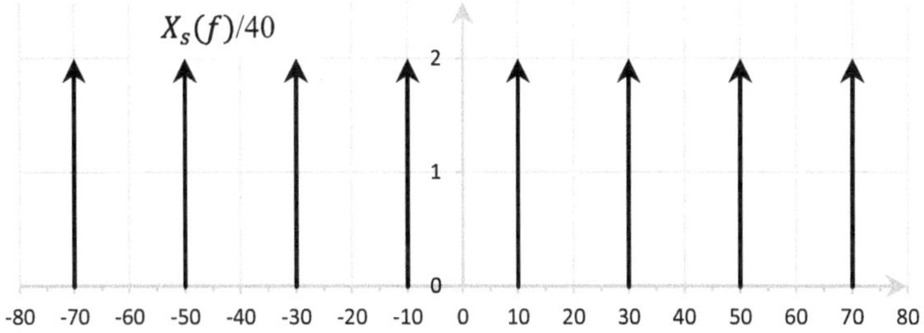

$y(t) = 4\cos(20\pi t) + 4\cos(60\pi t) + 4\cos(100\pi t)$.

(**b**) Aliasing;

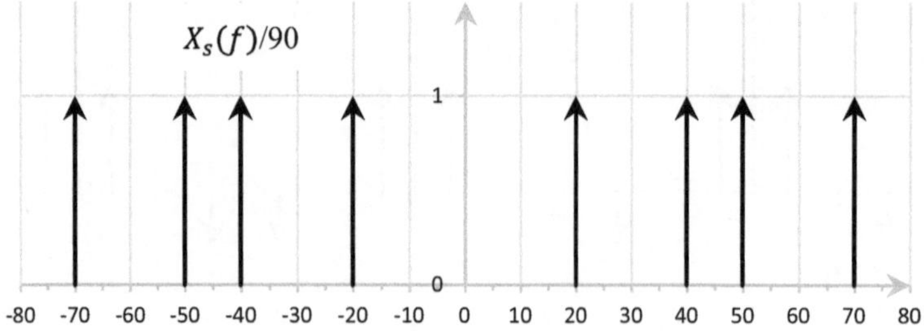

$y(t) = 2\cos(40\pi t) + 2\cos(80\pi t) + 2\cos(100\pi t)$.

(**c**) No aliasing;

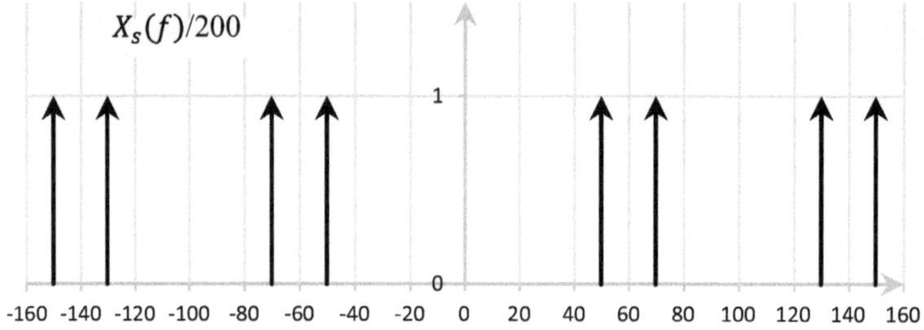

$y(t) = 2\cos(100\pi t)$.

6.4. The sampling frequency is close to the fundamental frequency 10 Hz, Nyquist condition is not fulfilled—we observe the stroboscopic effect.

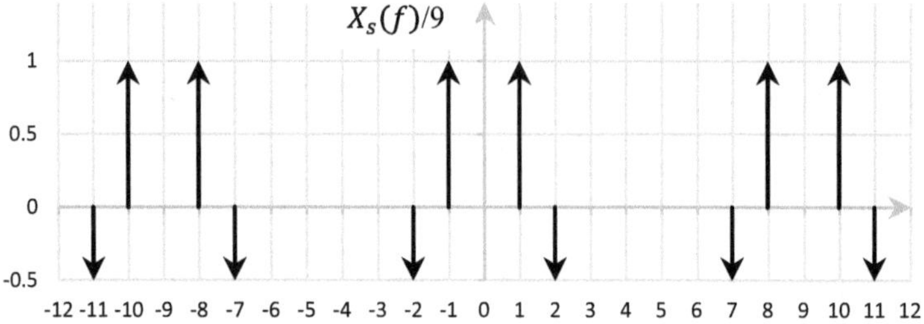

$y(t) = 2\cos(2\pi t) - \cos(4\pi t)$.

6.5. **(a)** $X_s(f) = 3\sum_{n\in\mathbb{Z}} \Pi\left(\frac{f-3n}{4}\right)$; aliasing;

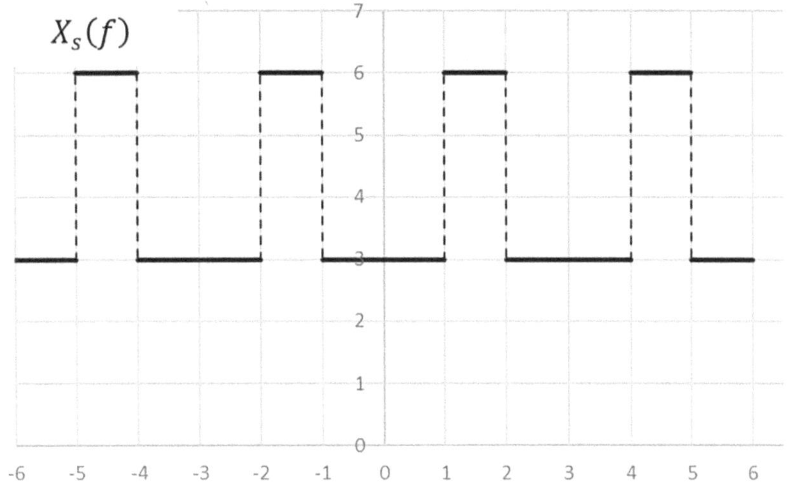

$y(t) = 4 \cdot \mathrm{Sa}(4\pi t) + 2 \cdot \mathrm{Sa}(\pi t)\cos(3\pi t)$.

(b) $X_s(f) = 3\sum_{n\in\mathbb{Z}} \Lambda\left(\frac{f-3n}{2}\right)$; aliasing;

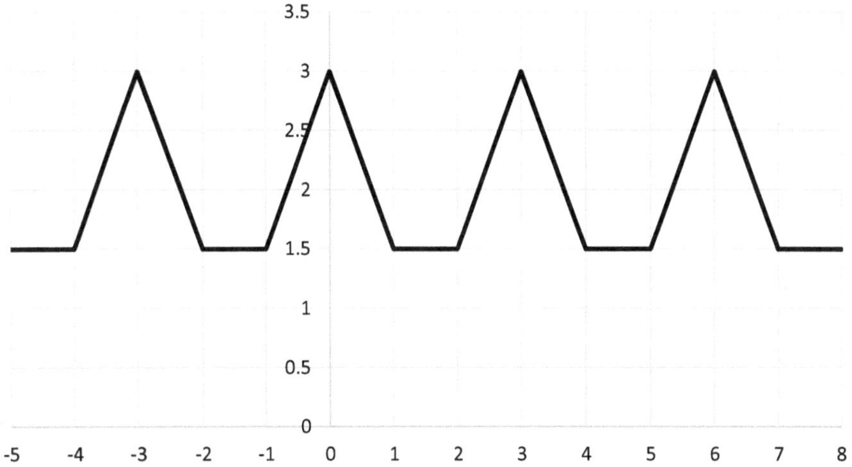

$y(t) = 4 \cdot \mathrm{Sa}(4\pi t) + \mathrm{Sa}^2(\pi t)$.

6.6. $X_s(f) = 10\sum_{n\in\mathbb{Z}} \Lambda\left(\frac{f-10n}{5}\right)$; no aliasing;

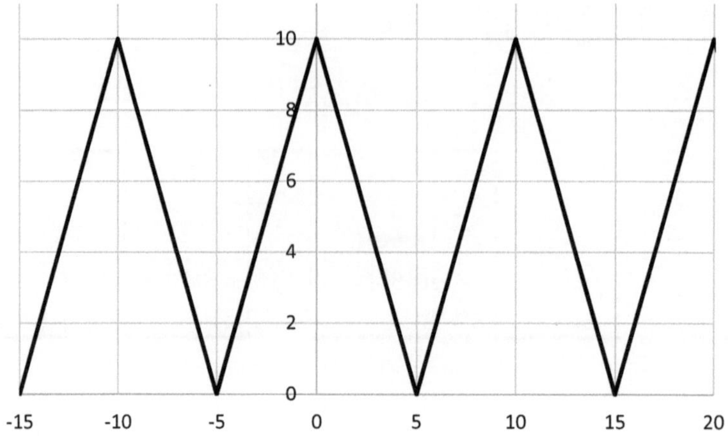

$$H_1(f) = \begin{cases} \dfrac{1}{10 - 2|f|}, & |f| \leq 5 \\ 0, & \text{elsewhere.} \end{cases}$$

6.7. $X_s(f) = \frac{3}{4} \sum_{n \in \mathbb{Z}} \Pi\left(f - \frac{3}{4}n\right).$

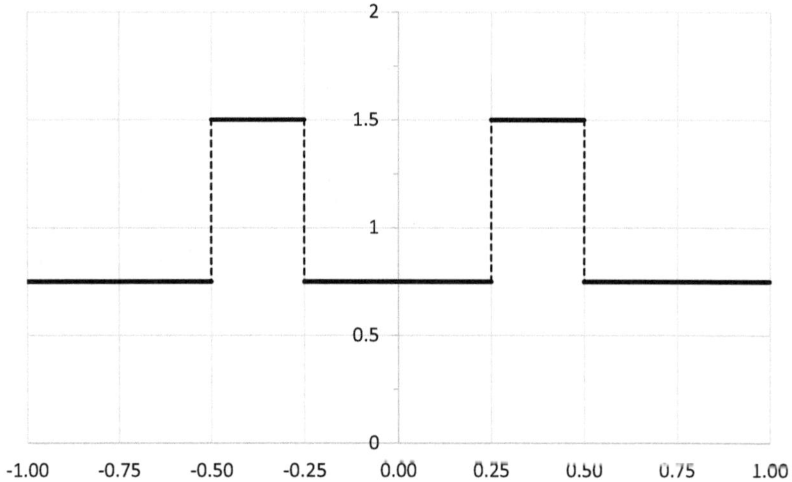

(a) $Y(f) = \Pi(f) + \Pi\left(\frac{f - 3/8}{1/4}\right) + \Pi\left(\frac{f + 3/8}{1/4}\right);$
$y(t) = \mathrm{Sa}(\pi t) + \frac{1}{2} \cdot \mathrm{Sa}\left(\frac{\pi}{4}t\right) \cos\left(\frac{3\pi}{4}t\right).$

(b) $Y(f) = \Pi\left(\frac{f}{1/2}\right);\ y(t) = \frac{1}{2} \cdot \mathrm{Sa}\left(\frac{\pi}{2}t\right).$

6.8. $X(f) = \frac{1}{2}\Pi(f-1) + \frac{1}{2}\Pi(f+1); f_N = 3$ Hz;
$X_s(f) = \frac{3}{4} \sum_{n \in \mathbb{Z}} \left[\Pi\left(f - 1 - \frac{3}{2}n\right) + \Pi\left(f + 1 - \frac{3}{2}n\right)\right];$

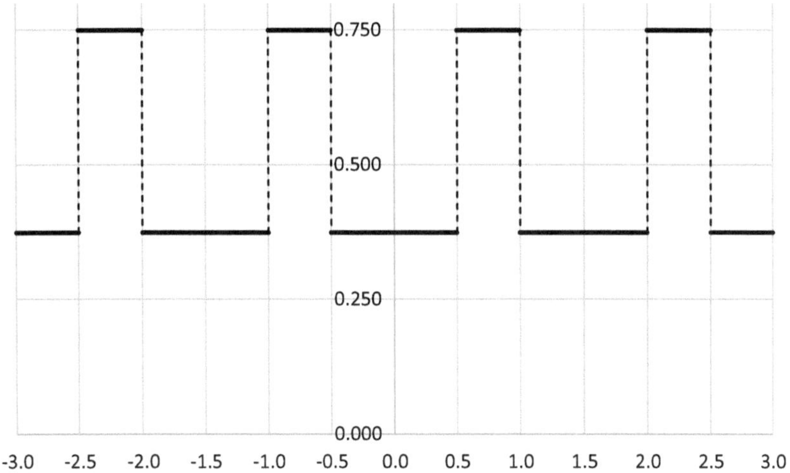

$y(t) = \frac{3}{4} \cdot \mathrm{Sa}(2\pi t) + \frac{3}{8} \cdot \mathrm{Sa}\left(\frac{\pi}{2} t\right) \cos\left(\frac{3\pi}{2} t\right).$

6.9. $X(f) = 2A \cdot \Lambda\left(\frac{f}{2f_0}\right) - A \cdot \Lambda\left(\frac{f}{f_0}\right);$

$x(t) = 4Af_0 \cdot \mathrm{Sa}^2(2\pi f_0 t) - Af_0 \cdot \mathrm{Sa}^2(\pi f_0 t).$

(a) $f_s = 4f_0;$

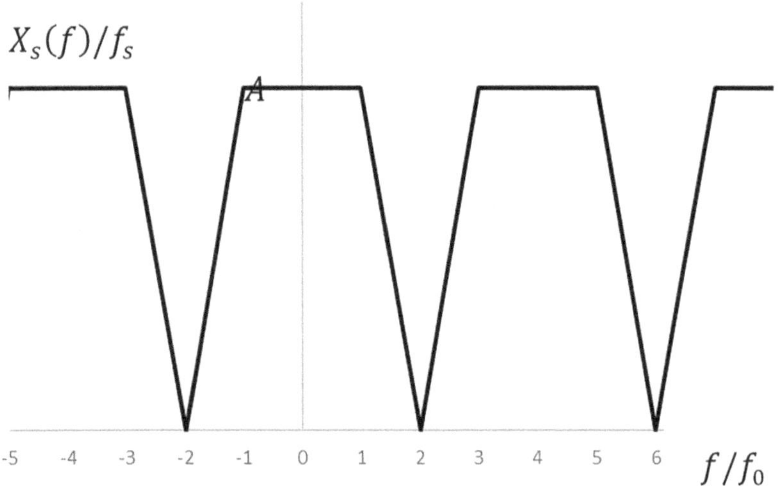

$y(t) = x(t).$

(b) $f_s = 3f_0;\ X_s(f) = Af_s, f \in \mathbb{R};$

$Y(f) = A\Pi\left(\frac{f}{4f_0}\right);\ y(t) = 4Af_0 \cdot \mathrm{Sa}(4\pi f_0 t).$

(c) $f_s = \frac{7}{2} f_0;$

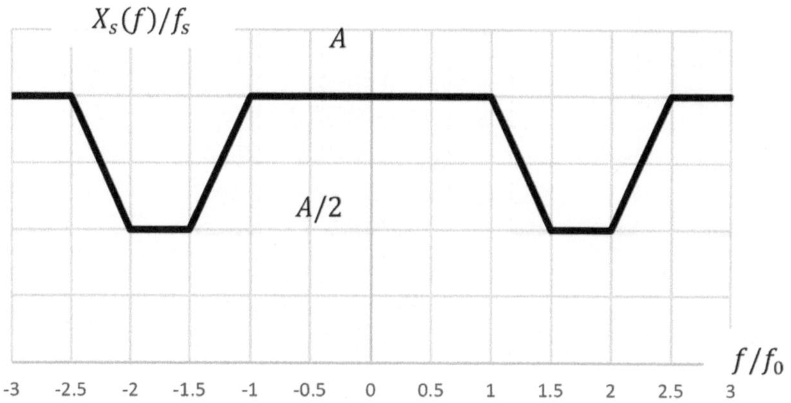

$Y(f) = \frac{A}{2}\Pi\left(\frac{f}{4f_0}\right) + A\cdot\Lambda\left(\frac{f}{\frac{3}{2}f_0}\right) - \frac{A}{2}\cdot\Lambda\left(\frac{f}{f_0}\right)$;

$y(t) = 2Af_0\cdot\mathrm{Sa}(4\pi f_0 t) + \frac{3}{2}Af_0\cdot\mathrm{Sa}^2\left(\frac{3}{2}\pi f_0 t\right) - \frac{A}{2}f_0\cdot\mathrm{Sa}^2(\pi f_0 t)$.

(d) $f_s = 2f_0$;

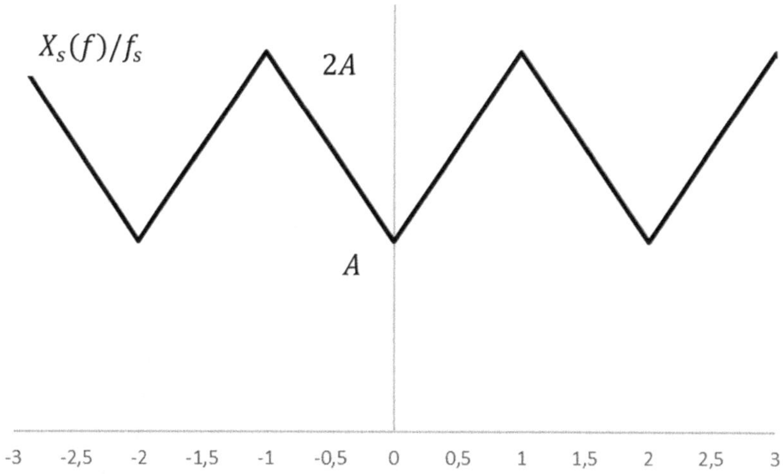

$Y(f) = A\Pi\left(\frac{f}{4f_0}\right) + A\cdot\Lambda\left(\frac{f-f_0}{f_0}\right) + A\cdot\Lambda\left(\frac{f+f_0}{f_0}\right)$;

$y(t) = 4Af_0\cdot\mathrm{Sa}(4\pi f_0 t) + 2Af_0\cdot\mathrm{Sa}^2(\pi f_0 t)\cos(2\pi f_0 t)$.

6.10. $X(f) = 3\Lambda(f/3)$; $W(f) = \Pi(f/4) + 2\Lambda(f/2)$; $W_s(f) = 8, f \in \mathbb{R}$;

$$H_{LP}(f) = \begin{cases} \frac{1}{8}, & |f| \leq 2 \\ 0, & \text{elsewhere.} \end{cases}$$

6.11.

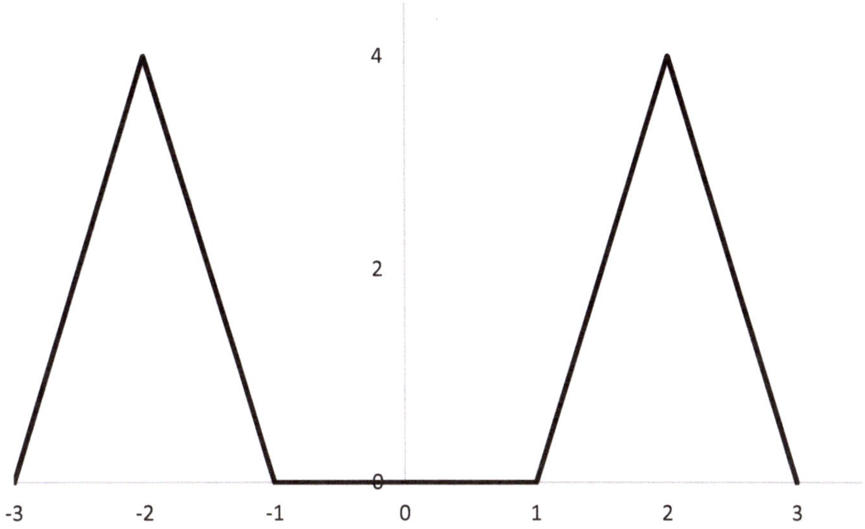

$Y(f) = \Lambda(f-2) + \Lambda(f+2)$; $y(t) = 2\mathrm{Sa}^2(\pi t)\cos(4\pi t)$.

6.12. $x(n) = \cos\left(\frac{\pi n}{4}\right)$; $x(0) = 1$; $x(1) = 1/\sqrt{2}$; $x(2) = 0$;$x(3) = -1/\sqrt{2}$; $x(4) = -1$; $x(5) = -1/\sqrt{2}$; $x(6) = 0$; $x(7) = 1/\sqrt{2}$;

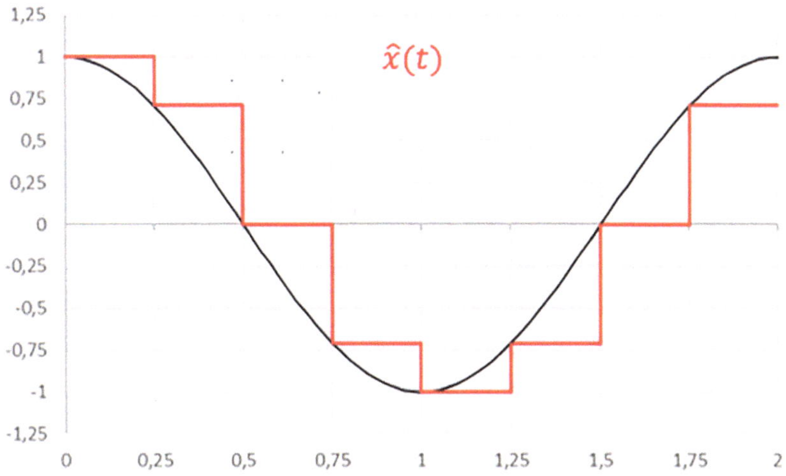

$X(f) = \frac{1}{2}\delta\left(f - \frac{1}{2}\right) + \frac{1}{2}\delta\left(f + \frac{1}{2}\right)$;

$$|Y(f)| = \frac{1}{2}\left|\mathrm{Sa}\left(\frac{\pi f}{4}\right)\right| \sum_{n\in\mathbb{Z}} \left[\delta\left(f - 4n - \frac{1}{2}\right) + \delta\left(f - 4n + \frac{1}{2}\right)\right].$$

n	$\frac{1}{2}\lvert\mathrm{Sa}(\pi(n+1/8))\rvert$	f [Hz]	$\frac{1}{2}\lvert\mathrm{Sa}(\pi(n-1/8))\rvert$	f [Hz]
−5	0,012	−19, 5	0, 012	−20, 5
−4	0,016	−15, 5	0,015	−16, 5
−3	0,021	−11, 5	0, 02	−12, 5
−2	0,032	−7, 5	0,029	−8, 5
−1	0, 07	−3, 5	0,054	−4, 5
0	0,487	0,5	0,487	−0, 5
1	0,054	4,5	0,07	3,5
2	0,029	8,5	0,032	7,5
3	0, 02	12,5	0,021	11,5
4	0,015	16,5	0,016	15,5
5	0,012	20,5	0,012	19,5

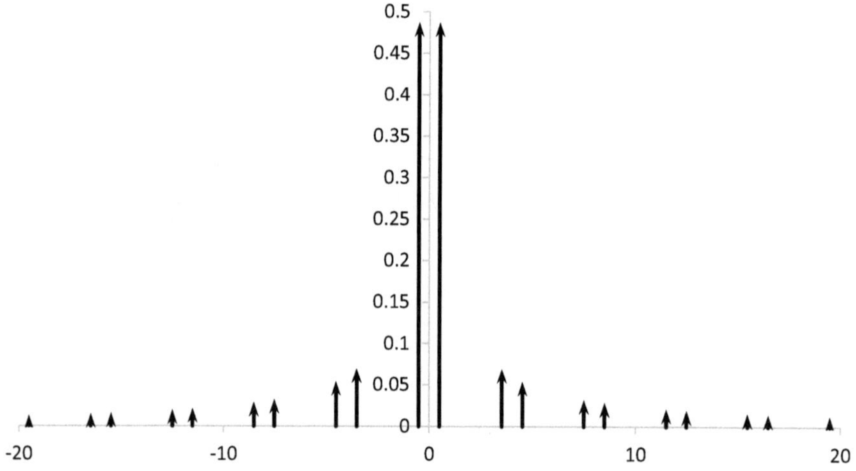

6.13. $E_x = 21$, $X(\mathrm{F}) = \frac{\sin(2,1\pi \mathrm{F})}{\sin(\pi \mathrm{F})}$; $|X(\mathrm{F})| = \left|\frac{\sin(2,1\pi \mathrm{F})}{\sin(\pi \mathrm{F})}\right|$;

$$\text{if } X(\mathrm{F}) < 0, \text{ then } \arg X(F) = \begin{cases} \pi, & \mathrm{F} \in \left[0, \frac{1}{2}\right) \\ -\pi, & \mathrm{F} \in \left(-\frac{1}{2}, 0\right) \end{cases}.$$

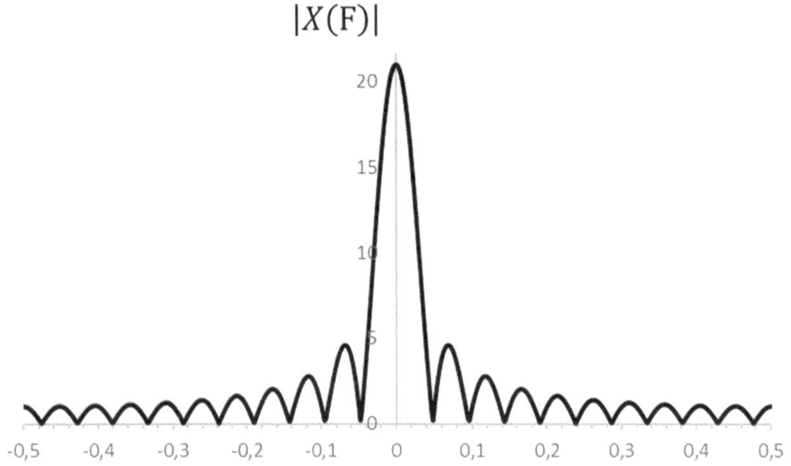

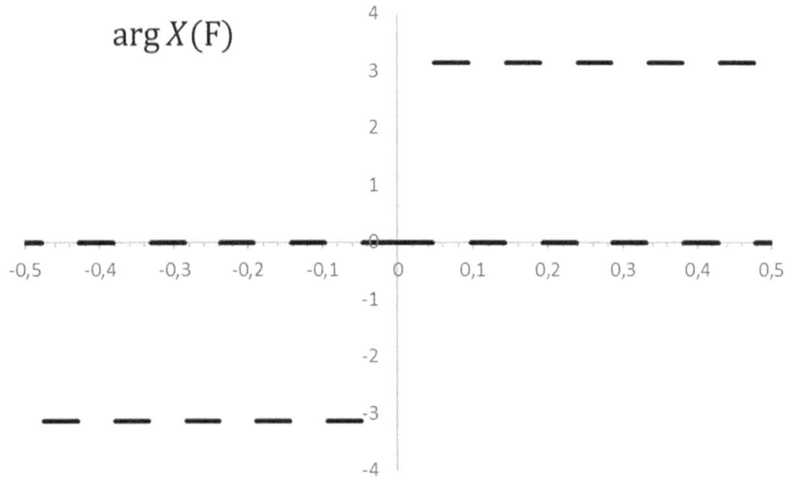

6.14. $E_x = 5/3$.
For $a = \frac{1}{2}$: $X(\mathrm{F}) = \frac{3}{5 - 4\cos(2\pi\mathrm{F})}$.

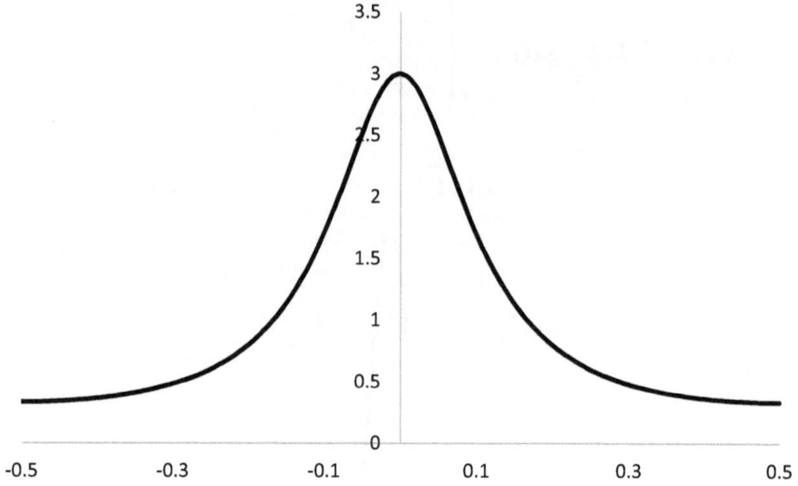

For $a = -\frac{1}{2}$: $X(\mathrm{F}) = \frac{3}{5+4\cos(2\pi\mathrm{F})}$.

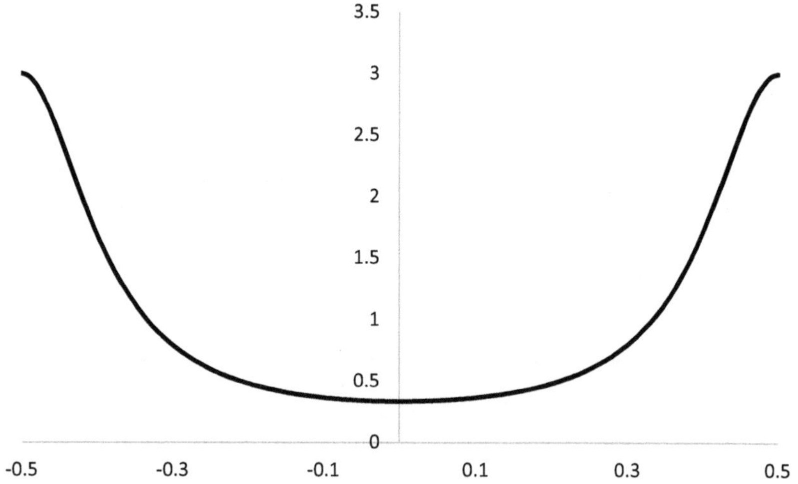

6.15. **(a)** $X(F) = 2\cos(4\pi F)$; $|X(F)| = 2|\cos(4\pi F)|$.

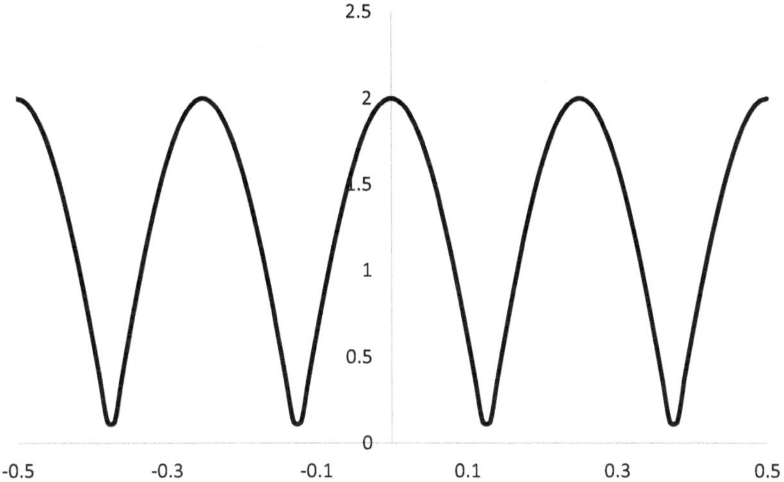

(b) $X(\mathrm{F}) = e^{j4\pi \mathrm{F}}(1 + e^{-j5\pi \mathrm{F}})$; $|X(\mathrm{F})| = 2|\cos(5\pi F)|$.

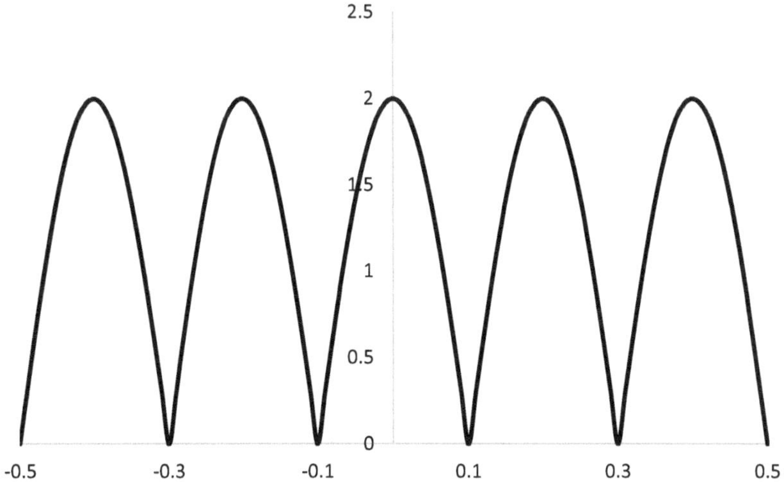

6.16. *Hint*: Since $x_1(n) = \delta(n) + \delta(n-1) + \delta(n-2) + \delta(n-3) + \delta(n-4)$ and $x_2(n) = \delta(n) + \frac{1}{2}\delta(n-1) + \frac{1}{4}\delta(n-2) + \frac{1}{8}\delta(n-3)$, it is easy to compute their DTFTs and apply the convolution property from Table 6.2.

6.17. $y(n) = \begin{cases} 2, & n < -1 \\ n+4, & |n| \le 1 \\ 7, & n > 1 \end{cases}$.

6.18. $z(n) = \{7,4,5,2\}$;

$X(k) = 2 + e^{-j\pi k} = \{3, 1, 3, 1\}$;
$Y(k) = 3 + 2e^{-j\frac{\pi}{2}k} + e^{-j\pi k} = \{6, 2 - 2j, 2, 2 + 2j\}$.

6.19. *Hint*: Compute the circular convolution of spectra $X(k)$ and $Y(k)$ from Exercise 6.18 and then calculate the inverse DFT.
$x(n)y(n) = \{6, 0, 1, 0\}$.

6.20. **(a)** $X_1(k) = \{8, 0, 0, 0, 0, 0, 0, 0\}$; **(b)**$X_2(k) = \{1, 1, 1, 1, 1, 1, 1, 1\}$

6.21. $X(k) = \{2, 0, 2, 0, 2, 0, 2, 0\}$. *Hint:* Notice that $y(n)$ can be expressed as the sum: $y(n) = \frac{1}{2}x(n) + \frac{1}{4}X(n)$, then $Y(k) = \frac{1}{2}X(k) + 2x(-k)$ (see the dual symmetry property of DFT—Table 6.3).

Chapter 7

7.1. $H(\Omega) = H_{\mathrm{LP}}(\Omega) + H_{\mathrm{LP}}(\Omega - \pi)$; BS ideal filter.

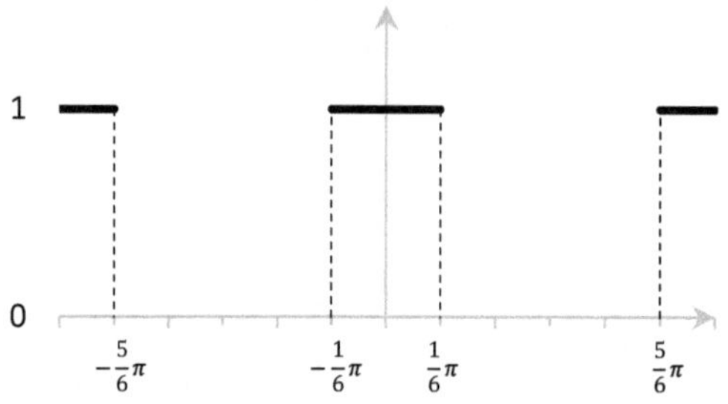

7.2. $h(n) = \delta(n)$.

7.3. **(a)** $y(n) = 0$; **(b)**$y(n) = \cos\left(\frac{3}{8}\pi n + \frac{\pi}{4}\right)$.

7.4. **(a)** $y(n) = \sin\left(\frac{\pi}{8}n\right)$; **(b)**$y(n) = 2\sin\left(\frac{\pi}{8}n\right) - 2\cos\left(\frac{\pi}{4}n\right)$;
(c) $y(n) = \frac{1}{6}\sin\left(\frac{\pi}{8}n\right) - \frac{1}{4}\sin\left(\frac{\pi}{4}n\right)$.

7.5. **(a)** $y(n) = \left(\frac{1}{2}\right)^{n+1}\left[1 + \cos\left(\frac{\pi}{2}n\right) + \sin\left(\frac{\pi}{2}n\right)\right] \cdot \mathbf{1}(n)$;
(b) $y(n) = \frac{4}{3}\cos\left(\frac{\pi}{2}n\right)$.

7.6. $H(\Omega) = \frac{1}{1 - \frac{1}{6}e^{-j\Omega} - \frac{1}{6}e^{-j2\Omega}}$; $h(n) = \left[\frac{3}{5}\left(\frac{1}{2}\right)^n + \frac{2}{5}\left(-\frac{1}{3}\right)^n\right] \cdot \mathbf{1}(n)$.

7.7. $y(n) = \frac{3}{4}y(n-1) - \frac{1}{8}y(n-2) + \frac{5}{3}x(n) - \frac{1}{2}x(n-1)$;
$H(\Omega) = \frac{4}{3}\frac{10 - 3e^{-j\Omega}}{(e^{-j\Omega} - 2)(e^{-j\Omega} - 4)}$; $h(n) = \left[\frac{4}{3}\left(\frac{1}{2}\right)^n + \frac{1}{3}\left(\frac{1}{4}\right)^n\right] \cdot \mathbf{1}(n)$.

7.8. $H(\Omega) = 1 - e^{-j2\Omega}$; $h(n) = \delta(n) - \delta(n-2)$; $A(\mathrm{F}) = 2|\sin(2\pi \mathrm{F})|$.

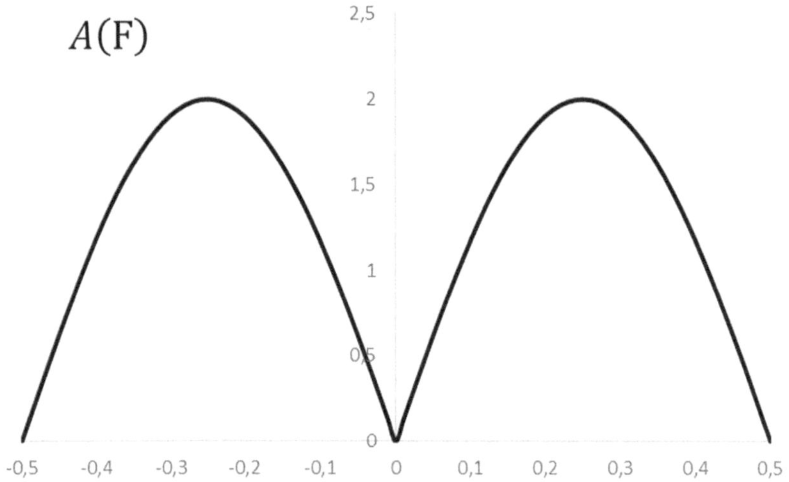

7.9. $y(n) = \frac{1}{2}y(n-1) + y(n-2) - \frac{1}{2}y(n-3) + 2x(n)$;
$H(\Omega) = \frac{2}{(1-e^{-j2\Omega})\left(1-\frac{1}{2}e^{-j\Omega}\right)}$; $X(\Omega) = 1 - e^{-j2\Omega}$; $Y(\Omega) = \frac{2}{1-\frac{1}{2}e^{-j\Omega}}$;
$y(n) = 2\left(\frac{1}{2}\right)^n \cdot \mathbf{1}(n)$; $|Y(\Omega)| = \frac{4}{\sqrt{5-4\cos\Omega}}$; $|Y(\mathrm{F})| = \frac{4}{\sqrt{5-4\cos(2\pi \mathrm{F})}}$.

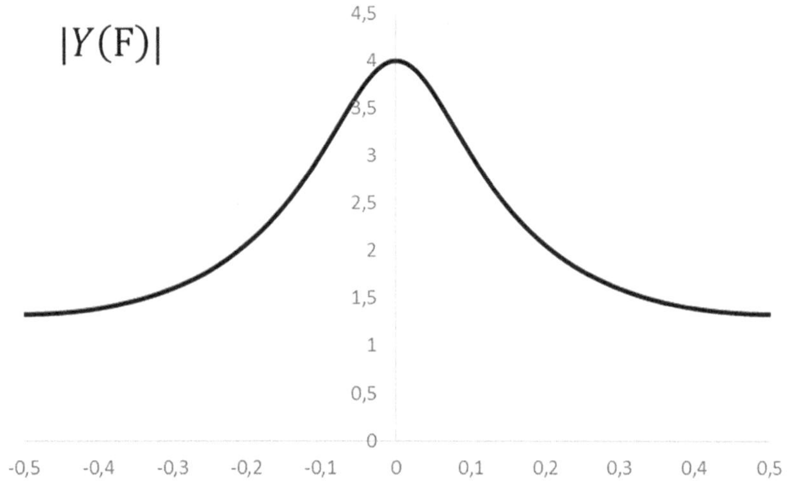

7.10. $H(\Omega) = \frac{1}{1+\frac{1}{2}e^{-j2\Omega}}$; $A(\mathrm{F}) = \frac{4}{\sqrt{5+4\cos(4\pi \mathrm{F})}}$.

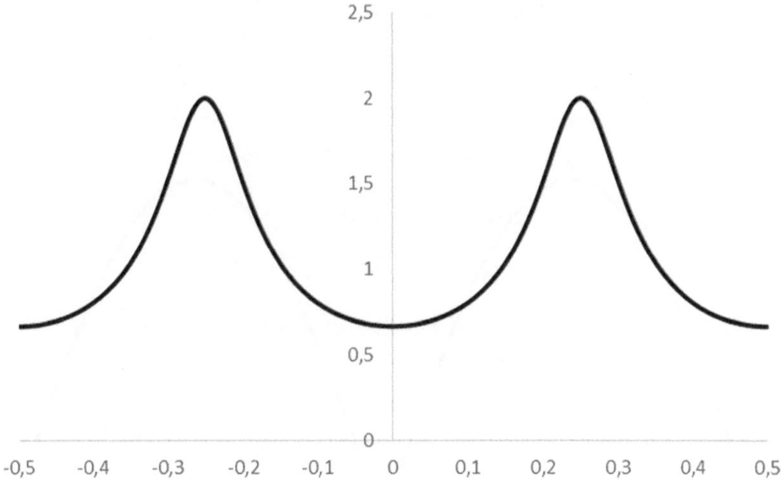

Hint: The 3-dB bandwidth is defined as $[\mathrm{F}_1, \mathrm{F}_2]$, where F_1, F_2 are solutions of the equation $\cos(4\pi\mathrm{F}) = -\frac{3}{4}$, that is, $\mathrm{F}_1 = \frac{\pi - \operatorname{acos}(3/4)}{4\pi} \cong 0{,}192$, $\mathrm{F}_2 = \frac{\pi + \operatorname{acos}(3/4)}{4\pi} \cong 0{,}308$.

7.11. $A(\mathrm{F}) = \frac{5}{\sqrt{34 - 30\cos(2\pi\mathrm{F})}}$.

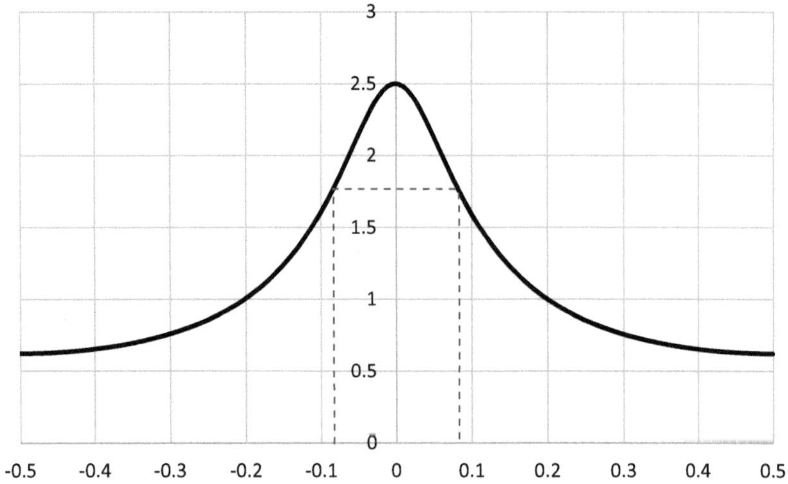

Hint: The 3-dB bandwidth is defined as $[-\mathrm{F}_c, \mathrm{F}_c]$, where F_c is the solution of the equation $\cos(2\pi\mathrm{F}) = \frac{13}{15}$, that is, $\mathrm{F}_c \cong 0{,}083$.

$y(n) = \left[-2\left(\frac{2}{5}\right)^n + 3\left(\frac{3}{5}\right)^n\right] \cdot \mathbf{1}(n)$.

7.12. $b = -a$; $y(n) = \left(-\frac{1}{2}\right)^n\left(-\frac{3}{2}n + \frac{1}{2}\right) \cdot \mathbf{1}(n)$.

7.13. $H(\Omega) = \frac{-\frac{1}{4} + e^{-j2\Omega}}{1 - \frac{1}{4}e^{-j2\Omega}}$; $A(\Omega) = 1$ – all-pass filter;

$y(n) = \frac{1}{4}y(n-2) - \frac{1}{4}x(n) + x(n-2).$

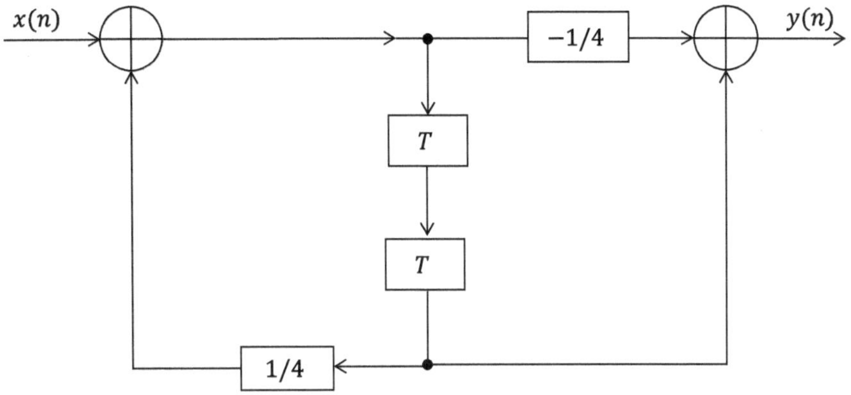

7.14. *Hint*: $h(n) = \delta(n) + 2\delta(n-2)$; $H(\Omega) = 1 + 2e^{-j2\Omega}$;
$y(n) = x(n) + 2x(n-2)$ – FIR filter; $A(\mathrm{F}) = \sqrt{5 + 4\cos(4\pi \mathrm{F})}$.

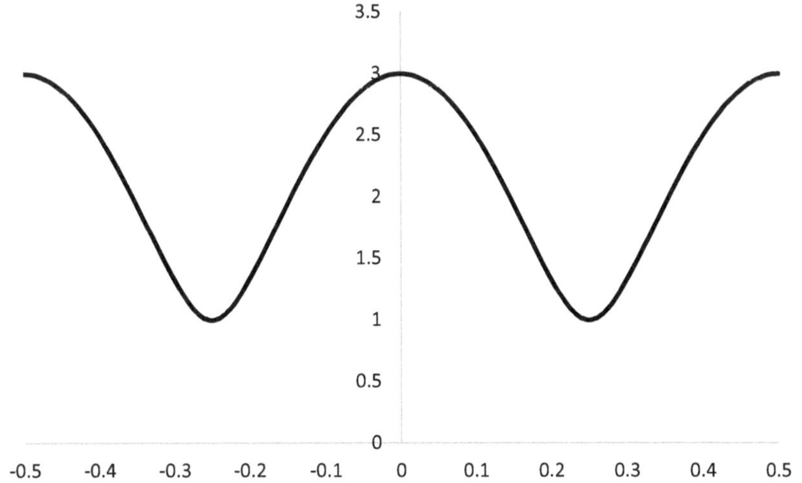

7.15. $\Omega_{c_1} = \mathrm{acos}\left(\frac{3}{4}\right) \cong 0{,}115$;
Hint: Use formulas from Problem 7.6 (a) to calculate $\alpha = -2/5$;
$H_{\mathrm{LP}}(\Omega) = \frac{10+4e^{-j\Omega}}{8-e^{-j\Omega}}$; $A_{\mathrm{LP}}(\mathrm{F}) = \sqrt{\frac{116+80\cos(2\pi \mathrm{F})}{65-16\cos(2\pi \mathrm{F})}}$.

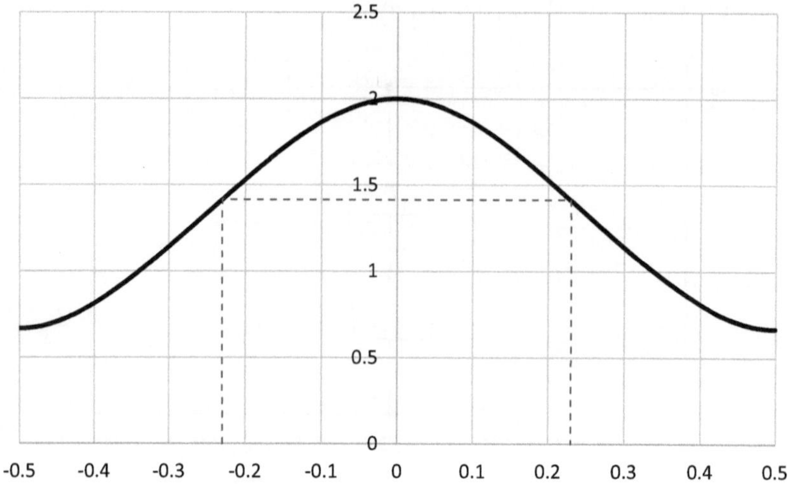

7.16. $\Omega_{c_1} = \text{acos}\left(\frac{23}{24}\right) \cong 0,29;$

Hint: Use formulas from Problem 7.6 (b) to calculate $\alpha = -11/12$;

$H_{\text{HP}}(\Omega) = \frac{48 - 44e^{-j\Omega}}{15 - 8e^{-j\Omega}};$

$A_{\text{HP}}(\text{F}) = \sqrt{\frac{4240 - 4224\cos(2\pi\text{F})}{289 - 240\cos(2\pi\text{F})}}.$

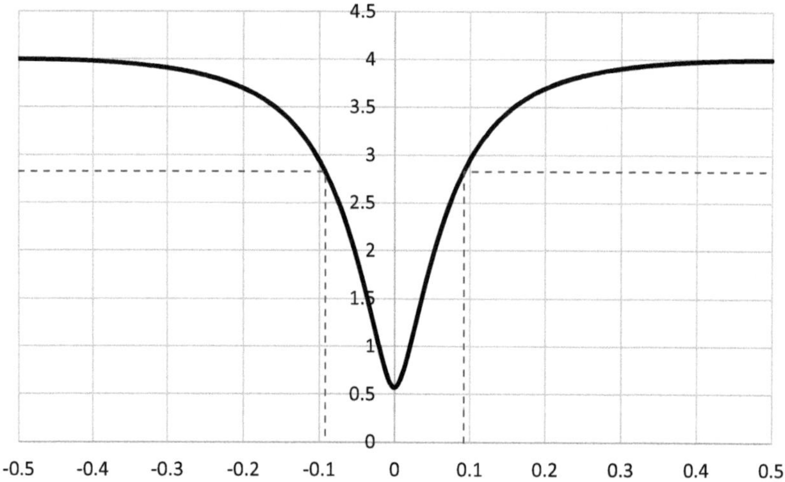

Chapter 8

8.1. $y(n)=\frac{7}{4}\left(\frac{1}{4}\right)^n\cdot\mathbf{1}(n)-\frac{1}{8}\left(\frac{1}{4}\right)^{n-1}\cdot\mathbf{1}(n-1)$.

8.2. $y(n)=\left[-4\left(-\frac{1}{2}\right)^n+2\left(-\frac{1}{4}\right)^n\right]\cdot\mathbf{1}(n)$.

8.3. $y_{0s}(n)=\left[\frac{2}{3}+\frac{1}{3}\left(\frac{1}{4}\right)^n\right]\cdot\mathbf{1}(n)$; $y_{0i}(n)=\left(\frac{1}{4}\right)^{n+1}\cdot\mathbf{1}(n)$.

8.4. $y_{0s}(n)=\left[\frac{1}{6}\left(\frac{1}{4}\right)^n+\frac{1}{3}\left(-\frac{1}{2}\right)^n\right]\cdot\mathbf{1}(n)$; $y_{0i}(n)=-\left(-\frac{1}{2}\right)^n\cdot\mathbf{1}(n)$.

8.5. $H(z)=\frac{z^2}{(z-1)^2}$; $h(n)=(n+1)\cdot\mathbf{1}(n)$;

Hint: One method of computing $k(n)$ is solving the recursive equation: $k(n)=h(n)+k(n-1)$. The second method uses the formula $K(z)=\frac{z}{z-1}H(z)=\frac{z^3}{(z-1)^3}=\frac{z^2+z+1}{(z-1)^2}+\frac{1}{(z-1)^3}=\frac{z^2}{(z-1)^2}+\frac{z}{(z-1)^2}+\frac{1}{(z-1)^2}+\frac{1}{(z-1)^3}$;

$k(n)=(2n+1)\cdot\mathbf{1}(n)+\frac{n(n-1)}{2}\cdot\mathbf{1}(n-1)$; $y(n)=\left[2n+\left(\frac{1}{2}\right)^n\right]\cdot\mathbf{1}(n)$.

8.6. $H(z)=\frac{1/4}{1-\frac{1}{4}z^{-2}}$; $h(n)=\frac{1}{8}\left[\left(-\frac{1}{2}\right)^n+\left(\frac{1}{2}\right)^n\right]\cdot\mathbf{1}(n)$;

$A(\mathrm{F})=\frac{1}{\sqrt{17-8\cos(4\pi\mathrm{F})}}$.

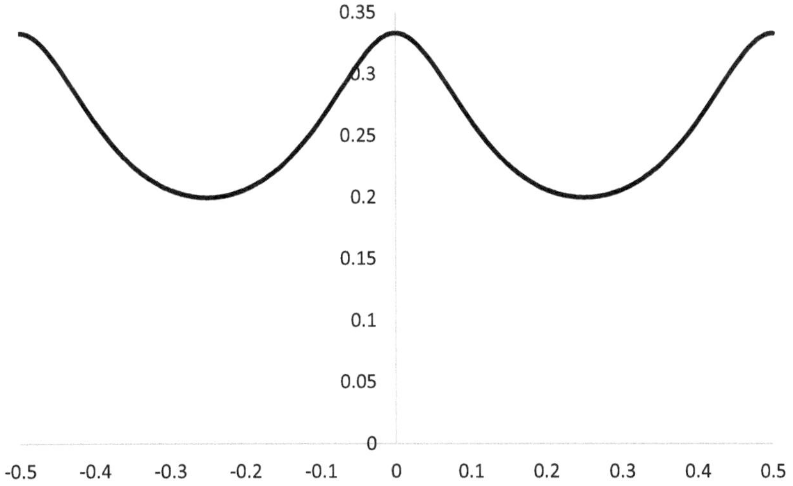

8.7. System equation: $y(n)=\frac{1}{4}y(n-1)-\frac{1}{8}x(n-1)+\frac{3}{2}x(n)$;

Hint: From the equation $y(n)=\frac{1}{2}x(n)+w(n)$, we get $y(-1)=1$. We recommend computing the system response $y(n)$ via $Y(z)=\frac{7}{4}\frac{z}{z-\frac{1}{4}}-\frac{1}{8}\frac{1}{z-\frac{1}{4}}$;

$y(n)=\frac{7}{4}\left(\frac{1}{4}\right)^n\cdot\mathbf{1}(n)-\frac{1}{8}\left(\frac{1}{4}\right)^{n-1}\cdot\mathbf{1}(n-1)$; $H(z)=\frac{-\frac{1}{8}z^{-1}+\frac{3}{2}}{1-\frac{1}{4}z^{-1}}$;

$A(\mathrm{F})=\frac{1}{2}\sqrt{\frac{145-24\cos(2\pi\mathrm{F})}{17-8\cos(2\pi\mathrm{F})}}$.

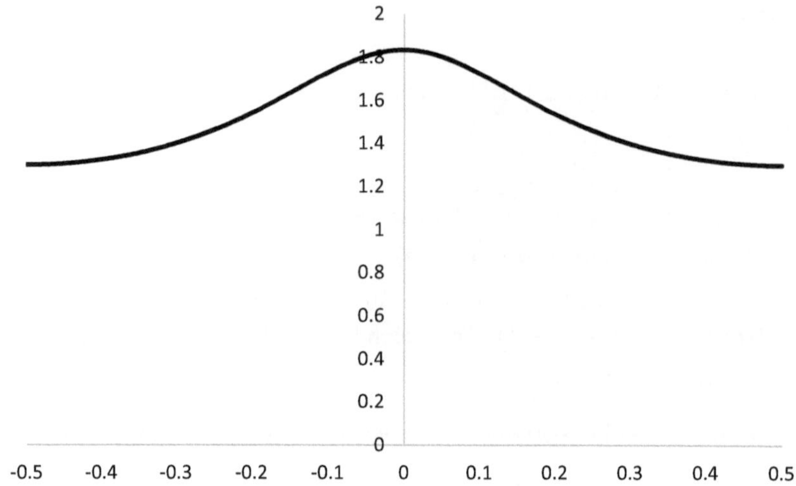

8.8. $H(z) = \frac{1}{1-\frac{1}{4}z^{-2}}$; system equation: $y(n) - \frac{1}{4}y(n-2) = x(n)$;

$A(\mathrm{F}) = \frac{4}{\sqrt{17 - 8\cos(4\pi \mathrm{F})}}$ (for the plot see answer to Ex. 8.6).

8.9. $\alpha = 1/4$; $\beta = 1$; $h(n) = \left[2\left(\frac{1}{2}\right)^n - \left(\frac{1}{4}\right)^n\right] \cdot \mathbf{1}(n)$.

8.10. **(a)** $H_1(z) = \frac{1}{1-0,5z^{-1}}$; $H_2(z) = 1 - 1,5z^{-1}$.

(b) $H_1(z) = \frac{1}{1-0,5z^{-1}}$; $H_2(z) = \frac{-1,5z^{-1}}{1-0,5z^{-1}}$;

$h(n) = \left(\frac{1}{2}\right)^n \cdot \mathbf{1}(n) - \frac{3}{2}\left(\frac{1}{2}\right)^{n-1} \cdot \mathbf{1}(n-1)$.

8.11. $H(z) = \frac{1+\frac{1}{2}z^{-1}}{1-\frac{1}{3}z^{-1}}$; $h(n) = \left(\frac{1}{3}\right)^n \cdot \mathbf{1}(n) + \frac{1}{2}\left(\frac{1}{3}\right)^{n-1} \cdot \mathbf{1}(n-1)$; system equation: $y(n) = \frac{1}{3}y(n-1) + \frac{1}{2}x(n-1) + x(n)$.

8.12. System equation: $y(n) = 5x(n) - x(n-1) + y(n-1) - y(n-2)$;

$H(z) = \frac{z(5z-1)}{z^2-z+1}$; zeros: 0, 1/5; poles: $\frac{1}{2} \pm \frac{\sqrt{3}}{2}j$; system nonstable.

Hint: In order to calculate $y(n)$, determine $Y(z) = \frac{z^2(5z-1)}{\left(z+\frac{1}{2}\right)(z^2-z+1)}$ and decompose $Y(z)/z$ into partial fractions and get $Y(z) = \frac{z}{z+\frac{1}{2}} + 4\frac{z\left(z-\frac{1}{2}\right)}{z^2-z+1}$. Then apply the Table 8.1 and get

$y(n) = \left[\left(-\frac{1}{2}\right)^n + 4\cos\left(\frac{\pi}{3}n\right)\right] \cdot \mathbf{1}(n)$.

8.13. System equation: $y(n) = x(n) - y(n-1) - \frac{1}{2}y(n-2)$;

$H(z) = \frac{2z^2}{2z^2+2z+1}$; zero: 0; poles: $-\frac{1}{2} \pm \frac{1}{2}j$; system stable;

$A(\mathrm{F}) = \frac{2}{\sqrt{9+12\cos(2\pi \mathrm{F})+4\cos(4\pi \mathrm{F})}}$.

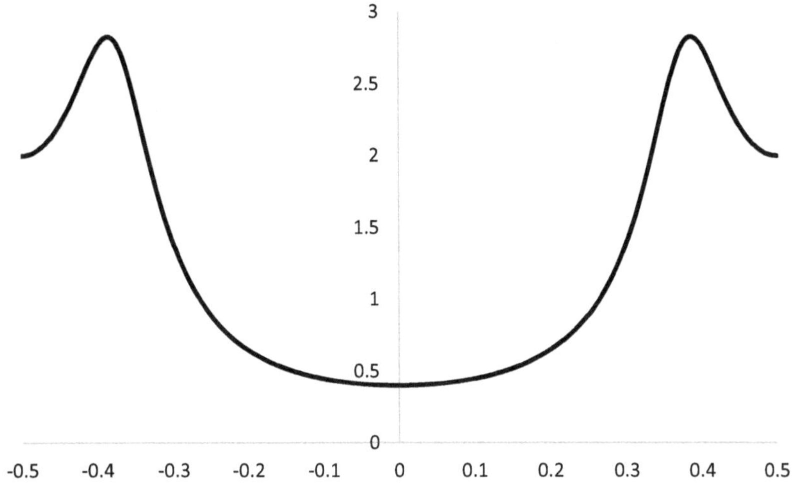

8.14. **(a)** System stable; **(b)** system stable.
8.15. *Hint:* Follow the procedure described in Problem 8.8.

Chapter 9

9.1. $y(n) = \frac{20}{21}y(n-1) + \frac{1}{21}x(n);$

$$y_a(t) = \left(1 - e^{-\frac{1}{2}t}\right) \cdot \mathbf{1}(t);$$

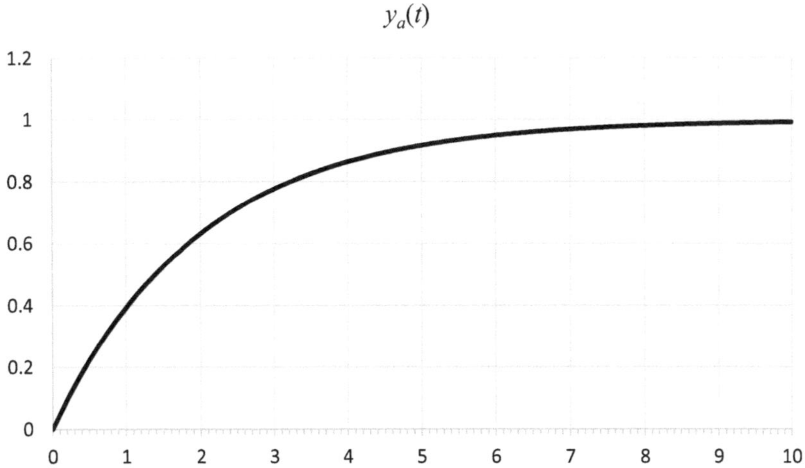

$y_d(n) = \left[1 - \left(\frac{20}{21}\right)^n\right] \cdot \mathbf{1}(n).$

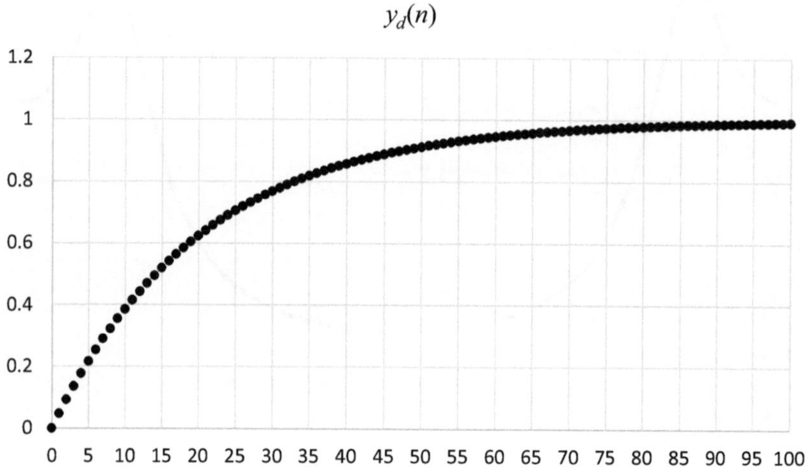

9.2. $\mathbf{S}_a$:$H(s) = \frac{1}{(s+2)^2}$;

$y(t) = \frac{1}{4}x(t) - \frac{1}{4}\frac{d^2y(t)}{dt^2} - \frac{dy(t)}{dt}$;

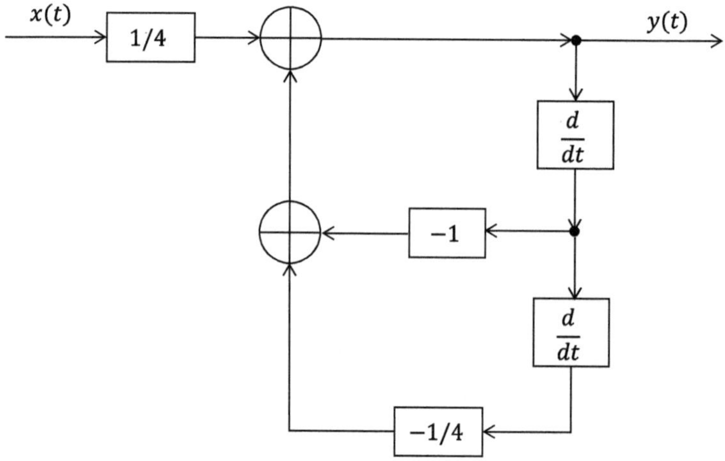

$A(\omega) = \frac{1}{4+\omega^2}$;

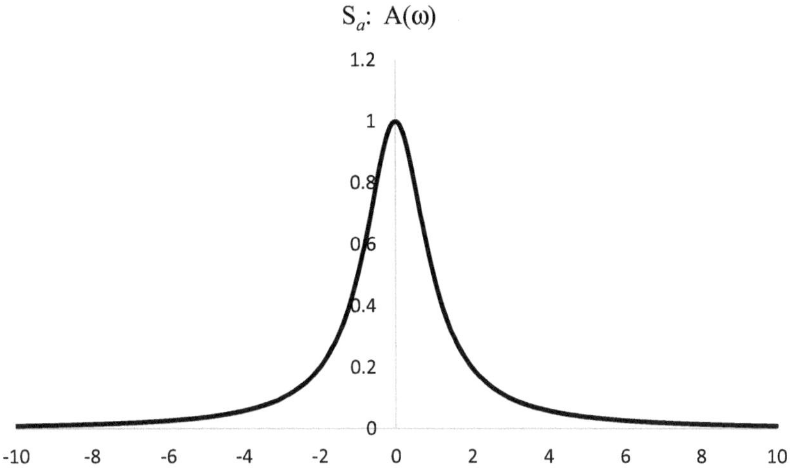

$\mathbf{S_d}$: $y(n) = \frac{1}{144}x(n) + \frac{5}{3}y(n-1) - \frac{25}{36}y(n-2);$

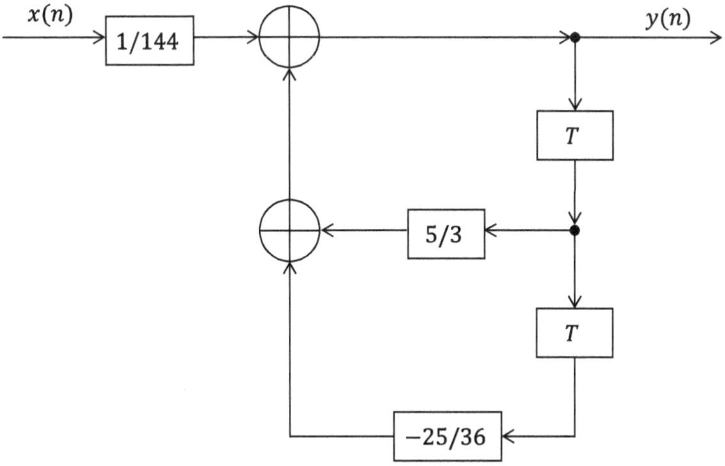

$H(z) = \frac{1}{144}\,\frac{1}{\left(1-\frac{5}{6}z^{-1}\right)^2};\ A(\mathrm{F}) = \frac{1}{4[61-60\cos(2\pi \mathrm{F})]}.$

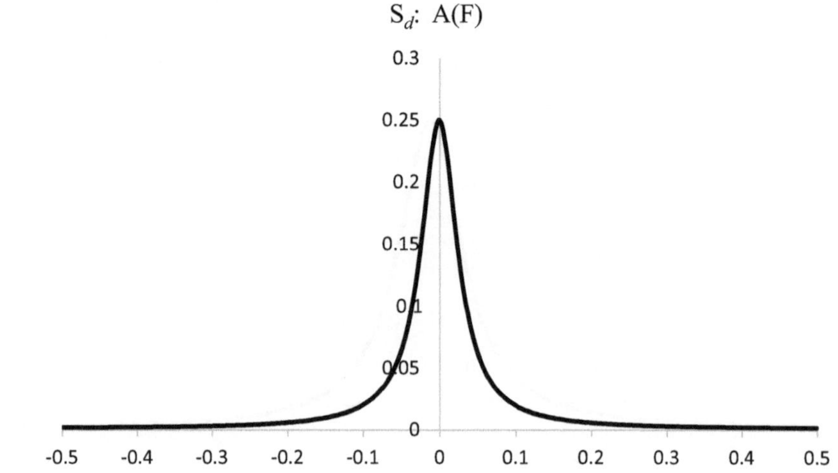

9.3. $h_a(t) = 0,5e^{-0,5t} \cdot \mathbf{1}(t)$; $h_d(n) = 0,05e^{-0,05n} \cdot \mathbf{1}(n)$;
$H(z) = 0,05\frac{1}{1-e^{-0,05}z^{-1}}$; $y(n) - e^{-0,05}y(n-1) = 0,05x(n)$;
$Y(z) = 0,05\frac{z^2}{(z-1)(z-e^{-0,05})}$.
Response of S_d:
$y(n) = 0,05[A + Be^{-0,05n)}] \cdot \mathbf{1}(n)$ where $= \frac{1}{1-e^{-0,05}}$, $B = \frac{e^{-0,05}}{e^{-0,05}-1}$.

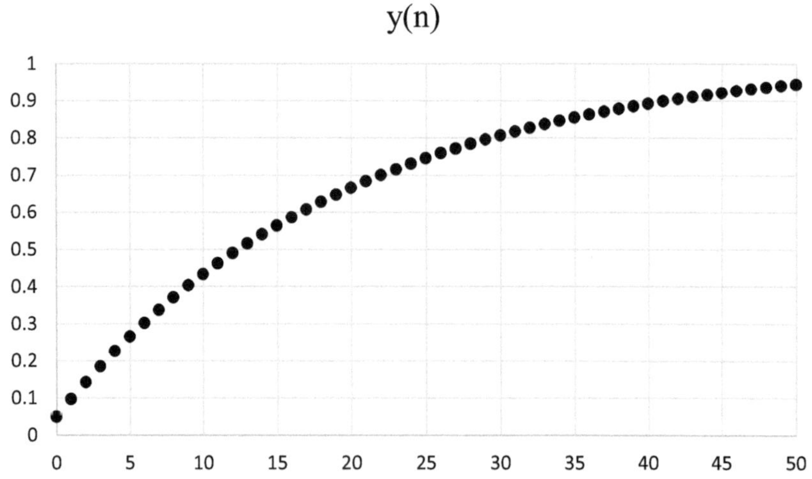

Magnitude response of S_d: $A(\mathrm{F}) = \frac{0,05}{1+e^{-0,1} - 2e^{-0,05}\cos(2\pi\mathrm{F})}$.

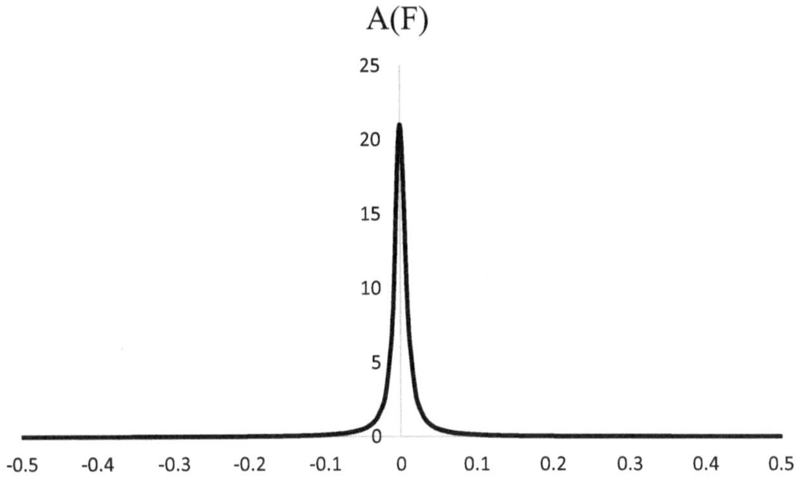

9.4. $h_a(t) = te^{-2t} \cdot \mathbf{1}(t)$; $h_d(n) = 0,01ne^{-0,2n} \cdot \mathbf{1}(n)$;

$H(z) = 0,01\frac{e^{-0,2}z}{(z-e^{-0,2})^2}$;

$y(n) - 2e^{-0,2}y(n-1) + e^{-0,4}y(n-2) = 0,01e^{-0,2}x(n-1)$;

$Y(z) = 0,01\frac{e^{-0,2}z^2}{(z-1)(z-e^{-0,2})^2}$.

Response of S_d:

$y(n) = 0,01[A + (B + Ce^{0,2}(n-1))e^{-0,2(n-1)}] \cdot \mathbf{1}(n-1)$, where $=$ $\frac{e^{-0,2}}{(1-e^{-0,2})^2}$, $C = \frac{e^{-0,6}}{e^{-0,2}-1}$, $B = e^{0,2}C - e^{-0,2}A$.

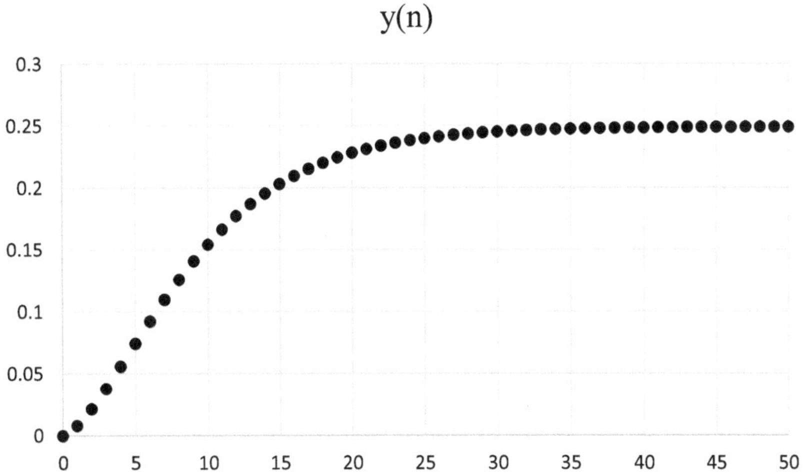

Magnitude response of S_d: $A(\mathrm{F}) = 0,01\frac{e^{-0,2}}{1+e^{-0,4}-2e^{-0,2}\cos(2\pi\mathrm{F})}$.

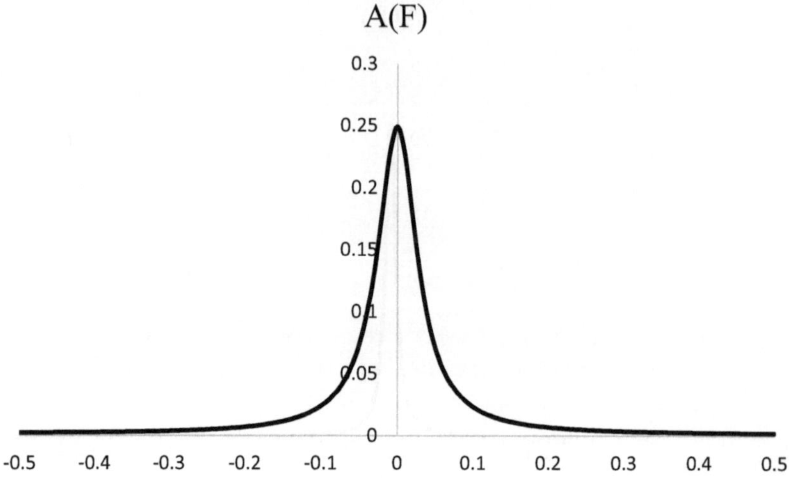

9.5. $H(z) = \frac{T_s}{2}\frac{z+1}{z-1} = \frac{T_s}{2}\frac{1+z^{-1}}{1-z^{-1}} = 0,05\frac{1+z^{-1}}{1-z^{-1}}$;
$y(n) = 0,05[x(n) + x(n-1)] + y(n-1)$.

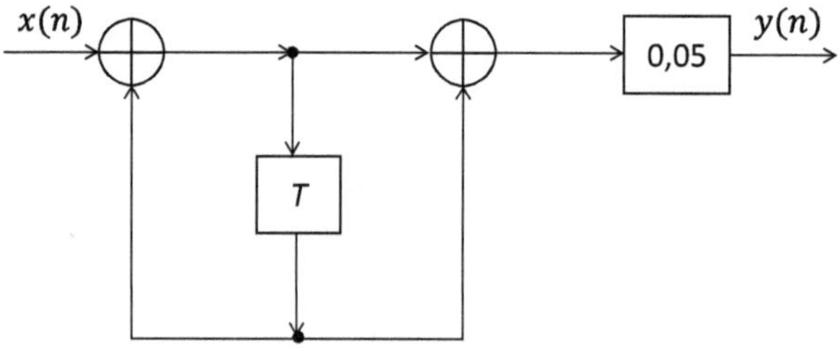

9.6. $\mathbf{S}_a$:$H(s) = \frac{1}{1+2s}$; $y_a(t) = \left(e^{-\frac{1}{2}t} - e^{-t}\right) \cdot \mathbf{1}(t)$;

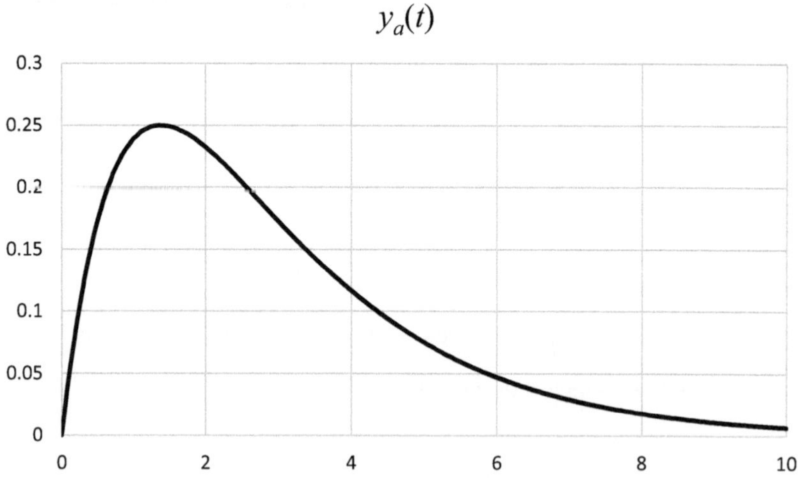

$\mathbf{S}_d$: $H(z)=\frac{1+z^{-1}}{41-39z^{-1}}$; $y(n)=\frac{1}{41}[x(n)+x(n-1)]+\frac{39}{41}y(n-1)$;

Hint: The input signal: $x(n)=e^{-0{,}2n}\cdot\mathbf{1}(n)\leftrightarrow X(z)=\frac{1}{1-e^{-0{,}2}z^{-1}}$. The output can be computed by solving the above equation in a recursive way or via $Y(z)=\frac{1+z^{-1}}{(41-39z^{-1})(1-e^{-0{,}2}z^{-1})}$.

$y_d(n)=\left[\frac{A}{41}\left(\frac{39}{41}\right)^n+Be^{-0{,}2n}\right]\cdot\mathbf{1}(n)$.

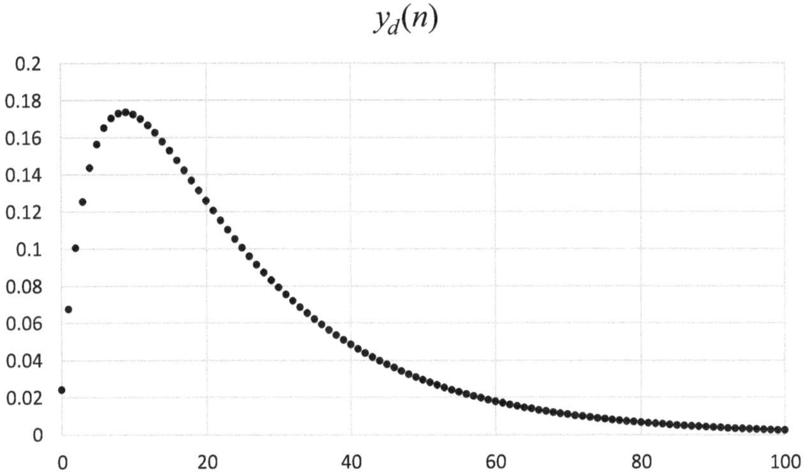

9.7. $H(z)=\frac{(z+1)^2}{(22z-18)^2}$;

$484y(n)-792y(n-1)+324y(n-2)=x(n)+2x(n-1)+x(n-2)$;

$A(\mathrm{F})=\frac{1+\cos(2\pi\mathrm{F})}{404-396\cos(2\pi\mathrm{F})}$.

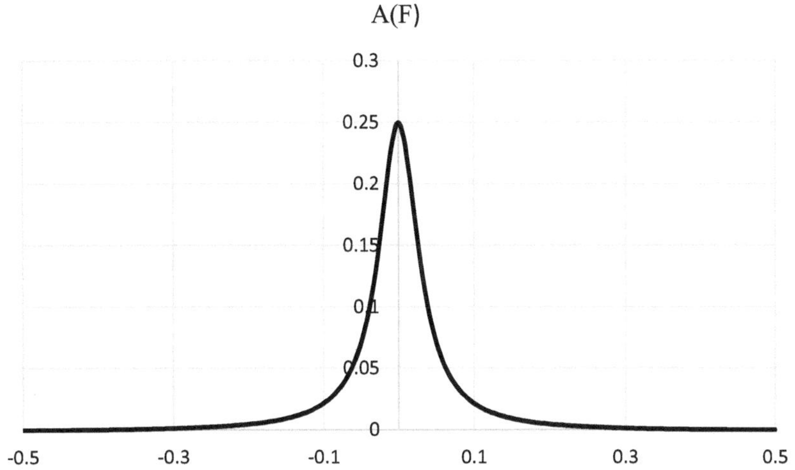

References

1. Y. Bistritz, Zero location with respect to the unit circle of discrete-time linear system polynomials. **Proc. IEEE, 72**(9), 1131–1142
2. C.T. Chen, System and Signal Analysis, 3rd, New York/Oxford: Oxfor University Press, 2004
3. J.W. Cooley, J.W. Tuckey, An algorithm for the machine calculation of complex Fourier series. Math. Comput. **19**, 76–79 (1965)
4. P. Denbigh, *System Analysis & Signal Processing* (Addison-Wesley, Boston, 1998)
5. R.A. Gabel, R.A. Roberts, *Signals and Linear Systems*, 3rd edn. (Wiley, 2009)
6. O. Gazi, *Principles of Signals and Systems* (Springer, Berlin, 2022)
7. J.H. Harter, W.D. Beitzel, *Mathematics Applied to Electronics* (Pearson, 2000)
8. L.B. Jackson, *Digital Filters and Signal Processing*, 3rd edn. (Springer Science+Business Media, New York, 1995)
9. A. Jeffrey, D. Zwillinger, *Table of Integrals, Series, and Products*, 6th edn. (Academic, 2000)
10. E.J. Jury, A modified stability table for linear discrete systems. Proc. IEEE **124**, 184–185 (1964)
11. G.A. Korn, T.M. Korn, *Mathematical Handbook for Scientists and Engineers* (Dover Publications, 2013)
12. H. Kwakernaak, R. Sivan, *Modern Signals and System Analysis* (Prentice Hall, Englewood Cliffs, 1991)
13. A. Mertins, *Signal Analysis – Wavelets, Filter Banks, Time-Frequency Transforms and Applications* (Wiley, 1999)
14. A.V. Oppenheim, R.W. Schafer, *Discrete-Time Signal Processing*, 3rd edn. (Pearson Education, 2013)
15. A.V. Oppenheim, A.S. Willsky, *Signals and Systems*, 2nd edn. (Prentice-Hall, Upper Saddle River, 1996)
16. A. Papoulis, *Circuits and Systems – A Modern Approach* (Oxford University Press, 1995)
17. A. Papoulis, *The Handbook of Formulas and Tables for Signal Processing* (CRC Press, 1998)
18. A.D. Poularikas (ed.), *The Transforms and Applications Handbook*, 2nd edn. (CRC Press, IEEE Press, 2000)
19. A.D. Poularikas, S. Seely, *Signals and Systems* (PWS Publishers, Boston, 1985)
20. M.S. Roden, *Analog and Digital Communication Systems*, 5th edn. (Discovery Press, 2003)
21. H. Stark, F.B. Tuteur, *Modern Electrical Communications. Theory and Systems* (Prentice-Hall, Inc, Englewood Cliffs, 1979)
22. J.L. Walsh, A closed set of normal orthogonal functions. Am. J. Math. **55**, 5–24 (1921)

K. M. Snopek, *Hands-on Signals and Systems Theory*,
https://doi.org/10.1007/978-3-031-56081-1

Index

K. M. Snopek, *Hands-on Signals and Systems Theory*,
https://doi.org/10.1007/978-3-031-56081-1

Z